100th Anniversary of Brillouin Scattering

100th Anniversary of Brillouin Scattering

Guest Editor

Seiji Kojima

MDPI

Basel • Beijing • Wuhan • Barcelona • Belgrade • Novi Sad • Cluj • Manchester

Guest Editor
Seiji Kojima
Department of Materials
Science
University of Tsukuba
Tsukuba
Japan

Editorial Office
MDPI AG
Grosspeteranlage 5
4052 Basel, Switzerland

This is a reprint of the Special Issue, published open access by the journal *Materials* (ISSN 1996-1944), freely accessible at: https://www.mdpi.com/journal/materials/special_issues/100th_Anniversary_Brillouin_Scattering.

For citation purposes, cite each article independently as indicated on the article page online and as indicated below:

Lastname, A.A.; Lastname, B.B. Article Title. *Journal Name* **Year**, *Volume Number*, Page Range.

ISBN 978-3-7258-4107-3 (Hbk)
ISBN 978-3-7258-4108-0 (PDF)
https://doi.org/10.3390/books978-3-7258-4108-0

Contents

Preface

The year 2022 marked one hundred years since the discovery of Brillouin scattering by L. Brillouin in 1922. Nowadays, the term Brillouin scattering is extensively used for the inelastic light scattering in the gigahertz range. Owing to the non-contact method, phase transition, liquid–glass transition, and polymerization have been studied by changing temperature, pressure, and external fields such as electric and magnetic fields and stress. Investigations of various excitations, not only acoustic phonons, but also magnons, for example, provide new insights into physics, chemistry, mineralogy, biology, pharmacy, medical science, and engineering.

The history of experimental methods and applications of Brillouin scattering spectroscopy in materials science has been reviewed from 1922 to the present. The historical development of spin wave studies on magnetic materials has also been reviewed. Brillouin scattering spectroscopy with diamond anvil cells was reported for the pressure dependence of the sound-relevant polymer material. Characterization of three-dimensional direct laser writing technology by Brillouin light scattering has been reported. The studies on structural phase transitions on various perovskite oxides and halides have been reported concerning relaxor ferroelectrics, optoelectronics, and photovoltaic devices.

Seiji Kojima
Guest Editor

materials

Review

100th Anniversary of Brillouin Scattering: Impact on Materials Science

Seiji Kojima

Division of Materials Science, University of Tsukuba, Tsukuba 305-8573, Japan; kojima@ims.tsukuba.ac.jp; Tel.: +81-29-874-7728

Abstract: L. Brillouin predicted inelastic light scattering by thermally excited sound waves in 1922. Brillouin scattering is a non-contact and non-destructive method to measure sound velocity and attenuation. It is possible to investigate the elastic properties of gases, liquids, glasses, and crystals. Various kinds of phase transitions, i.e., liquid–glass transitions, crystallization, polymorphism, and denaturation have been studied by changing the temperature, pressure, time, and external fields such as the electric, magnetic, and stress fields. Nowadays, Brillouin scattering is extensively used to measure various elementary excitations and quasi-elastic scattering in the gigahertz range between 0.1 and 1000 GHz. A brief history, spectroscopic methods, and Brillouin scattering studies in materials science on ferroelectric materials, glasses, and proteins are reviewed.

Keywords: Brillouin scattering; elastic properties; acoustic phonon; central peak; crystal; glass; liquid

Citation: Kojima, S. 100th Anniversary of Brillouin Scattering: Impact on Materials Science. *Materials* **2022**, *15*, 3518. https://doi.org/10.3390/ma15103518

Academic Editor: Alexandros Lappas

Received: 3 March 2022
Accepted: 11 May 2022
Published: 13 May 2022

1. Introduction

The well-known method to measure sound velocity is the ultrasonic method. Piezoelectricity was discovered by Jacques and Pierre Curie in 1880 [1]. Piezoelectric quartz transducers enable the detection of sound waves. However, the ultrasonic method cannot detect thermally excited sound waves due to their very weak intensity. Louis Brillouin predicted inelastic light scattering by thermally excited sound waves in 1922 [2]. It was two years before the discovery of the Bose-Einstein statistics in 1924 [3] which defined the population of thermally excited acoustic phonons. The inelastic light scattering by sound waves is called Brillouin scattering and it is a powerful tool to investigate the elastic properties of materials as a non-contact and non-destructive method.

The binding of materials involves the attractive and repulsive interactions between atoms and molecules. These interactions play the dominant roles of stability in polymorphism, amorphization, non-equilibrium states, and phase transitions. Elasticity is closely related to binding and the experimental study of elastic constants has given new insights into materials science. Various methods were developed to measure elastic constants. The time-domain and frequency-domain ultrasonic methods have been extensively used. The former ultrasonic pulse-echo method [4] measures the delay time of an ultrasonic pulse and determines the sound velocity which is related to the elastic constant and density. The latter frequency-domain resonant ultrasonic spectroscopy (RUS) measures a series of acoustic resonance frequencies of a sample [5,6]. The operating frequency of these methods is in the kilohertz to megahertz range and to measure a small sample of less than 1 mm is difficult. The mechanical contact between a sample and an ultrasonic transducer is necessary.

In contrast, Brillouin scattering is inelastic light scattering without any contact to a sample that is observed. The nondestructive measurement is possible due to the relatively weak power of the incident light. Consequently, Brillouin scattering spectroscopy is a powerful tool to study the elastic properties of various materials. Nowadays, the term Brillouin scattering is extensively used as the inelastic light scattering in the gigahertz range between 0.1 and 1000 GHz. Due to the non-contact method, phase transition, liquid–glass transition, and polymerization have been studied by changing the temperature, pressure, time, and external fields such as the electric, magnetic, and stress fields. Recently, the exciting light source was extended to UV and X-rays, which covers the terahertz-nanometer range. Brillouin scattering imaging or Brillouin microscopy enables the detection of elastic heterogeneity, which is useful for in-vivo diagnostics. The investigations of various excitations, not only acoustic phonon in the bulk and surfaces, but also magnon in magnetic materials, fracton in fractal structures, etc., give new insights into physics, chemistry, materials science, mineralogy, biology, pharmacy, medical science, and engineering.

In this review paper, a brief history of Brillouin scattering and overview of Brillouin scattering spectroscopy are described in Section 2. The various applications for materials science are discussed in Section 3, and the summary in Section 4.

2. Brief History and Overview of Brillouin Scattering Spectroscopy

The experimental observation of Brillouin scattering was reported eight years after the theoretical study by L. Brillouin in 1922 [2]. In 1930, E. Gross observed the inelastic light scattering from acoustic oscillations of highly scattering liquids, such as toluene and benzene at right scattering geometry using 4358 A monochromatic light from a mercury lamp and high-resolution 30 steps echelon grating [7]. The history timeline related to Brillouin scattering in the early days is listed in Table 1.

Table 1. History timeline related to Brillouin scattering in the early days.

Year	Event	Ref.
1899	A. Perot and C. Fabry developed the Fabry-Perot interferometer.	[8,9]
1912	P. Debye discussed acoustic oscillations in a box.	[10]
1922	L. Brillouin predicted inelastic light scattering by thermally excited acoustic waves.	[2]
1923	H. Nagaoka and T. Mishima developed the method to measure high resolution spectra using an echelon grating.	[11]
1926	L. Mandelstam independently predicted light scattering from thermally excited acoustic waves.	[12]
1930	E. Gross observed the inelastic light scattering from acoustic oscillations of liquids.	[7]

The Brillouin scattering process among the incident and scattered light and the periodic modulation of density by sound waves in a medium is shown in Figure 1. Sound waves are thermally excited and their density fluctuations cause the periodic modulation of the refractive index. An incident monochromatic light is scattered by moving periodic modulations in opposite directions.

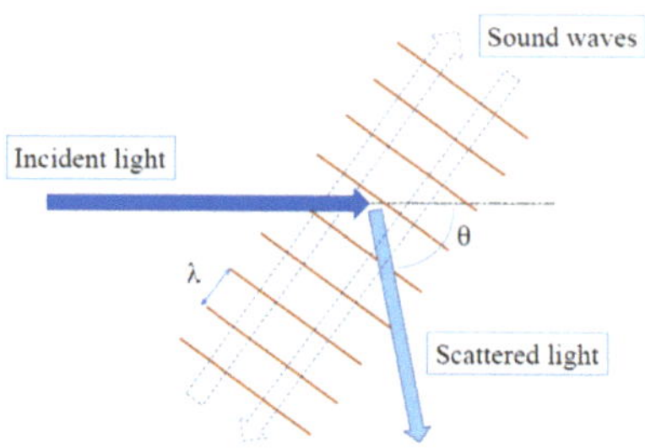

Figure 1. Brillouin scattering process among the incident and scattered light and periodically modulation of the density by thermally excited sound waves, where λ and θ are the wavelength of sound waves and scattering angle, respectively.

According to the energy and momentum conservations, it holds that

$$\nu_B = \nu_i - \nu_s,\ \vec{k}_i = \vec{k}_s \pm \vec{q} \tag{1}$$

where $\vec{k}_i$, $\vec{k}_s$, and $\vec{q}$, ν_B are the wavevectors of the incident photon, scattered photon, and acoustic phonon, respectively. ν_i, ν_s, and ν_B are the frequency of incident light, scattered light, and the acoustic phonon, respectively.

The Brillouin scattering spectrum of the liquid phase of propylene glycol observed by a Fabry-Perot interferometer (FPI) [8,9] is shown in Figure 2. A central intense peak corresponds to Rayleigh scattering and the doublet peaks correspond to Brillouin scattering by sound waves.

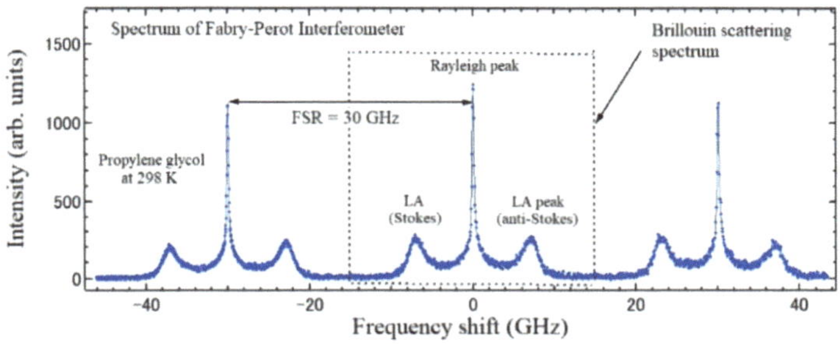

Figure 2. The spectrum of an angular dispersive Fabry-Perot interferometer of propylene glycol in a liquid phase at 298 K with the free spectral range of 30 GHz.

The differential cross section of the Brillouin scattering of a fluid is given by

$$\frac{d\sigma}{d\Omega} \propto \beta_S \rho^2 \left(\frac{\partial \chi}{\partial \rho}\right)_S^2 + \frac{T}{\rho C_p}\left(\frac{\partial \chi}{\partial \rho}\right)_p^2, \tag{2}$$

where β_S is the adiabatic compressibility, ρ is the density,χ is the susceptibility, and C_p is the heat capacity at a constant pressure. S, T, p are the entropy, temperature, and pressure, respectively [13]. The first term on the right side corresponds to the propagating density fluctuations of the sound waves at a constant entropy and it is related to a doublet of Brillouin shifts by sound waves. The second term corresponds to non-propagating entropy fluctuations at a constant pressure, and it is related to a Rayleigh peak. In a fluid, only the longitudinal acoustic (LA) mode is observed, while in a solid, the transverse acoustic (TA) mode is also observed due to the existence of a shear modulus. Brillouin scattering of KCl, RbCl, and KI crystals with a cubic system was studied [14]. The scattering intensity of a cubic crystal by the LA and TA modes, which propagate along the [100] axis is given by

$$I_{VV} \propto \frac{(p_{12})^2}{c_{11}},\ I_{VH} \propto \frac{(p_{44})^2}{2c_{44}}, \tag{3}$$

respectively. VV denotes that the polarization of both the incident and scattered light is vertical and VH denotes that the polarization of the incident light is vertical while that of the scattered light is horizontal. The p_{ij} and c_{ij} are the elasto-optic and elastic stiffness constants, respectively. The displacements of the LA and TA modes are [100] and [001] directions, respectively. The method for the determination of the elastic and photoelastic constants was analyzed for all the crystal systems, except for the low-symmetry systems [15].

In the visible light excitation, the frequency of the scattered acoustic phonon (Brillouin shift) is usually from 10 to 200 GHz. To measure the Brillouin scattering spectra, a spectrometer with a high resolution of 0.1 GHz or 0.01 A is necessary. Therefore, it is difficult to observe by using a conventional grating monochromator for Raman scattering spectroscopy.

For the first observation of Brillouin scattering, the higher order diffraction of an echelon grating was used [7], while in most experiments, a Fabry-Perot interferometer (FPI) [8,9] was used. The FPI consists of two parallel partially reflecting surfaces with the distance d to observe the interference by the multi-reflection between reflecting surfaces. The free spectral range (FSR) of FPI is $1/2d$, for example, FSR = 30 GHz for d = 5 mm. Figure 2 shows the FPI spectrum of propylene glycol with FSR = 30 GHz [16]. On the horizontal axis, the wavelength of the scattered light is converted to the frequency shift from the frequency of the incident monochromatic light. The zero-frequency shift is adjusted to a Rayleigh peak and the cyclic Brillouin scattering spectra appears with the period of 30 GHz. This spectrum was measured by an angular dispersive FPI (ADFPI), in which the finess = 100 is possible using a solid etalon [17,18]. In ADFPI, the angular dispersion of a solid etalon was used, and the one-dimensional output signals were detected by a highly sensitive CCD detector.

During the early stage, the output wavelength of the FPI was scanned by the change of in the refractive index induced by the release of air into a sealed FPI chamber in a vacuum. The scattered light was analyzed using a pressure-scanned Fabry-Perot interferometer (FPI) and "photon-counting" electronics. For crystals with intense elastic scattering, a tandem FPI was used to measure an improper ferroelectric β-$Gd_2(MoO_4)_3$ crystal [19] and a double-pass FPI was used to measure a proper ferroelastic $LiNH_4C_4H_4O_6$-H_2O crystal [20]. After scanning an etalon spacing by piezoelectric elements was developed, the long acquisition time of weak signals using a multi-pass FPI became possible. For the broadband spectra with a high contrast, the tandem multi-pass FPI was developed to measure a wide frequency range up to 1000 GHz using two FPIs with slightly different FSRs [21]. By the connection to a Raman scattering spectrum measured by a grating-spectrometer, a very broadband spectrum is available between 1 GHz and 120 THz including both acoustic and optical phonons.

A typical broadband inelastic light scattering spectrum of a ferroelectric $Pb(Sc_{1/2}Ta_{1/2})O_3$ crystal with the perovskite structure between 1 GHz (0.033 cm^{-1}) and 36 THz (1200 cm^{-1}) is shown in Figure 3 [22]. It consists of a Brillouin scattering spectrum measured by a 3 + 3 pass tandem Fabry-Perot interferometer and a Raman scattering spectrum measured by a triple grating spectrometer. In the low-frequency range below 300 GHz (10 cm^{-1}), the peak of an acoustic phonon is observed, while in the higher frequency above 10 cm^{-1}, several peaks of optical phonons are observed. In addition, a broad Rayleigh wing of quasi-elastic scattering by the relaxation mode related to polarization fluctuations appears below the lowest-frequency optical phonon. It is called a central peak (CP) and it plays an important role in the dynamical properties in order-disorder phase transitions and liquid–glass transitions.

The double monochromator double pass (DMDP) spectrometer allows the simultaneous measurement of Brillouin scattering and Raman scattering by using a high-resolution echelle grating. Since the radiation effectively travels a distance of 32 m, a DMDP spectrometer is large and delicate [23].

Usually, the measurements of FPI with a single channel detector, such as photomultiplier, or photodiode, require a long acquisition time. For a rapid measurement, the angular dispersive FPI (ADFPI) was developed by using a solid etalon and a highly sensitive multichannel detector for the rapid spectrum acquisition. The acquisition time of 1 s or less is possible, which is much shorter than that of a scanning FPI [17]. Using a solid etalon with a high reflectivity (99.5%), the high resolution and high contrast spectrum was achieved. The finesse, the inverse of the resolution, reached 100 [18]. A virtually imaged phased array (VIPA) is another type of angular dispersive FPI [24]. VIPA coupled to a CCD detector enables Brillouin scattering imaging, and the in situ biomechanical measurement of the crystalline lens in a mouse eye using Brillouin shifts. [25]. The high performance of Brillouin microscopy was reported using microwave-induced acoustic fields in a piezoelectric $LiNbO_3$ [26]. The possibility was shown to detect the microwave-induced bulk as well as surface acoustic waves.

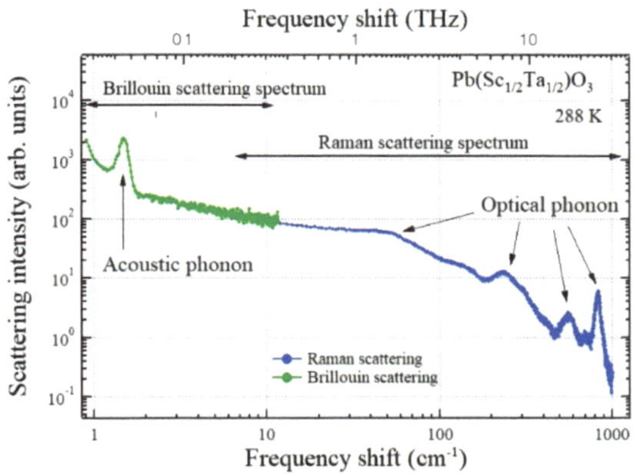

Figure 3. Broadband inelastic light scattering spectrum of a ferroelectric $Pb(Sc_{1/2}Ta_{1/2})O_3$ crystal. Brillouin and Raman scattering spectra were measured by a tandem multi-pass FPI and triple-grating spectrometer, respectively [22].

As an ultrasonic method, scanning acoustic microscopy (SAM) was developed using a finely focused ultrasonic beam in the GHz range [27]. SAM is an imaging system with a resolution comparable to an optical microscope. Another advantage of the SAM is the determination of the phase velocity of surface acoustic waves by measurement of the periodicity of a $V(z)$ curve, where V is the output signal of an acoustic lens and z is the distance between a sample surface and a lens [28]. By using the ultrasonic frequency of 8 GHz and liquid helium as the coupling medium at 0.2 K, the resolution of 200 A was reported [29]. The improvement of the special resolution better than an optical microscopy may be expected for Brillouin imaging.

The stimulated Brillouin scattering (SBS) was developed by Chiao, Townes, and Stoicheff [30]. SBS is a stimulated process of light scattering that occurs when the intensity of the light field itself affects the propagating medium. The rate of Brillouin scattering is markedly enhanced by the powerful coherent laser light. SBS has recently become relevant in the optical fiber industry [31,32]. The time-domain Brillouin scattering (TDBS) has been used for the detection of inhomogeneity [33].

Nowadays, Brillouin scattering is extensively used in the studies of various kinds of materials such as gases [31], colloids [34], multilayers [35], ceramics [36], amino acids [37], biological tissues [38,39], insects [40], and fossils [41]. Brillouin scattering has been also used as inelastic light scattering in the gigahertz range for the investigations of various elementary excitations, not only for the acoustic phonon, but also the magnon [42,43], fracton [44], roton [45], ripplon [46], and soliton [31] have been studied. This knowledge offers new insights into physics, chemistry, materials science, mineralogy, biology, pharmacy, medical science, and engineering.

3. Brillouin Scattering Spectroscopy in Materials Science

The combination of a tandem multi-pass FPI with an optical microscope enables the measurements of a very small material or a very small area of a sample as shown in Figure 4 [47]. The monochromatic light beam from a single frequency CW laser is focused into a specimen to be observed through the objective lens of a reflection optical microscope. The scattered light from a specimen is collected by the objective lens at backward scattering geometry. The scattered light is collimated and focused on the input pinhole of a Sandercock-type 3 + 3 pass tandem multi-pass FPI and detected by a photon-counting system. A compact IR furnace is connected to measure a sample at high temperatures [48]. For the measurements at low temperatures, a heating/cooling stage [49] or cryostat [50] is used. For the measurements at high pressures, diamond anvils are very convenient and powerful. The size of a sample is 0.1 mm or less as it is possible to observe Brillouin scattering using a finely focused laser beam.

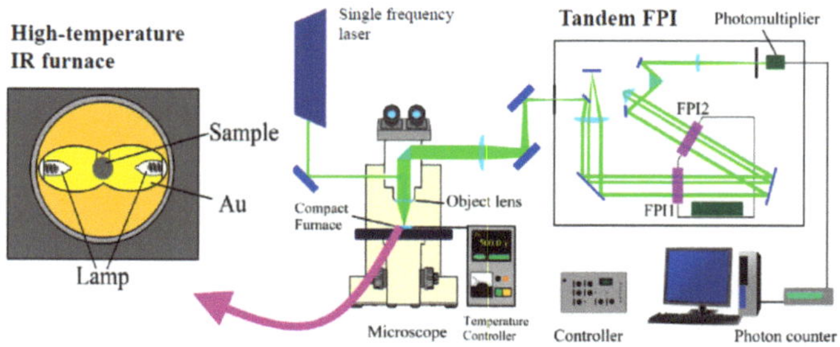

Figure 4. Experimental setup of micro-Brillouin scattering using a tandem multi-pass Fabry-Perot interferometer [48].

For the measurements under high pressure, diamond anvils are convenient as shown in Figure 5a [51]. By shortening the distance between a pair of diamond anvils, a high pressure is generated in the sample chamber consisting of the sample and pressure transmitting medium. Figure 5b shows the pressure dependences of compressibility and Poisson's ratio of Baltic amber determined by the LA and TA shifts up to 12 GPa [52]. Baltic amber is a natural organic glass with the high fragility index of $m = 90$. As the pressure becomes very high, the sample size becomes very small and the ultrasonic methods become difficult to measure its elastic properties.

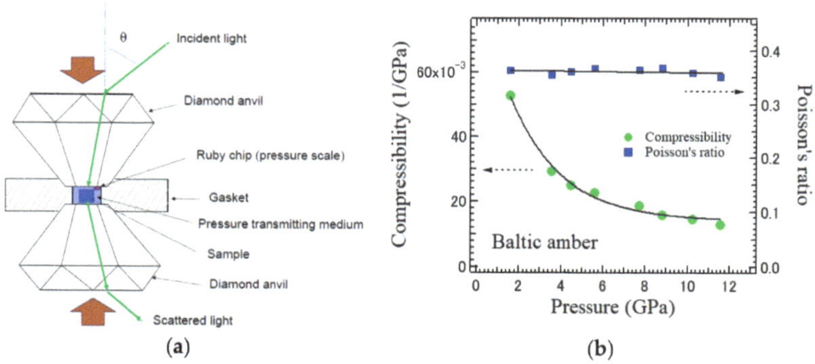

Figure 5. (**a**) Brillouin scattering experiment at high pressures using a diamond anvil cell. (**b**) Pressure dependences of compressibility and Poisson's ratio of the Baltic amber measured by Brillouin scattering [51].

Brillouin scattering spectroscopy is a powerful tool to investigate elastic properties and dynamical properties in the gigahertz range, and various materials have been studied. It is possible to measure a sample under extreme conditions such as high temperature up to 2000 °C [53] and high pressure up to 100 GPa [54]. In this section, a few topics regarding the Brillouin scattering studies of functional ferroelectric materials, glass-forming materials, and proteins are reviewed.

3.1. Ferroelectric Materials

Ferroelectricity is defined by the existence of spontaneous polarization and its switching by external electric fields. Ferroelectricity was first found in Rochel salt in 1921 by Balasec [55]. Ferroelectric crystals have no center of symmetry and their physical properties of odd order tensors, such as pyroelectric, piezoelectric [1], electro-optic, nonlinear-optic properties, and polaritons, are technologically very important [56]. They have been extensively used as functional materials. A ferroelectric material undergoes a paraelectric-ferroelectric phase transition at its Curie temperature upon heating. The highest Curie temperature is 1342 °C for $Sr_2Nb_2O_7$ with a perovskite-slab structure [57]. The high tunability of the Curie temperature from such a high temperature down to −107 °C is possible by the control of the composition of $Sr_2(Nb_{1-x}Ta_x)_2O_7$. It is also technologically important [58].

Most ferroelectric materials are oxygen octahedra ferroelectrics such as perovskite, layered perovskite, perovskite slab, and tungsten bronze (TB) structures. In this section, normal and relaxor ferroelectrics with the TB structure are described. The projection on the (001)-plane of the oxygen octahedral framework of the TB structure is shown in Figure 6 [59]. There are three types of A1, A2, and C sites. B1 and B2 are the central sites of the oxygen octahedra. A tetragonal unit cell includes two A1 sites, four A2 sites, four C sites, two B1 sites, eight B2 sites and thirty oxygens. The site occupancy formula of the tetragonal TB structure is $(A1)_2(A2)_4(C)_4(B1)_2(B2)_8O_{30}$. The smallest triangular C sites are occupied only by Li ions. The ferroelectrics with filled TB, such as $Ba_2NaNb_5O_{15}$ (BNN), undergo a normal ferroelectric phase transition of which dielectric constant along the *c*-axis obeys the Curie-Weiss law. However, the ferroelectrics with an unfilled TB structure, such as $Sr_{1-x}Ba_xNb_2O_6$ (SBN100x), show relaxor behaviors with a broad dielectric anomaly and remarkable frequency dispersion [60].

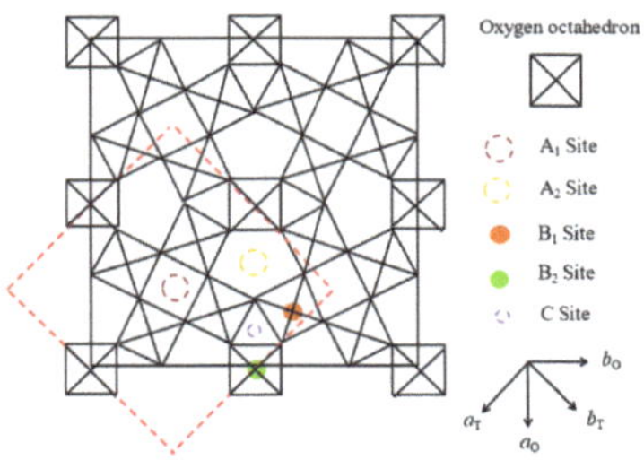

Figure 6. Projection of tungsten bronze structure on the *c*-plane. Tetragonal and orthorhombic unit cells are shown by dotted and solid lines, respectively [60].

BNN is important in the field of telecommunications due to its superior electro-optical and nonlinear optical properties such as second harmonic generation, and its resistance to optical damage is high. The A1 and A2 sites are fully filled by Na and Ba, respectively, and there is no vacancy. The B1 and B2 sites are fully occupied by Nb, while the C sites are vacant. BNN is called a filled TB structure with no charge disorder at A1 and A2 sites. BNN undergoes successive phase transitions at 575 and 290 °C. The higher one is associated with the ferroelectric Curie temperature at 575 °C and its crystal symmetry changes from prototypic *4/mmm* to ferroelectric tetragonal *4 mm* with a spontaneous polarization P_3 along the *c*-axis. Upon the symmetry changes from tetragonal to incommensurate (IC) orthorhombic *mmm* systems at 290 °C, the *a* and *b* axes are rotated 45° along the *c*-axis as shown in Figure 6. The modulation direction of the IC wave vectors is along the *a* and *b* axes of the orthorhombic coordinate [59].

3.1.1. Elastic Anomaly

The sound velocity V is determined by the frequency shift ν_B in a Brillouin scattering spectrum using the following equation:

$$V = \frac{\lambda_i v_B}{2\, n \sin\frac{\theta}{2}} \tag{4}$$

where, λ_i, θ, and n are the wavelength of the incident beam, the scattering angle, and the refractive index of the sample, respectively. The TA and LA velocities are determined from the TA and LA frequency shifts, respectively.

The attenuation α is determined by,

$$\alpha = \frac{\pi\Gamma}{V} \tag{5}$$

where Γ is FWHM of the TA or LA peak.

The temperature dependence of the LA peaks observed at backward scattering using a *c*-plate is shown in Figure 7a [61]. The wave vector of the observed LA mode is parallel to the c-axis. The LA peaks shows softening from 800 °C to T_C = 575 °C. The temperature dependences of the LA velocity and attenuation determined from the LA shift and its FWHM using Equations (1) and (2), respectively, are shown in Figure 7b.

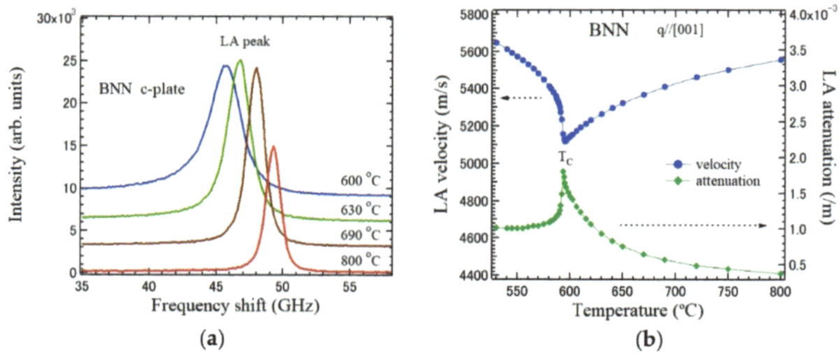

Figure 7. Temperature dependences of (**a**) Brillouin scattering spectra of a *c*-plate of BNN in a paraelectric phase, and (**b**) sound velocity and attenuation of the LA mode along the *c*-axis of BNN [61].

Since the polarization fluctuations along the *c*-axis couple with the strain along the *c*-axis by the electrostrictive coupling, remarkable changes were observed in the temperature dependences of the sound velocity and attenuation of the LA mode, which propagate along the *c*-axis. Such a sharp elastic anomaly in the vicinity of the Curie temperature is the typical nature of a normal ferroelectric phase transition. In contrast, the diffusive changes were observed in a relaxor ferroelectric phase transition with random fields [22,62]. Using the power law behaviors of the sound velocity and attenuation above the Curie temperature, the dimensionality of fluctuations of the order parameters and the order-disorder nature were discussed [63,64].

3.1.2. Critical Slowing Down

Ferroelectrics are divided into two types, namely, the displacive type and the order-disorder type according to the difference in the mechanism upon the appearance of a spontaneous polarization. In the displacive type, an infrared active soft-optic mode exists and a spontaneous polarization appears by the freezing of a soft-mode displacement. However, in the order-disorder type, the relaxation time of the polarization fluctuations along a ferroelectric axis diverges at the Curie temperature, and the critical slowing down towards the Curie temperature is observed. No soft optic mode was observed in the ferroelectrics with the TB structure, however, an intense central peak (CP) of BNN was observed in the vicinity of the Curie temperature in the broadband Brillouin scattering spectra as shown in Figure 8a [61].

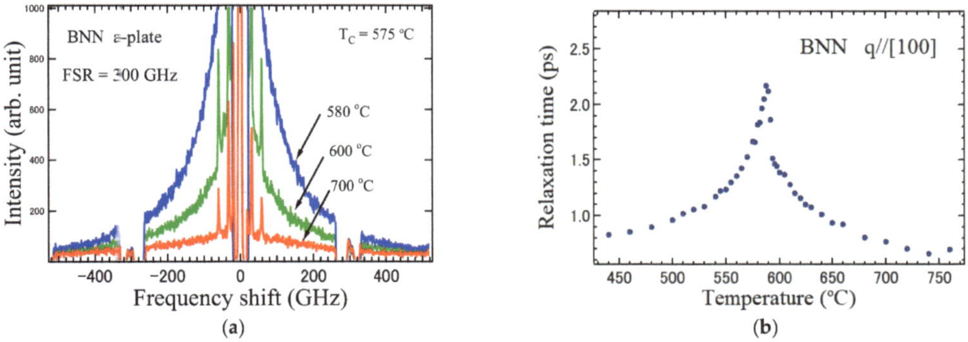

Figure 8. (**a**) Broadband Brillouin scattering spectra observed at backward scattering in a paraelectric phase of BNN [61]. A broad central peak appears in the vicinity of the Curie temperature. (**b**) Temperature dependence of relaxation time determined by the width of a broad central peak.

The temperature dependence of the relaxation time determined by the width of a broad central peak observed at backward scattering in the *a*-plate of BNN is shown in Figure 8b [61]. The discontinuity of the relaxation time at the Curie temperature indicates the first order phase transition. The critical slowing down towards the Curie temperature was clearly observed. The divergence of the relaxation time of the polarization fluctuations along the ferroelectric *c*-axis is evidence for the order-disorder nature of the ferroelectric phase transition of BNN. A similar critical slowing down was also observed in ferroelectric $LiTaO_3$ [65] and $BaTi_2O_5$ [66]. However, in $LiTaO_3$, the soft optic mode was also observed far below the Curie temperature. Therefore, the crossover from the displacive type to order-disorder one occurs, and it is related to the single minimum potential of Nb ions and double minimum potential of Li ions along the ferroelectric *c*-axis. In contrast, in $Sr_2Nb_2O_7$, the soft optic mode was clearly observed at all the temperatures, and it is evidence of the pure displacive type [58].

3.2. Relaxor Ferroelectrics

In normal ferroelectrics, the dielectric constant along a ferroelectric axis shows a sharp peak at the Curie temperature with no frequency dispersion. However, the dielectric constant of relaxor ferroelectrics shows a broad diffusive peak with a remarkable frequency dispersion [67]. The origin of the relaxor natures is the structural disorders, such as the random occupancy of heterovalent cations, vacancies, and frustration of competing interactions. Typical relaxor ferroelectrics are $Pb(B_{1/3}B'_{2/3})O_3$ [62], $Pb(B_{1/2}B'_{1/2})O_3$ [22], with the perovskite structure and $M_{1-x}Ba_xNb_2O_6$ (M = Ca, Sr) with the unfilled TB structure [64]. Relaxor ferroelectrics (REFs) have attracted much attention due to their colossal piezoelectric effect and their wide variety of applications in piezoelectric devices. Pb-based multidimensional REFs with the perovskite structure have been extensively studied [22,62], whereas the understanding of uniaxial relaxors is still unresolved.

3.2.1. Characteristic Temperatures of Relaxor Ferroelectrics

Regarding the physical properties of the REFs, the polar nanoregions (PNRs) with a polar structure play an important role. Figure 9 shows the sequence of the characteristic temperatures of a relaxor ferroelectric phase transition, $T_B \Rightarrow T^* \Rightarrow T_C \Rightarrow T_f$ [22,62]. When a sample is cooled from a very high temperature, at the Burns temperature, T_B, dynamic PNRs nucleate in a paraelectric phase with a center of symmetry. The T_B is typically a few hundred degrees above the ferroelectric Curie temperature, T_C. With further cooling from T_B, a dynamic central peak in an inelastic scattering spectrum is observed by thermally fluctuated flipping of the local polarizations in the PNRs. An elastic anomaly is observed in the vicinity of T_C by the local piezoelectric coupling in the PNRs. In the temperature range between T_B and T_C, a dynamic-to-static transition of the PNRs occurs at the intermediate temperature T^*. All the PNRs are completely frozen at the freezing temperature, T_f, which is the Vogel-Fulcher temperature [22]. At T_C, the polarization fluctuations of the PNRs are frozen into nanodomain structures, because the growth into macrodomains is blocked by the random fields (RFs). Consequently, the nanodomain structures are nonequilibrium states, and gradually change into equilibrium macrodomain structures by aging with a long relaxation time. For the investigation of the REFs, Brillouin scattering spectroscopy is a powerful tool to investigate the elastic anomaly, relaxation processes, and aging versus their temperature, pressure, time, and electric field dependences [22,62].

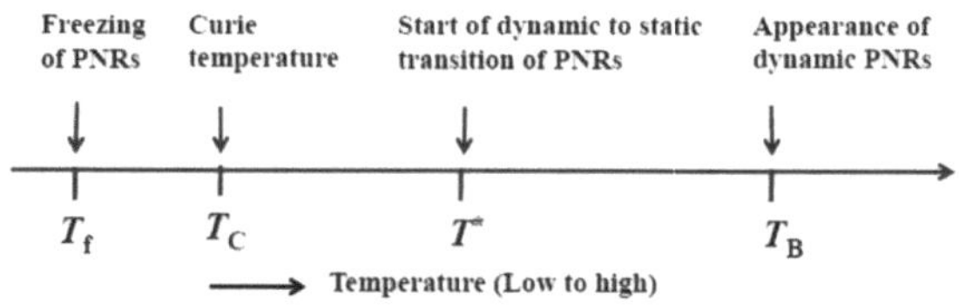

Figure 9. Characteristic temperatures of relaxor ferroelectrics. Polar nanoregions (PNRs) play a main role.

In the unfilled TB structure, there are vacancies in the A1 and A2 sites, which cause the charge disorder. This is due to the RFs of the relaxor nature. $Sr_xBa_{1-x}Nb_2O_6$ (SBN100x), is one of the technologically important uniaxial REFs due to its very high electro-optic and pyroelectric coefficients, making it useful for applications such as sensors and data storage. In the SBN, the largest A2 sites are occupied by the divalent Ba and Sr ions, while the A1 sites are occupied only by the smaller Sr ions. One-sixth of all the A sites (A1 + A2 sites) are vacancies, which are the main origin of the RFs. The relaxor nature of SBN was described by the random field Ising Model [68]. With the increasing Sr/Ba ratio, SBN shows a crossover from normal to relaxor ferroelectrics by the enhancement of the charge disorder. $Ca_xBa_{1-x}Nb_2O_6$ (CBN100x) has also attracted attention by its similar relaxor nature and high Curie temperatures, which are important for application in a high-temperature environment [69]. The crossover from normal to relaxor ferroelectrics has been studied in SBN100x and CBN100x. In SBN26 with weak RFs, a sharp elastic anomaly was observed in the vicinity of the Curie temperature similar to the anomaly of BNN. However, in SBN80 with strong RFs, a broad and diffuse elastic anomaly was observed [63].

For the optical applications of the SBN, in most cases, the congruent melt SBN61 has been used, because large and high-quality single crystals are available by the Czochralski method. The relaxor ferroelectric SBN61 with a strong charge disorder undergoes a diffuse phase transition upon cooling from a prototypic tetragonal *4/mmm* phase to a ferroelectric tetragonal *4mm* phase with a spontaneous polarization along the *c*-axis [60]. The acoustic phonons are sensitive to the three characteristic temperatures due to the scattering caused by the PNRs. The temperature dependence of the LA attenuation of SBN61 is related to the characteristic temperatures as shown in Figure 10 [70]. The increase in the LA width was observed near T_B. Below T^*, the scattering of the LA mode by the static PNRs causes a significant increase in the LA width. At the T_C, the increase in the scattering by the PNRs was stopped by the freezing of the PNRs into ferroelectric nanodomain structures [71]. The three characteristic temperatures, T_B = 350 °C, T^* = 190 °C, and T_C = 72 °C, were determined. T_B and T^* are very close to those of SBN75, therefore, T_B and T^* may be common to all the compositions of the relaxor SBN100x.

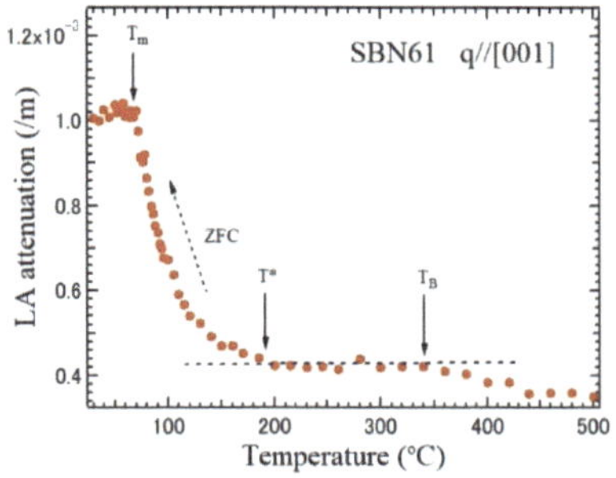

Figure 10. Temperature dependence of LA attenuation of SBN61 observed on zero-field cooling (ZFC), where $T_B = 350$ °C, $T^* = 190$ °C, and $T_m = 72$ °C [70].

3.2.2. Electric Field Effects on Relaxor Ferroelectrics

The electric field dependence of the LA velocity along the *c*-axis of SBN61 after zero field cooling (ZFC) is shown in Figure 11 [70]. The electric field along the ferroelectric *c*-axis was first applied to a ZFC crystal in the nonequilibrium nanodomain state, "Nano". The initial LA velocity of 5270 m/s of "Nano" gradually increased up to 5430 m/s at Es = 1.7 kV/cm, then the splitting of a LA peak occurred up to Es = 2.4 kV/cm. Another high frequency peak produced a higher velocity of 5760 m/s in the equilibrium macrodomain state, "Macro", which is a nearly single domain state and only macrodomains exist. In the field range between Es = 1.7 kV/cm and Ef = 2.4 kV/cm, the "Nano" and "Macro" states coexist, and it is called the "Mix" state. For a further increase in the electric field, the splitting of the LA peaks disappeared, and only a single higher frequency peak appeared. Above Ef, the high LA velocity of 5760 m/s showed only a "Macro" state up to 3.6 kV/cm. When the electric field decreased from E = 3.6 kV/cm, the velocity kept the high value and a "Macro" state remained down to E = 0 kV/cm.

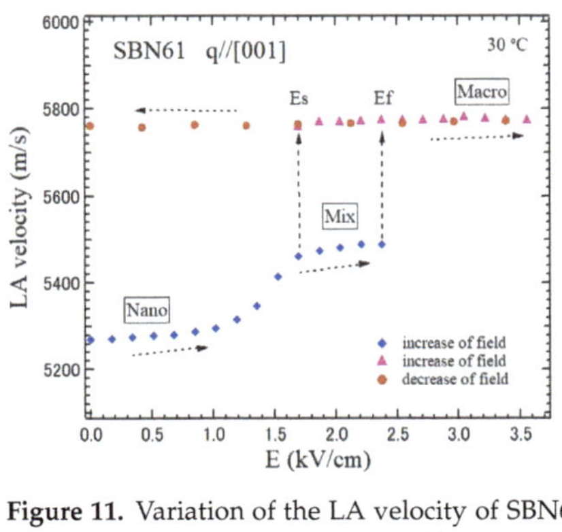

Figure 11. Variation of the LA velocity of SBN61 by the application of the electric field along the *c*-axis at 30 °C. By the application of the electric field along the *c*-axis to a (Nano) nanodomain state, (Mix) mixed, and (Macro) macrodomain states appeared [70].

In order to clarify the temperature dependence of the three different states, "Nano", "Mix", and "Macro" in Figure 11, heating and cooling with and without an electric field were investigated. The LA velocity was first measured during field heating (FH) from the "Macro" state at room temperature up to T^* as shown in Figure 12. A large thermal hysteresis was then observed below T_C during the ZFC. A similar hysteresis was also reported for the Pb-based REFs [72]. At room temperature, it was found that the LA velocity after the ZFC is much lower than after FH, and the "Nano" state appears. Next, after annealing for a month at room temperature, the velocity increased by about 2%, and the "Mix" state appeared as the intermediate state with a mixture of nano- and macrodomain states. The slow aging from the "Nano" to "Macro" states was also investigated at room temperature. For the complete recovery to the "Macro" macrodomain state at room temperature, it is necessary to keep a ZFC crystal for more than 10 years. The temperature, electric field dependences, and aging were also reported for SBN40 and SBN70 [73,74].

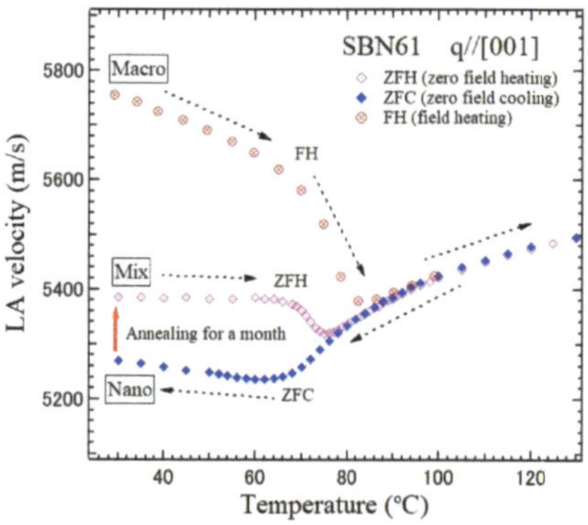

Figure 12. Variation of the LA velocity of SBN61 in the "Nano" state during ZFC without poling, the "Mix" state during zero field heating (ZFH) after a month at room temperature, and the "Macro" state during FH [70].

3.3. Liquid, Glass, and Crystalline States

3.3.1. Liquid–Glass Transition

When a liquid is cooled from a high temperature, a simple liquid is solidified into a crystalline state at its melting temperature, T_m. However, a complex liquid changes into a supercooled liquid state, and with further cooling, it undergoes a liquid–glass transition into a glassy state at the glass transition temperature T_g. Figure 13 shows the temperature dependence of enthalpy for a liquid (AB), supercooled liquid (BD), glassy (DE), and crystalline (CG) states, where T_g, and T_K are the glass transition and Kauzmann temperatures, respectively. The enthalpy of a liquid crosses to that of a crystal at the point, F. The temperature at F is the Kauzmann temperature, T_K, which is a static ideal glass transition temperature and close to the Vogel–Fulcher temperature, the dynamic ideal glass transition temperature, T_{VF}.

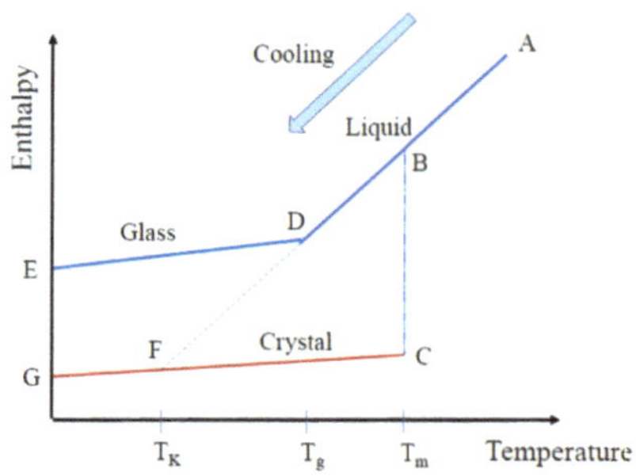

Figure 13. Temperature dependence of the enthalpy for the liquid (AB), supercooled liquid (BD), glass (DE), and crystalline (CG) states, where T_m, T_g, and T_K are the melting, glass transition, and Kauzmann temperatures, respectively.

When a liquid is cooled to a liquid–glass transition, various types of dynamical properties, such as a main α-relaxation, slow and fast β-processes and a boson peak were observed in the high frequency range from the mHz to THz ranges [47]. The relaxation time of the main α-structural relaxation process in a fragile liquid, such as ortho-terphenyl (OTP), diverges at T_g, and obeys the Vogel–Tammann–Fulcher law reflecting the cooperative motion in a cage. However, in a strong liquid, such as a silica glass, the relaxation time obeys the Arrhenius law and no remarkable change is observed at T_g.

Glycerol undergoes a liquid–glass transition at about T_g = 187 K. It is one of the typical glass-forming materials and well-known cryoprotectants due to its strong glass-forming tendency [75]. According to strong-fragile classification, it is the intermediate liquid with the fragility index, m = 53 [76]. Figure 14 shows several relaxation processes and vibrational dynamics in glycerol. A low-frequency Raman scattering spectrum covers a boson peak [77]. A broadband Brillouin scattering spectrum covers the high-frequency part of the α-relaxation, the γ-relaxation, the fast β-relaxation in a cage, and the low-frequency part of a boson peak [78]. The relaxation times of the α- and slow β-processes were observed by dielectric spectroscopy, which detects polar motions [79]. Heat capacity spectroscopy detects polar and nonpolar motions [80]. For understanding of the dynamical features of a liquid–glass transition, it is very important to draw a complete picture of an ultra-broadband relaxation map, which covers from very slow to very fast processes, and a boson peak.

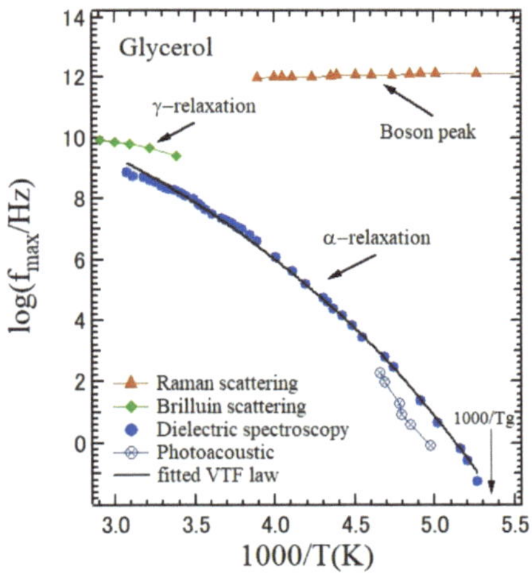

Figure 14. Relaxation processes and boson peaks of liquid and supercooled liquid states of glycerol. The main structural relaxation, α-relaxation, obeys the Vogel–Tammann–Fulcher (VTF) law [77–80].

The polarized and depolarized broadband spectra between 1 GHz and 10 THz measured by a tandem multi-pass FPI and a Raman spectrometer were studied in silica, OTP and ethanol [81]. It was found that the amplitude of this excess quasi-elastic scattering in polarized scattering is proportional to the ratio of the integral intensities of the Brillouin line and the boson peak, while the relaxation mechanisms are the same for both the polarized and depolarized spectra.

For comparison with the dynamic susceptibility predicted by the idealized mode coupling theory of a liquid–glass transition, the composite depolarized spectra were extensively studied in various glass-forming liquids [82]. In this study, the problem of a "knee" in the fragile $Ca_{0.4}K_{0.6}(NO_3)_{1.4}$ (CKN) was reported as an experimental artifact, which can be eliminated by the addition of a narrow-band dielectric filter to the Fabry-Perot optics [83]. However, for T that was above but close to the critical temperature of the mode coupling theory (MCT), T_C, the susceptibility spectra were found to obey the MCT predictions for all the materials such as salol, propylene carbonate, glycerol, iso-propyl benzene, OTP, and CKN.

For the liquid–glass transition of symmetric molecules, a rapid cooling is necessary to avoid crystallization in a supercooled liquid state. The liquid–glass transitions of trimethylene glycol, deuterated ethylene glycol-d_4, and poly(propylene glycol) diglycidyl ether were studied by a rapid cooling using the short acquisition time of ADFPI [17,84,85].

When a liquid is cooled at a constant pressure, the free volume decreases and a liquid–glass transition occurs. When the hydrostatic pressure is applied to a liquid at a constant temperature, the free volume decreases and a pressure-induced liquid–glass transition occurs. The pressure induced liquid–glass transitions of a few lower alcohols were studied using a DAC at room temperature and the change in the elastic properties was observed at the glass transition pressure [86,87].

3.3.2. Drug Materials

The polymorphic nature of pharmaceutical materials is attributed to their good glass-forming tendency. Since nonequilibrium glassy/amorphous states have a solubility and dissolution rate that is higher than the equilibrium crystalline states, the presence of the glassy/amorphous states have the advantage in pharmaceutical materials. The liquid, glassy, and crystalline states of various drug materials, such as aspirin [88,89], ibuprofen [90,91], ketoprofen [92], and indomethacin [34,93] were studied by Brillouin scattering.

Indomethacin (IMC) [1-(p-chlorobenzoyl)-5-methoxy-2-methy lindole-3-acetic acid] is one of the non-steroidal anti-inflammatory drugs and is used for the treatment of fever, pain, and swelling [94]. The liquid, supercooled liquid, and crystalline states of IMC were studied by Brillouin scattering at the backward scattering and reflection induced ΘA scattering geometries [95]. The temperature dependences of the Brillouin scattering spectra are shown in Figure 15. The temperature dependences of the Brillouin shifts of the LA and TA modes of the liquid, glass, and γ-form crystalline IMC are shown in Figure 16. Upon heating of the glassy IMC, the elastic anomaly at the liquid–glass transition temperature T_g = 315 K was clearly observed. With further heating of the supercooled liquid state, crystallization at T = 410 K and melting at T_m = 443 K into an equilibrium liquid phase were also observed.

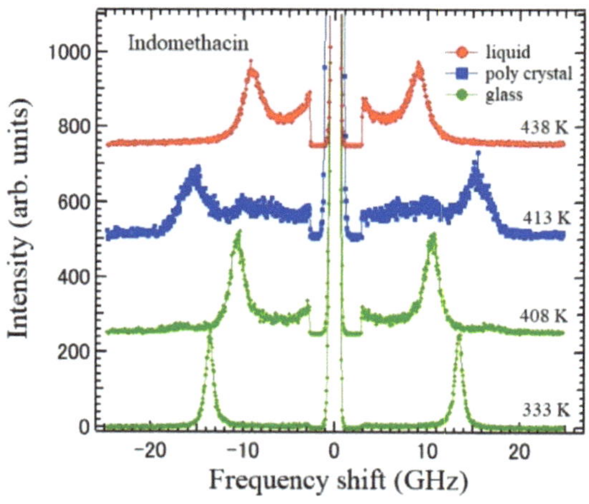

Figure 15. Temperature dependence of Brillouin scattering spectra of the liquid, glass, and polycrystalline states of indomethacin. The glassy states at 333 and 408 K, polycrystalline phase at 413 K after crystallization from glass, and liquid states at 438 K [95].

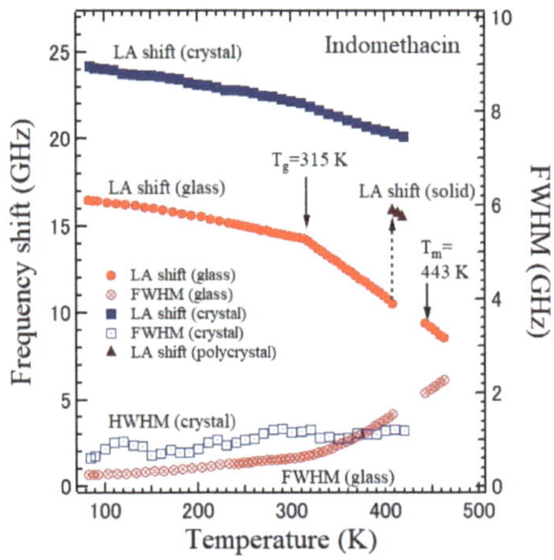

Figure 16. Temperature dependences of the LA frequency shift and FWHM of a γ-form indomethacin crystal with a triclinic structure, glass, and liquid states. The polycrystalline state created by the crystallization from a glass state at 410 K is also plotted [95].

The TA mode of a glassy state was measured using reflection-induced ΘA scattering [95] as shown in Figure 17a. The observed spectrum is shown in Figure 17b. From the LA and TA Brillouin shifts, the bulk modulus $K = 5.41$ GPa and shear modulus $G = 1.33$ GPa were determined.

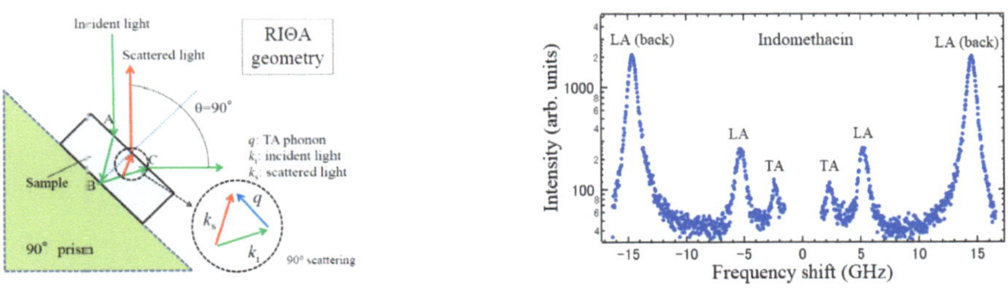

(a) (b)

Figure 17. (**a**) Scattering geometry of the reflection-induced ΘA scattering to observe TA modes at the backward scattering geometry. By the reflection of an incident light at the point B, right angle scattering is observed. (**b**) The Brillouin scattering spectrum of the reflection induced ΘA scattering of a glassy IMC. The LA peaks of backward scattering were also observed [95].

The temperature dependence of the LA and TA mode frequency of a γ-form IMC crystal was measured at the backward scattering geometry. The elastic constants at 296 K are shown in Table 2 [95].

Table 2. Elastic constants, c_{ij}, of the γ-form IMC crystal at 296 K. Those of similar drug materials, aspirin, ibuprofen, are also shown for comparison [95].

Elastic Constant (GPa)	C_{11}	C_{22}	C_{33}	C_{44}	C_{55}	C_{66}
Indomethacin			17.89	5.44	3.39	
Aspirin [82]	11.29	12.10	10.67	3.67	2.68	4.41
Ibuprofen [83]	6.80				1.71	3.25

3.4. Polymorphism, Dehydration, and Denaturation of Proteins

Brillouin scattering is a convenient method to investigate the elastic properties of protein crystals and their phase transitions or transformations. Regarding the study of tetragonal lysozyme crystals grown by the hanging-drop vapor-diffusion method, an elastic anomaly was observed at 34 °C [96]. This fact indicated a structural phase transition. The elastic properties of a tetragonal hen egg-white lysozyme (HEWL) crystal were studied in a cryoprotective solution over a wide temperature range [97]. The dehydration process of tetragonal and monoclinic lysozyme crystals grown by the two-liquids interface method [98] was studied by exposure of the crystals to open air using Brillouin scattering [99]. The LA velocity markedly increased due to the increase in intermolecular interaction between the lysozyme molecules, while the LA attenuation monotonically decreased due to the decrease in the friction generated by the mobile water. The time dependence of the LA velocity of the tetragonal HEWL at a constant temperature was measured. At 40 °C, the LA velocity markedly increased after 20 min and approached the horizontal asymptote. In contrast, the LA velocity at 30 °C moderately increased and reached a constant value that is slightly lower than that at 40 °C.

The time dependence of the LA velocity was analyzed by the Avrami–Erofe'ev model [100]. The normalized dehydration fraction is defined by

$$r(t) = \frac{V(t) - V(0)}{V(\infty) - V(o)} \tag{6}$$

For the Avrami–Erofe'ev type reaction, it holds that

$$-\ln\{1 - r(t)\} = kt^m, \tag{7}$$

where k is the dehydration rate constant, t is the time, and m is the dimension number of the reaction. Figure 18 shows the timed dependence of $r(t)$ during dehydration at 30 °C and 40 °C. The dimension of this process is different between a slow two-dimensional process (m = 2) at 30 °C and a fast three-dimensional process (m = 3) at 40 °C in accordance with the AE model. However, for a monoclinic HEWL crystal, the $r(t)$ was three-dimensional at 32 °C. These findings may indicate that the protein–water interaction is sensitive to the dehydration rate and the rate depends on the crystal structure.

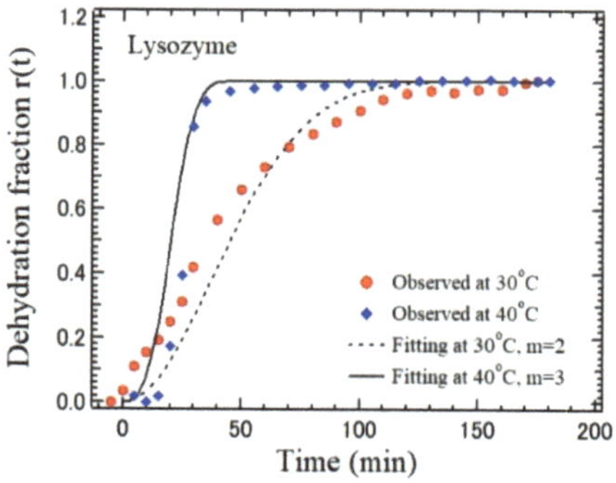

Figure 18. Time dependence of the normalized fraction of a tetragonal HEWL crystals at 30 and 40 °C [98].

The lysozyme and denaturing guanidine-hydrochloride solutions were studied by Brillouin scattering [95,101]. The Brillouin spectrum of the lysozyme unfolded in 6M guanidine hydrochloride contained narrow and broad quasi-elastic scattering peaks. The narrow peak corresponds to interactions of the denaturing agent with the polypeptide chain of the lysozyme. However, the broad peak is attributed to the interaction among the guanidinium (Gdm^+) $[CH_6N_3]^+$ cations.

4. Conclusions

Brillouin scattering is the inelastic light scattering in the gigahertz range from 0.1 to 1000 GHz with a high resolution. It allows the measurement of not only the elastic properties but also various elementary excitations and relaxation processes. Since its measurement is noncontact and nondestructive, it is possible to investigate the dependences of various dynamical properties on the temperature, pressure, time, and external fields. The sharp elastic anomaly and critical slowing down of a normal ferroelectric phase transition are shown. The electric field effect and aging of the non-equilibrium nanodomain states into equilibrium macrodomain states are discussed in relaxor ferroelectrics. The liquid, glass, and crystalline states of drug materials are clarified by their temperature dependence on the elastic properties. The polymorphism and denaturation of proteins are discussed by the variation in the elastic properties and quasi-elastic scattering.

Funding: This research was funded in part by JSPS KAKENHI, Grant No. JP17K05030.

Acknowledgments: The author is very thankful to following collaborators: T. Nakamura; K. Ohi; S. Itoh; M. Ahart; J.-H. Ko; S. Lushnikov; M. Maczka,; A. Yoshihara; Y. Takagi; V.N. Novikov; M.S. Jang; Q.R. Yin; M. Kodama; Z.-G. Ye; W. Kleemann; J. Dec; I.S. Park; S. Tsukada; A. Hushur; S. Islam; M. Rahaman; M. Helal; M. Aftebuzzaman; Y. Ike; F. Jiang; Y. Matsuda; S. Anna; A.M. Pugachev; A.I. Fedoseev; V. Sivasubramanian; D.-H. Kim; G. Shabbir; T.H. Kim; and G. Matsui.

Conflicts of Interest: The author declares no conflict of interest.

References

1. Curie, J.; Curie, P. Développement par compression de l'électricité polaire dans les cristaux hémiédres à faces inclinées. *Bull. Minéral.* **1880**, *3–4*, 90–93. [CrossRef]
2. Brillouin, L. Diffusion de la lumière et des rayons X par un corps transparent homogène. *Ann. Phys.* **1922**, *17*, 88–122. [CrossRef]
3. Bose, S.N. Plancks Gesetz und Lichtquantenhypothese. *Z. Für Phys.* **1924**, *26*, 178–181. (In German) [CrossRef]
4. McSkimin, H.J. Pulse superposition method for measuring ultrasonic wave velocities in solids. *J. Acoust. Soc. Am.* **1961**, *33*, 12–16. [CrossRef]
5. Ohno, I. Free vibration of a rectangular parallelpiped crystal and its application to determination of elastic constants of orthorhombic crystals. *J. Phys. Earth* **1976**, *24*, 355–379. [CrossRef]
6. Yasunaga, M.; Funatsu, Y.; Kojima, S.; Otsuka, K.; Suzuki, T. Measurement of elastic constants. *Scr. Metall.* **1983**, *17*, 1091–1094. [CrossRef]
7. Gross, E. Change of wave-length of light due to elastic heat waves at scattering in liquids. *Nature* **1930**, *126*, 201–202. [CrossRef]
8. Fabry, C.; Pérot, A. Théorie et applications d'une nouvelle méthode de spectroscopie interférentielle. *Ann. Chim. Phys.* **1899**, *16*, 115–144. (In French)
9. Perot, A.; Fabry, C. On the application of interference phenomena to the solution of various problems of spectroscopy and metrology. *Astrophys. J.* **1899**, *9*, 87–115. [CrossRef]
10. Debye, P. Zur Theorie der spezifischen Warme. *Ann. Physik* **1912**, *39*, 789–839. (In German) [CrossRef]
11. Nagaoka, H.; Mishima, T. A combination of a concave grating with a Lummer-Gehrcke plate or an echelon grating for examining fine structure of spectral lines. *Astrophys. J.* **1923**, *57*, 92–97. [CrossRef]
12. Mandelstam, L.I Light scattering by inhomogeneous media. *Zh. Russ. Fiz. Khim. Ova.* **1926**, *58*, 381–391. (In Russia)
13. Hayes, W.; Loudon, R. *Scattering of Light by Crystals*; Dover Publishing, Inc.: Minesota, NY, USA, 1978.
14. Benedek, G.B.; Fritsch, K. Brillouin scattering in cubic crystals. *Phys. Rev.* **1966**, *149*, 647–662. [CrossRef]
15. Vacher, R.; Boyer, L. Brillouin scattering: A tool for the measurement of elastic and photoelastic constants. *Phys. Rev. B* **1972**, *6*, 639–673. [CrossRef]
16. Itoh, S.; Yamana, Y.; Kojima, S. Quick measurement of brillouin spectra of glass-forming material trimethylene glycol by angular dispersion-type Fabry-Perot interferometer system. *Jpn. J. Appl. Phys.* **1996**, *35*, 2879–2881. [CrossRef]
17. Ko, J.-H.; Kojima, S. Nonscanning Brillouin spectroscopy applied to solid materials. *Rev. Sci. Instrum.* **2002**, *73*, 4390–4392. [CrossRef]
18. Ike, Y.; Tsukada, S.; Kojima, S. High-resolution Brillouin spectroscopy with angular dispersion-type Fabry-Perot interferometer and its application to a quartz crystal. *Rev. Sci. Instrum.* **2007**, *78*, 076104. [CrossRef]
19. Itoh, S.; Nakamura, T. Brillouin scattering study of $Gd_2(MoO_4)_3$ using a double Fabry-Perot-Interferometer. *Ferroelectrics* **1974**, *8*, 589–590. [CrossRef]
20. Sawada, A.; Udagawa, M.; Nakamura, T. Proper ferroelastic transition in piezoelectric lithium ammonium tartrate. *Phys. Rev. Lett.* **1977**, *39*, 829–832. [CrossRef]
21. Mock, R.; Hillebrands, B.; Sandercock, R. Construction and performance of a Brillouin scattering set-up using a triple-pass tandem Fabry-Perot interferometer. *J. Phys. E Sci. Instrum.* **1987**, *20*, 656–659. [CrossRef]

22. Kojima, S.; Muhtar, A.; Sivasburmanian, V.; Vokov, A.A.; Ye, J.C. Precursor dynamics of $Pb(B_{1/2}B'_{1/2})O_3$-type relaxor ferroelectrics studied by micro-Brillouin scattering. *J. Adv. Dielec.* **2012**, *2*, 1241004. [CrossRef]
23. Sharma, L.P.; Gowd, T.N. Raman and Brillouin scattering studies on liquids and crystals using an ultra high resolution spectrometer. *Ind. J. Pure Appl. Sci.* **2000**, *38*, 335–343.
24. Shirasaki, M. Large angular dispersion by a virtually imaged phased array and its application to a wavelength demultiplexer. *Opt. Lett.* **1996**, *21*, 366–368. [CrossRef] [PubMed]
25. Scarcelli, G.; Polacheck, W.J.; Nia, H.T.; Patel, K.; Grodzinsky, A.J.; Kamm, R.D.; Yun, S.H. Noncontact three-dimensional mapping of intracellular hydromechanical properties by Brillouin microscopy. *Nat. Meth.* **2015**, *12*, 1132–1136. [CrossRef]
26. Vincent, B.; Kruger, J.K.; Elmazria, O.; Bouvot, L.; Mainka, J.; Sanctuary, R.; Rouxel, D.; Alnot, P. Imaging of microwave-induced acoustic fields in $LiNbO_3$ by high-performance Brillouin microscopy. *J. Phys. D Appl. Phys.* **2005**, *38*, 2026–2030. [CrossRef]
27. Lemons, R.A.; Quate, C.F. Acoustic microscope—Scanning version. *Appl. Phys. Lett.* **1974**, *24*, 163–165. [CrossRef]
28. Kojima, S. Reflection acoustic microscopy to crystalline materials. *Int. J. Eng. Sci.* **1991**, *29*, 309–314. [CrossRef]
29. Foster, J. High resolution acoustic microscopy in superfluid helium. *Phys. B C* **1984**, *126*, 199–205. [CrossRef]
30. Chiao, R.Y.; Townes, C.H.; Stoicheff, B.P. Stimulated Brillouin scattering and coherent generation of intense hypersonic waves. *Phys. Rev. Lett.* **1964**, *12*, 592–594. [CrossRef]
31. Damzen, M.J.; Vlad, V.I.; Babin, V.; Mocofanescu, A. *Stimulated Brillouin Scattering*; Institute of Physics Publishing: London, UK, 2003.
32. Mizuno, Y.; Nakamura, K. Brillouin scattering in polymer optical fibers: Fundamental properties and potential use in sensors. *Polymers* **2011**, *3*, 886. [CrossRef]
33. Gusev, V.E.; Ruello, P. Advances in applications of time-domain Brillouin scattering for nanoscale imaging. *Appl. Phys. Rev.* **2018**, *5*, 031101. [CrossRef]
34. Still, T. *Frequency Acoustics in Colloid-Based Meso-and Nanostructures by Spontaneous Brillouin Light Scattering*; Springer: Berlin/Heiderberg, Germany, 2010.
35. Carlotti, G. Elastic characterization of transparent and opaque films, multilayers and acoustic resonators by surface brillouin scattering: A review. *Appl. Sci.* **2018**, *8*, 124. [CrossRef]
36. Ko, J.-H.; Kim, D.H.; Kojima, S.; Kim, J.-H.; Choo, W.K. Brillouin scattering study on polycrystalline relaxor ferroelectrics. *Jpn. J. Appl. Phys.* **2003**, *42*, 3076–3079. [CrossRef]
37. Kang, D.H.; Oh, S.H.; Ko, J.-H.; Lee, K.-S.; Kojima, S. Elastic properties of taurine single crystals studied by Brillouin spectroscopy. *Int. J. Mol. Sci.* **2021**, *22*, 7116. [CrossRef]
38. Mathieu, V.; Fukui, K.; Matsukawa, M.; Kawabe, M.; Vayron, R.; Soffer, E.; Anagnostou, F.; Haiat, G. Measurements in mature and newly formed bone tissue surrounding an implant. *J. Biomech. Eng.* **2011**, *133*, 021006. [CrossRef]
39. Randall, J.T.; Vaughan, J.M. The measurement and interpretation of Brillouin scattering in the lens of the eye. *Proc. R. Soc. Lond. B* **1982**, *214*, 449–470.
40. Yoshihara, A.; Miyazaki, A.; Maeda, T.; Imai, Y.; Itoh, T. Spectroscopic characterization of dragonfly wings common in Japan. *Vib. Spec.* **2012**, *61*, 85–93. [CrossRef]
41. Oh, K.H.; Ko, Y.-H.; Kojima, S.; Ko, J.-H. High-pressure elasticity of Baltic amber studied by Brillouin spectroscopy. *J. Korean Phys. Soc.* **2020**, *77*, 773–779. [CrossRef]
42. Camley, R.E.; Grünberg, P.; Mayr, C.M. Magnon: Stokes—Anti-stokes asymmetry in Brillouin scattering from magnons in thin ferromagnetic films. *Phys. Rev. B* **1982**, *26*, 2609–2614. [CrossRef]
43. Kargar, F.; Alexander, A.; Balandin, A.A. Advances in Brillouin–Mandelstam light-scattering spectroscopy. *Nat. Photonics* **2021**, *15*, 720–731. [CrossRef]
44. Courtens, E.; Pelous, J.; Phalippou, J.; Vacher, R.; Woignier, T. Brillouin-scattering measurements of phonon-fracton crossover in silica aerogels. *Phys. Rev. Lett.* **1987**, *58*, 128. [CrossRef] [PubMed]
45. Greytak, T.J.; Yan, J. Light scattering from rotons in liquid helium. *Phys. Rev. Lett.* **1969**, *22*, 987–990. [CrossRef]
46. Loudon, R. Theory of surface-ripple Brillouin scattering by solids. *Phys. Rev. Lett.* **1978**, *40*, 581–583. [CrossRef]
47. Kojima, S. Gigahertz acoustic spectroscopy by Micro-Brillouin scattering. *Jpn. J. Appl. Phys.* **2010**, *49*, 07HA01. [CrossRef]
48. Aramomi, S.; Matsuda, Y.; Kojima, S. High temperature Brillouin scattering study on elastic properties of lithium cesium borate glasses, crystals and melts. *J. Non Cryst. Solids* **2014**, *401*, 142–146. [CrossRef]
49. Jiang, F.M.; Kojima, S. Microheterogeneity and relaxation in $0.65Pb(Mg_{1/3}Nb_{2/3})O_3$-$0.35PbTiO_3$ relaxor single crystals. *Appl. Phys. Lett.* **2000**, *77*, 1271–1273. [CrossRef]
50. Shabbir, G.; Ko, J.-H.; Kojima, S. Observation of low-temperature acoustic anomaly in lanthanum lead zirconate titanate. *Appl. Phys. Lett.* **2005**, *86*, 012908. [CrossRef]
51. Kojima, S.; Nakamura, T.; Asaumi, K.; Takashige, M.; Minomura, S. Raman scattering studies on phase transition in barium sodium niobate at high-pressure. *Solid State Commun.* **1979**, *29*, 779–783. [CrossRef]
52. Tkachev, S.N.; Ahart, M.; Novikov, V.N.; Kojima, S. Pressure dependence of Poisson's ratio of glassy Baltic amber studied by Brillouin scattering spectroscopy. *Jpn. J. Appl. Phys.* **2021**, *60*, SDDA04. [CrossRef]
53. Polian, A.; Vo-Thanh, D.; Richet, P. Elastic properties of a-SiO_2 up to 2300 K from Brillouin scattering measurements. *Europhys. Lett.* **2002**, *57*, 375–381. [CrossRef]

54. Shi, W.; Sun, N.; Li, X.; Mao, Z.; Liu, J.; Prakapenka, V.B. Single-crystal elasticity of high-pressure ice up to 98 GP by Brillouin scattering. *Geophys. Res. Lett.* **2021**, *48*, e92514. [CrossRef]
55. Valasek, J. Piezoelectric and allied phenomena in Rochelle salt. *Phys. Rev.* **1920**, *15*, 537–538.
56. Kojima, S. Broadband terahertz spectroscopy of phonon-polariton dispersion in ferroelectrics. *Photonics* **2018**, *5*, 55. [CrossRef]
57. Nanamatsu, S.; Kimura, M.; Doi, K.; Takahashi, M. Ferroelectric properties of $Sr_2Nb_2O_7$ single crystal. *J. Phys. Soc. Jpn.* **1971**, *30*, 300–301. [CrossRef]
58. Kojima, S. Super-high curie temperature ferroelectrics and superior T_C tunability: Lattice instability and ferroelectric phase transitions. *Ferroelectrics* **2020**, *569*, 122–135. [CrossRef]
59. Christy, Y.; Matsumoto, K.; Kojima, S. Elastic anomaly and aging of new type of incommensurate phase transition in ferroelectric barium sodium niobate. *Jpn. J. Appl. Phys.* **2015**, *54*, 214112. [CrossRef]
60. Jiang, F.; Ko, J.-H.; Kojima, S. Central peaks and Brillouin scattering in uniaxial relaxor single crystals of $Sr_{0.61}Ba_{0.39}Nb_2O_6$. *Phys. Rev. B* **2002**, *66*, 184301. [CrossRef]
61. Ota, S.; Matsumoto, K.; Suzuki, K.; Kojima, S. Elastic anomaly and order-disorder nature of multiferroic barium sodium niobate studied by broadband Brillouin scattering. *IOP Conf. Ser. Mater. Sci. Eng.* **2014**, *54*, 012018. [CrossRef]
62. Kojima, S.; Ko, J.-H. Broadband micro-Brillouin scattering spectroscopy of $Pb(B_{1/3}B'_{2/3})O_3$-based relaxor ferroelectrics. *Curr. Appl. Phys.* **2011**, *11*, S22–S32. [CrossRef]
63. Kojima, S.; Aftabuzzaman, M.; Dec, J.; Kleemann, W. Ferroelectric phase transitions of uniaxial $Sr_{1-x}Ba_xNb_2O_6$ and their composition variation. *Jpn. J. Appl. Phys.* **2019**, *58*, SLLA02. [CrossRef]
64. Knauss, L.A.; Wang, X.M.; Toulouse, J. Polarization-strain coupling in the mixed ferroelectric $KTa_{1-x}Nb_xO_3$. *Phys. Rev. B* **1995**, *52*, 13261. [CrossRef] [PubMed]
65. Hushur, A.; Gvasaliya, S.N.; Roessli, B.; Lushnikov, S.G.; Kojima, S. Ferroelectric phase transition of stoichiometric lithium tantalate studied by Raman, Brillouin, and neutron scattering. *Phys. Rev. B* **2007**, *76*, 064104. [CrossRef]
66. Hushur, A.; Shigematsu, H.; Akishige, Y.; Kojima, S. Order-disorder nature of ferroelectric $BaTi_2O_5$. *Appl. Phys. Lett.* **2005**, *86*, 112903. [CrossRef]
67. Cross, L.E. Relaxor ferroelectrics. *Ferroelectrics* **1987**, *76*, 241–267. [CrossRef]
68. Kleemann, W.; Dec, J.; Lehnen, P.; Blinc, R.; Zalar, B.; Pankrath, R. Uniaxial relaxor ferroelectrics: The ferroic random-field Ising model materialized at last. *Europhys. Lett.* **2002**, *57*, 14–19. [CrossRef]
69. Suzuki, K.; Matsumoto, K.; Dec, J.; Łukasiewicz, T.; Kleemann, W.; Kojima, S. Critical slowing down and elastic anomaly of uniaxial ferroelectric $Ca_{0.28}Ba_{0.72}Nb_2O_6$ crystals with tungsten bronze structure. *Phys. Rev. B* **2014**, *90*, 064110. [CrossRef]
70. Matsumoto, K.; Kojima, S. Effect of electric field on uniaxial relaxor ferroelectric strontium barium niobate. *Jpn. J. Appl. Phys.* **2015**, *54*, 10NC04. [CrossRef]
71. Dul'kin, E.; Kojima, S.; Roth, M. Dielectric maximum temperature non-monotonic behavior in uniaxial $Sr_{0.75}Ba_{0.25}Nb_2O_6$ relaxor seen via acoustic emission. *J. Appl. Phys.* **2011**, *110*, 044106. [CrossRef]
72. Tsukada, S.; Kim, T.H.; Kojima, S. Large acoustic thermal hysteresis in relaxor ferroelectric $Pb(Zn_{1/3}Nb_{2/3})O_3$-$PbTiO_3$. *APL Mater.* **2013**, *1*, 032114. [CrossRef]
73. Aftabuzzaman, M.; Helal, M.A.; Paszkowski, R.; Dec, J.; Kleemann, W.; Kojima, S. Electric field and aging effects of uniaxial ferroelectrics $Sr_xBa_{1-x}Nb_2O_6$ probed by Brillouin scattering. *Sci. Rep.* **2017**, *7*, 11615. [CrossRef]
74. Aftabuzzaman, M.; Helal, M.A.; Dec, J.; Kleemann, W.; Kojima, S. Electric field effect on elastic properties of uniaxial relaxor $Sr_xBa_{1-x}Nb_2O_6$ single crystals with strong random fields. *Jpn. J. Appl. Phys.* **2017**, *56*, 10PC06. [CrossRef]
75. Polge, C.; Smith, A.; Parkes, A. Revival of spermatozoa after vitrification and dehydration. *Nature* **1949**, *164*, 666. [CrossRef] [PubMed]
76. Bohmer, R.; Ngai, K.L.; Angell, C.A.; Plazek, D.J. Nonexponential relaxations in strong and fragile glass formers. *J. Chem. Phys.* **1993**, *99*, 4201–4209. [CrossRef]
77. Kojima, S. Low-Frequency Raman Investigation of the Liquid-Glass Transition in Glycerol. *Phys. Rev. B* **1993**, *47*, 2924–2927. [CrossRef]
78. Kojima, S.; Takanashi, K.; Yoshihara, A. Broadband light scattering of glycerol. *J. Mol. Struct.* **1995**, *349*, 227–230. [CrossRef]
79. Park, I.S.; Saruta, K.; Kojima, S. Broadband dielectric spectroscopy of glass transition in supercooled alcohols. *J. Korean Phys. Soc.* **1998**, *32*, S817–S820.
80. Kojima, S.; Kashiwada, T.; Nakahira, S. Photoacoustic measurements of thermal dispersion in glass-transitions of glycerol and propylene glycol. *High Temp. High Press.* **1991**, *23*, 701–704.
81. Novikov, V.N.; Sourouvtsev, N.V.; Wierdersich, J.; Adichtchev, S.; Kojima, S.; Rossler, E. On the origin of quasi-elastic light scattering in glasses. *Europhys. Lett.* **2002**, *57*, 838–844. [CrossRef]
82. Li, G.; Du, W.M.; Chen, X.K.; Cummins, H.Z.; Tao, N.J. Testing mode-coupling predictions for α and β relaxation in $Ca_{0.4}K_{0.6}(NO_3)_{1.4}$ near the liquid-glass transition by light scattering. *Phys. Rev. A* **1992**, *45*, 3867–3879. [CrossRef]
83. Barshilia, H.C.; Li, G.; Shen, G.Q.; Cummins, H.Z. Depolarized light scattering spectroscopy of $Ca_{0.4}K_{0.6}(NO_3)_{1.4}$: A reexamination of the "knee". *Phys. Rev. E* **1999**, *59*, 5625–5628. [CrossRef]
84. Matsui, G.; Itoh, S.; Kojima, S. Rapid Brillouin Scattering Measurement of Fast Relaxation Process in Glass-Forming Intermediate Liquid. *Jpn. J. Appl. Phys.* **1998**, *37*, 2812–2814. [CrossRef]
85. Ike, Y.; Kojima, S. Brillouin scattering study of polymer dynamics. *Mater. Sci. Eng. A* **2006**, *442*, 383–386. [CrossRef]

86. Takagi, Y.; Yano, Y.; Kojima, S. The liquid-glass transition in n-propanol: The pressure dependence of the Brillouin spectra. *J. Phys. Conden. Matter.* **1997**, *9*, 6995–7002. [CrossRef]
87. Ahart, M.; Jiang, F.; Kojima, S. Brillouin scattering study of pressure induced glass transition in ethanol and methanol. *Jpn. J. Appl. Phys.* **1998**, *36*, 2803–2804. [CrossRef]
88. Ko, J.-H.; Lee, K.-S.; Ike, Y.; Kojima, S. Elastic properties of aspirin in its crystalline and glassy phases studied by micro-Brillouin scattering. *Chem. Phys. Lett.* **2008**, *465*, 36–39. [CrossRef]
89. Ko, J.-H.; Kim, T.H.; Lee, K.-S.; Kojima, S. Acoustic properties of aspirin in its various phases and transformation stages studied by Brillouin scattering. *J. Non Cryst. Solids* **2011**, *357*, 541–551. [CrossRef]
90. Ko, J.-H.; Kim, T.H.; Lee, K.-S.; Kojima, S. Brillouin scattering study on crystalline and glassy states of anti-inflammatory racemic S(+)–R(−) ibuprofen. *Chem. Phys. Lett.* **2011**, *515*, 221–225. [CrossRef]
91. Kim, T.H.; Shibata, T.; Kojima, S.; Shin, D.-M.; Hwang, Y.-H.; Ko, J.-H. Comparison of thermal and elastic properties of glassy racemic and enantiomorphic ibuprofen studied by Brillouin light scattering and modulated differential scanning calorimetry. *Curr. Appl. Phys.* **2014**, *14*, 965–969. [CrossRef]
92. Shibata, T.; Takayama, H.; Kim, T.H.; Kojima, S. Acoustic and thermal anomalies in a liquid–glass transition of racemic S(+)–R(−) ketoprofen. *Chem. Phys. Lett.* **2014**, *592*, 80–84. [CrossRef]
93. Shibata, T.; Kojima, S. Liquid, glass, and crystalline indomethacin studied by Brillouin scattering. *Jpn. J. Appl. Phys.* **2018**, *57*, 07LB03. [CrossRef]
94. Shibata, T.; Mori, T.; Kojima, S. Low-frequency vibrational properties of crystalline and glassy indomethacin probed by terahertz time-domain spectroscopy and low-frequency Raman scattering. *Spectrochim. Acta Part A* **2015**, *150*, 207–211. [CrossRef] [PubMed]
95. Krüger, J.K.; Embs, J.; Brierley, J.; Jiménez, R. A new Brillouin scattering technique for the investigation of acoustic and opto-acoustic properties: Application to polymers. *J. Phys. D* **1998**, *31*, 1913–1918. [CrossRef]
96. Svaniza, A.V.; Lushnikov, S.G.; Kojima, S. Anomalous temperature behavior of hypersonic acoustic phonons in a lisozyme crystal. *JETP Lett.* **2006**, *84*, 551–555.
97. Ike, Y.; Hashimoto, E.; Aoki, Y.; Kanazawa, H.; Kojima, S. Micro-Brillouin scattering study of low temperature elastic properties of protein crystals. *J. Mol. Struct.* **2009**, *924–926*, 157–160. [CrossRef]
98. Adachi, H.; Watanabe, T.; Yoshimura, M.; Mori, Y.; Sasaki, T. Promotion of large protein crystal growth with stirring solution. *Jpn. J. Appl. Phys.* **2002**, *41*, L1026–L1027. [CrossRef]
99. Hahsimoto, E.; Aoki, Y.; Seshimo, Y.; Sasanuma, K.; Ike, Y.; Kojima, S. Dehydration process of protein crystals by micro-Brillouin scattering. *Jpn. J. Appl. Phys.* **2008**, *47*, 3839–3842. [CrossRef]
100. Liu, H.-B.; Zhang, X.-C. Dehydration kinetics of D-glucose monohydrate studied using THz time-domain spectroscopy. *Chem. Phys. Lett.* **2006**, *429*, 229–233. [CrossRef]
101. Svanidze, A.V.; Lushnikov, S.G.; Romanov, V.P.; Kojima, S. Multicomponent Rayleigh scattering from guanidine hydrochloride solutions. *J. Raman Spec.* **2012**, *43*, 1510–1514. [CrossRef]

materials

MDPI

Article

Sound Velocity and Equation of State of Ballistic Gelatin by Brillouin Scattering

Muhtar Ahart [1] and Russell J. Hemley [1,2,3,*]

[1] Department of Physics, University of Illinois Chicago, Chicago, IL 60607, USA
[2] Department of Chemistry, University of Illinois Chicago, Chicago, IL 60607, USA
[3] Department of Earth and Environmental Sciences, University of Illinois Chicago, Chicago, IL 60607, USA
* Correspondence: rhemley@uic.edu

Abstract: Brillouin scattering spectroscopy with diamond anvil cells was used by measuring the pressure dependence of the sound-relevant polymer material, glass-forming liquid, and H_2O (water and ice VII) velocities of the material from ambient pressure to 12 GPa at room temperature. Measurements of 20%, 10%, and 4% gelatin solutions were performed. For comparison purposes, we also measured the pressure dependence of the sound velocity of animal tissue up to 10 GPa. We analyzed the Brillouin data using the Tait and Vinet equations of state. We discussed the possible influence of frequency dispersion on bulk modulus at low pressure. We compared the elastic moduli obtained for gelatin to those of several other polymers.

Keywords: Brillouin scattering; equation of state; ballistic gelatin; high-pressure

1. Introduction

External stimuli, such as temperatures and pressure, have a pronounced influence on the mechanical behavior of polymers. An important biopolymer is gelatin, a yellow-white protein-rich powder extracted by acid or base hydrolysis from collagen, the main matrix material of animal skin, bone, and connective tissue. Gelatin also refers to aqueous solutions of this powder. As an aggregate material, the molecular weight ranges between 17,000 and 300,000 Daltons. Thus, only the average molecular weight can be used to characterize the material [1]. Gelatin is one of the most versatile hydrocolloids and has a wide variety of industrial uses, including food, pharmaceuticals, and photography. Another use of this polymer gel is as the model of the effects of traumatic impacts on animal tissue, i.e., bullet penetration of so-called "ballistic gelatin" [1], for which the elastic behavior and equation of state (EOS) to gigapascal pressures are of interest.

Citation: Ahart, M.; Hemley, R.J. Sound Velocity and Equation of State of Ballistic Gelatin by Brillouin Scattering. *Materials* **2023**, *16*, 1279. https://doi.org/10.3390/ma16031279

Academic Editor: Seiji Kojima

Received: 20 November 2022
Revised: 14 January 2023
Accepted: 25 January 2023
Published: 2 February 2023

Despite numerous studies of polymer gels [2,3], proteins [4], and aqueous solutions [5,6] as a function of temperature, limited experiments have been performed to examine pressure effects on these materials. Some efforts have been made to investigate the glass transition and EOS of glass-forming liquids under pressure by Brillouin scattering [7,8]. Similar investigations of the EOS and elastic properties of various polymers under pressure have been conducted during the past decade. For example, the EOS and elastic properties of three elastomers, sylgard (a crosslinked polydimethylsiloxane), estane (a segmented polyester-polyurethane copolymer), and VCE (a crosslinked terpolymer poly-ethylene-vinyl acetate-vinyl alcohol), have been studied by high-pressure Brillouin scattering up to 10 GPa [9]; Kel-F 800, a binder for explosives, has been measured up to 85 GPa by Brillouin scattering, and the EOS and pressure-dependent elastic properties were obtained [10,11].

Here we present the results of a detailed study of gelatin water solution in diamond anvil cells (DACs) in conjunction with Brillouin scattering from ambient pressure to 12 GPa. Motivated the implications for traumatic impacts on animal tissue, we also measured the pressure dependence of the sound velocities of several animal tissues, i.e., lamb muscle and brain tissues. The results extend the EOS and provide additional analyses of experimental data presented previously for the material in an unpublished report [12]. Specifically, we assess the elastic parameters and compare related materials that provide insight into the microscopic structure of these materials.

2. Experimental Methods

2.1. Sample Preparation

The samples of gelatin (20%, 10%, and 4%) were prepared with the methods described in [1]. Images of as-made 4% and 20% gelatin solution are shown in Figure 1. Before loading, gelatin solutions were refrigerated for at least 24 h. Gelatin samples were then loaded into diamond anvil cells (DACs) (Figure 2). The DACs consist of two opposed diamond anvils on the piston and cylinder sides of a cell, with a metal gasket with a hole in its center used as a sample chamber sandwiched between the diamonds [13]. Preindented stainless gaskets with gasket holes of 120–150 μm were used. We used the standard ruby fluorescence method [14] to determine the pressure. We measured the pressure dependencies of velocities of 20%, 10%, and 4% gelatin solutions as well as fresh lamb muscle and brain tissues. These experiments did not involve the use of a separate pressure medium because the materials studied remain soft and serve as a hydrostatic or quasi-hydrostatic medium.

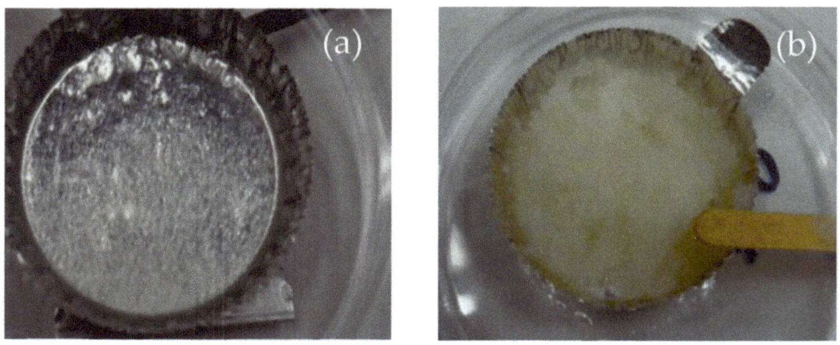

Figure 1. Images of gelatin solutions taken after the gelatin powders were mixed with water; (**a**) 4% solution; (**b**) 20% solution.

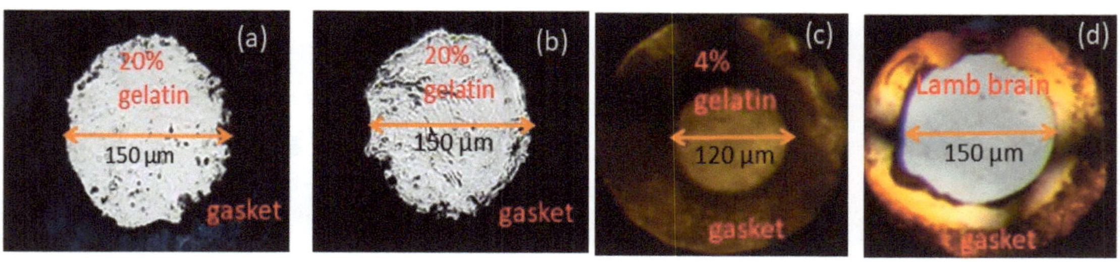

Figure 2. The photographs of gelatin solutions and animal tissue in DACs; (**a**) 20% gelatin solution at 2 GPa, 150 μm gasket hole; (**b**) 20% gelatin solution at 11 GPa, gasket hole is same as (**a**); (**c**) 4% gelatin solution at 1 GPa, 120 μm gasket hole; and (**d**) lamb brain tissue at 2 GPa, 150 μm gasket hole.

2.2. Brillouin Scattering

Brillouin light scattering is the inelastic scattering of an incident optical wave field by thermally excited elastic waves in a material [15,16]. The technique has been widely used in biology [17], chemistry [18], earth science [19], materials science [20], and polymer science [10]. In Brillouin scattering, incident light induces dynamic fluctuations in the strain field to bring about fluctuations in the dielectric constant; these in turn translate into fluctuations in the refractive index due to elasto-optic scattering. The phonons in the material move in thermal equilibrium with very small amplitudes and may be viewed as a moving diffraction grating by an incident light wave. The frequency shift of inelastically scattered light is given by

$$\Delta\nu = \frac{v}{\lambda}\sqrt{(n_i^2 + n_s^2 - 2n_i n_s \cos\theta)} \quad (1)$$

where $\Delta\nu$ is the Brillouin shift, λ is the wavelength of incident light, n_i is the refractive index in the direction of the incident light, n_s is the refractive index in the scattering direction, v is the velocity of acoustic phonons, and θ is the scattering angle.

In the present experiments, light from a single-mode Ar-ion laser was used as the excitation source with an average power of 100 mW. The DAC sample was placed symmetrically with respect to the incoming and collected light such that the difference vector of the incoming and detected light is in the plane of the sample. Scattering angles of 80° and 70° degrees were employed in these experiments (Figure 3). Since samples were placed symmetrically with respect to the incoming and collected light, the frequency shift of the incident light is independent of the refractive index of the samples. Thus, Equation (1) can be reduced to

$$\Delta\nu = (2v/\lambda)\sin(\theta/2) \quad (2)$$

where v is sound velocity, λ = 514.5 nm (laser wavelength), and θ is the scattering angle. The scattered light was collected by a lens and analyzed by a 3 + 3 tandem Fabry–Perot interferometer, detected by a photon-counting photomultiplier, and output to a multichannel scalar (Figure 3a). Spectra were collected for 10 to 120 min at each pressure at 295 K. Additional details about the Brillouin scattering system setup can be found in [21–23].

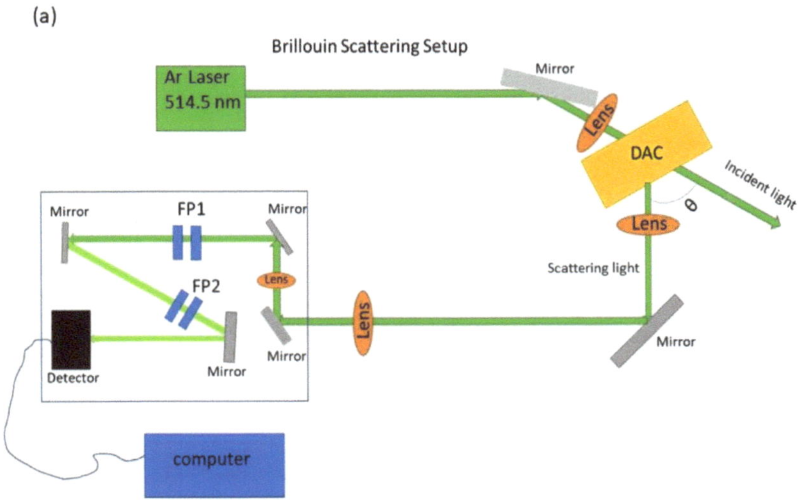

Figure 3. *Cont.*

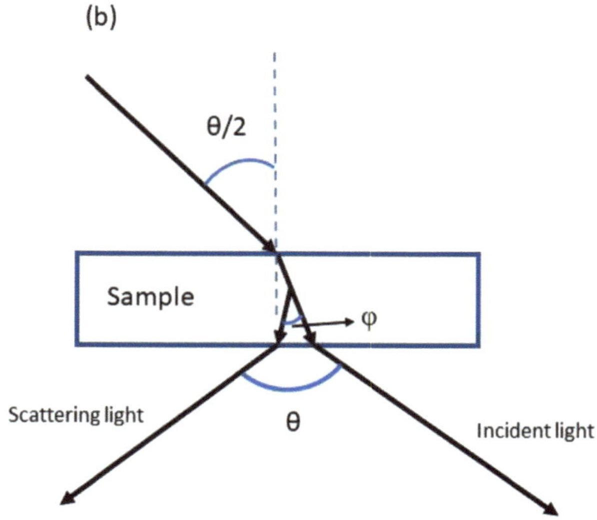

Figure 3. Schematic diagrams of Brillouin scattering setup and the symmetric scattering geometry. (**a**) Incident light is focused on the sample by the first lens, with scattered light collected by second lens, sent to the Fabry–Perot interferometer (FP1 and PP2), and measured by the detector. (**b**) The enlarged sample region inside of a DAC, showing the incident light entering the sample with the angle $\theta/2$. By Snell's law, $n_0\sin(\theta/2) = n\sin(\varphi/2)$, where n_0 is the refractive index of air and n is the refractive index of samples. Therefore, the Brillouin shifts have the following form: $\Delta\nu_\phi = \frac{2nv}{\lambda}\sin(\phi/2) = \frac{2v}{\lambda}\sin(\theta/2) = \Delta\nu_\theta$. The Brillouin shifts are uniquely defined by the scattering angle between the incident and scattering light, as the sound velocity of the sample (assumed to be isotropic) does not depend on the scattering angle.

3. Results and Analysis

3.1. Brillouin Scattering

A typical set of Brillouin spectra from 20% gelatin measured at room temperature is shown in Figure 4. There are two pairs of spectra peaks in each spectrum and a Rayleigh peak at zero frequency. One pair of peaks corresponds to the longitudinal acoustic (LA) mode; the other pair corresponds to the transverse acoustic (TA) mode.

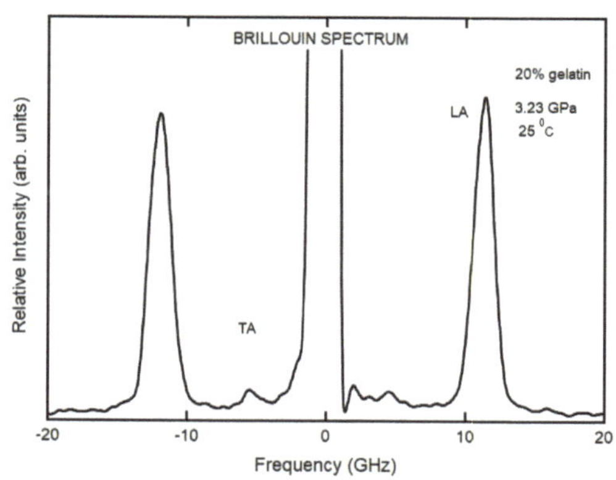

Figure 4. Representative Brillouin spectrum for 20% gelatin solutions. The elastic scattering (Rayleigh) line shown at zero frequency. The Brillouin peaks appear at around 5 and 12.5 GHz for transverse acoustic (TA) and longitudinal acoustic (LA) modes, respectively.

Figure 5 shows the pressure dependence of sound velocities for the 20% gelatin solutions. The results reveal that pressure dependencies of sound velocities from different experimental runs are consistent. On the other hand, the results for 4% and 10% gelatin differ from run to run (Figure 6). We attribute these differences to sample heterogeneity of the samples as evident from Figure 1.

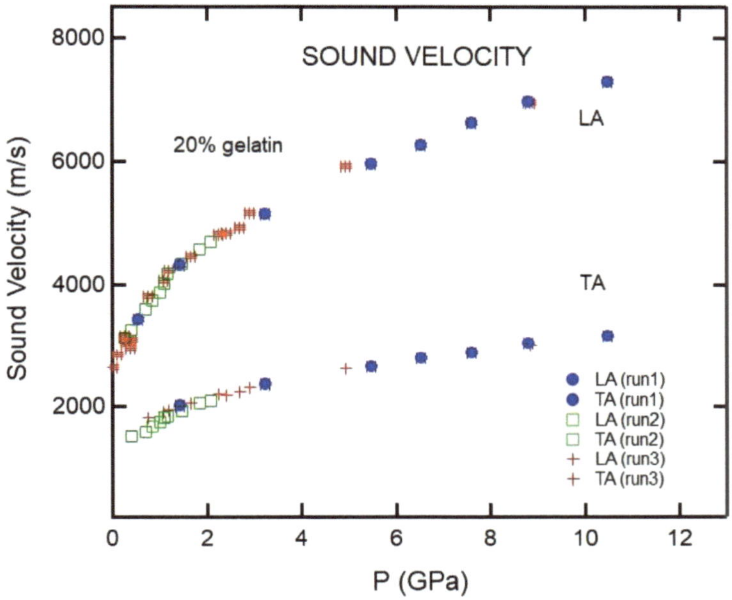

Figure 5. Pressure dependencies of sound velocities of 20% gelatin solutions. The reproducibility was confirmed in preparing samples over different time periods. In each case, the error bars are comparable to the symbol size.

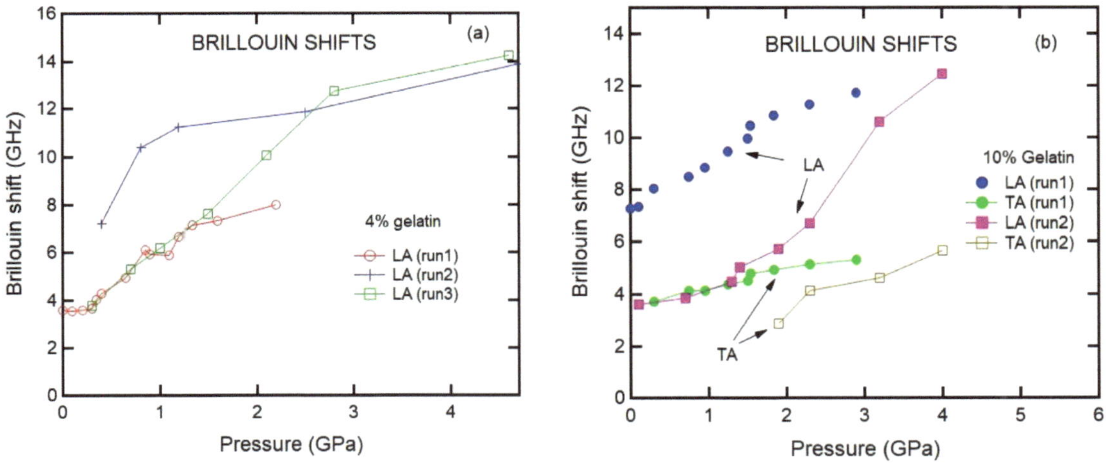

Figure 6. Pressure dependencies of the Brillouin shifts of 4% and 10% gelatin solutions. (**a**) The three experiments were performed for 4% gelatin samples at different time periods; (**b**) two experiments were performed for 10% gelatin solutions.

3.2. Density Calculations

Given that the gelatin samples are isotropic, we can calculate the pressure dependence of the density to obtain a pressure–density (pressure–volume, *P-V*) equation of state (EOS) from

$$\rho - \rho_0 = \int_{P_0}^{P} \frac{\gamma}{v_B^2} dP, \tag{3}$$

where ρ and ρ_0 are the densities at pressure P and P_0, respectively, v_B is the bulk sound velocity, and $v_B^2 = v_L^2 - \frac{4}{3}v_T^2$, and v_L and v_T are LA and TA velocities, respectively; $\gamma = \frac{C_P}{C_V}$ ([2,3,5]) is the ratio of heat capacities at constant pressure and constant volume. To fit Equation (3) to the experimental data, we note the definition of the bulk modulus $K = -V\left(\frac{dP}{dV}\right) = \rho\left(\frac{dP}{d\rho}\right)$, which can be measured under isothermal (K_T) or adiabatic (K_S) conditions. The latter is directly related to the bulk sound velocity by $K_s = \rho v_B^2$, and $K_T = \gamma K_S$. Replacing K_T with $\gamma \rho v_B^2$, the right side of Equation (3) is integrated to obtain the change in density with pressure. Further, from thermodynamics, we have the following relations between the two heat capacities:

$$C_P - C_V = VT\frac{\alpha^2}{\beta_T}, \text{ and } \frac{C_P}{C_V} = \frac{\beta_T}{\beta_S} = \frac{K_S}{K_T}, \tag{4}$$

where $\alpha = \frac{1}{V}\left(\frac{\partial V}{\partial T}\right)_P$ is the thermal expansion coefficient, $\beta_T = -\frac{1}{V}\left(\frac{\partial V}{\partial P}\right)_T = \frac{1}{K_T}$ is the isothermal compressibility, and $\beta_S = -\frac{1}{V}\left(\frac{\partial V}{\partial P}\right)_S = \frac{1}{K_S}$ is the adiabatic compressibility. Thus, the relationship between K_T and K_S is

$$K_T - K_S = \frac{T\alpha^2}{C_P} \tag{5}$$

One can estimate the difference between K_T and K_S if the C_P and α are known or vice versa. For example, for crystalline MgO, K_T = 162 GPa and K_S = 160 GPa; thus, γ = 1.01 [24]. For gelatin, the γ is slightly larger than 1 [25]. Given that $\gamma = \frac{C_P}{C_V} \approx 1$ in materials under pressure, we assume a limiting value of unity in the density calculations (Equation (3)).

The density can be determined by integrating the pressure dependence of the square of inversed bulk sound velocity (Equation (3)). Note that the pressure dependence of squared inverse bulk sound velocity shows a drastic change below 1 GPa for gelatin in contrast to, for crystalline materials, the pressure dependence of the square of inversed bulk sound velocity; e.g., see the almost linear behavior for MgO [20] (Figure 7 inset). To reduce the error in the density calculation, we measured the pressure dependence of the sound velocity in small pressure increments below 2 GPa (Figure 7).

To compare with gelatin, we also measured the pressure dependence of the velocity of lamb muscle and brain tissues. Figure 8 shows the Brillouin spectra from lamb brain tissue at selected pressures. The broad Brillouin peaks at low pressures indicate that the relaxation processes couple strongly with the acoustic modes. The Brillouin shifts from the animal tissue behaved similarly under pressure (Figure 9). Measurements performed on tissue from different animal sources yielded similar results. The TA peaks from the brain tissue sample were found to be extremely weak and were only observed above 4 GPa. On the other hand, the TA peaks were readily observed for the muscle tissue samples (Figure 9).

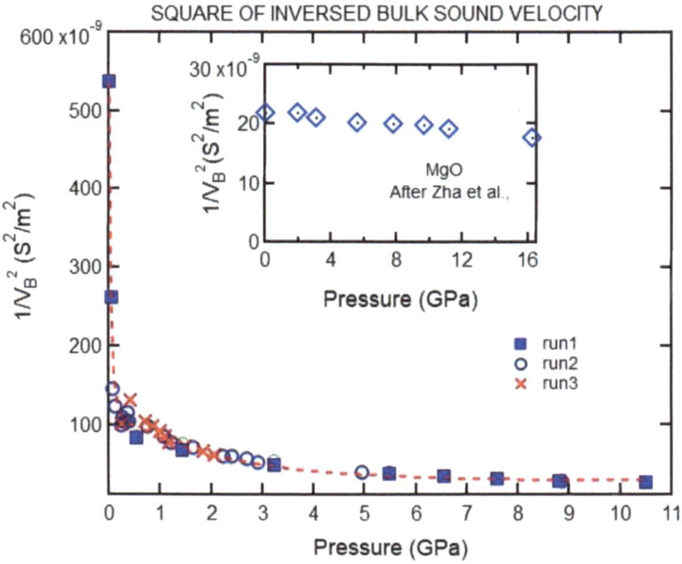

Figure 7. Pressure dependence of the square of inversed bulk velocity; different symbols correspond to the experiments performed at different times. Area under curve gives the density. More data points are obtained at lower pressures to improve the calculation of density. Dashed line is guide for eyes. Inset shows the pressure dependence of the square of inverse bulk sound velocity for crystalline MgO [20].

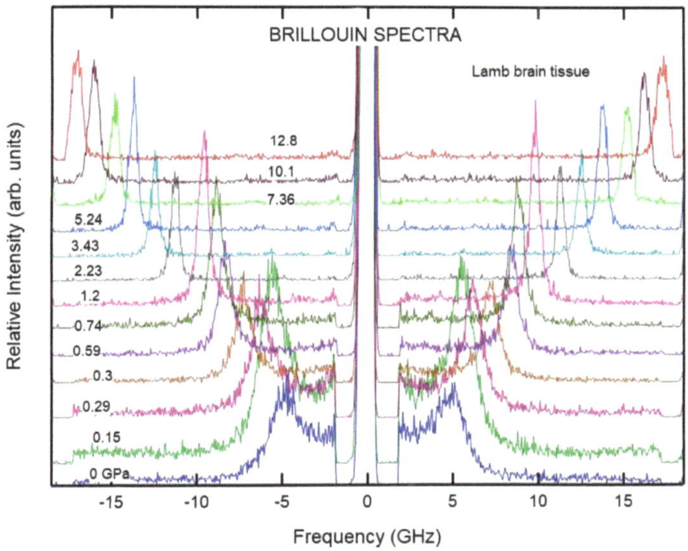

Figure 8. Brillouin spectra of lamb brain tissue at selected pressures. Note that the broad Brillouin peaks at low pressure indicate strong coupling between acoustic modes and concomitant relaxation effects.

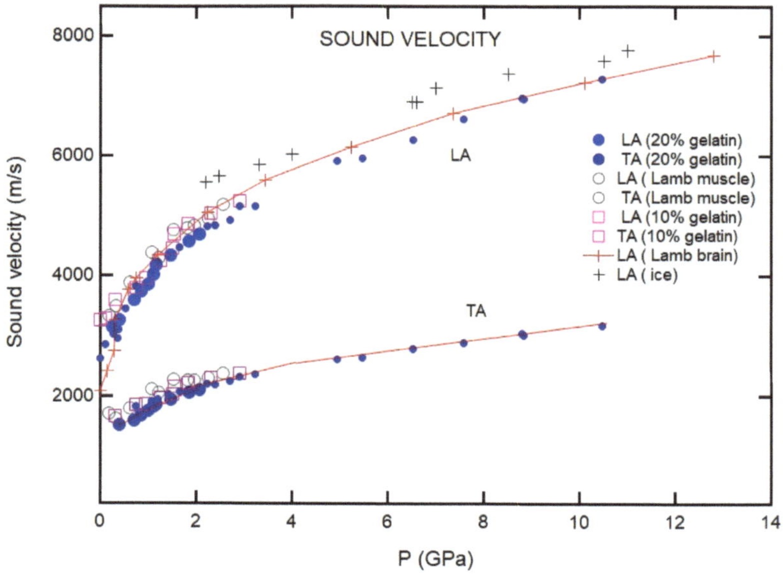

Figure 9. Pressure dependencies of the sound velocities of gelatin solutions, lamb tissues, and H_2O. Solid circles correspond to the longitudinal and transverse acoustic mode of 20% gelatin sample described in legend; open circles correspond the acoustic modes from lamb muscle tissue, whereas open squares correspond 10% gelatin samples; and crosshair with cross marks correspond to the longitudinal LA mode of the lamb brain tissue. Lines are guides for the eye.

3.3. *Equation of State Analysis*

The *P-V* EOS of gelatin is useful for understanding the behavior of this complex material and can provide information on effective intermolecular interactions. To begin, we analyze the *P-V* data with the Tait EOS [26], which was developed for liquids at modest pressures,

$$P = Bexp\left[\frac{1}{C}\left(1 - \frac{V}{V_0}\right)\right] - B \tag{6}$$

where V/V_0 is the relative volume at P, and B and C are empirical parameters. The isothermal bulk modulus at ambient pressure $K_0 = B/C$ [26]. We also use the Vinet EOS [27]

$$P = 3K_0\left[\frac{1-(V/V_0)^{1/3}}{(V/V_0)^{2/3}}\right]exp[\frac{3}{2}(K_0' - 1)\left(1 - \left(V/V_0\right)^{1/3}\right)] \tag{7}$$

where K_0 is the bulk modulus, $K_0{}'$ is its derivative at zero pressure, and V_0 is the initial volume at room temperature.

The Tait EOS fit to the data below 0.5 GPa yields a bulk modulus of 1.7 (±0.2) GPa (inset of Figure 10). However, it cannot be used to describe the compression over the entire pressure range or up to 12 GPa. We use Vinet EOS to fit *P-V* data up to 12 GPa. We also show that the Vinet EOS fits just the lower pressure range (Figure 10) and obtain a very similar bulk modulus. Given that the Tait EOS was developed for modest compressions (e.g., of fluids) and the Vinet EOS was developed for crystalline solids over a broad range of compression, the results are not surprising. We also include the shock Hugoniot for the 20% gelatin solution [27]. As described in [27], these shock compressions corresponded to extremely low frequency velocity.

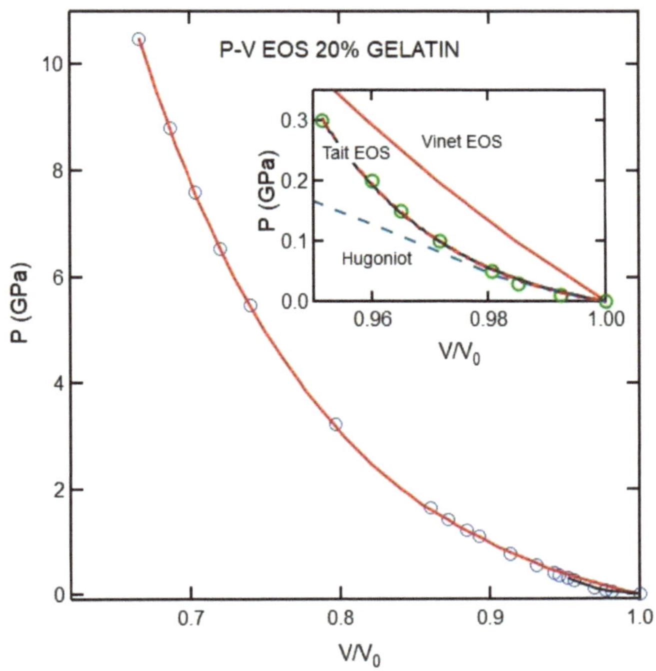

Figure 10. *P-V* EOS of 20% gelatin. Open circles are the data obtained from Equation (3). The black and red curves are fits to the Tait and Vinet EOS, respectively. The Tait EOS was fit to data only below 0.5 GPa; a Vinet EOS for data over this range was also performed (see inset); fitting curves are similar for both EOSs. The Tait EOS and Vinet EOS fit to the lower pressure data give similar bulk moduli of 1.7 (±0.2) and 1.6 (±0.2), respectively. Both EOS forms fail to represent the data over the entire pressure range studied. The dashed line is the Hugoniot reported for 20% gelatin [28].

3.4. Elastic Moduli

Elastic moduli C_{11} and C_{12} (Voigt notation) are directly related to the acoustic velocities through the following:

$$\rho v_L^2 = C_{11} \tag{8}$$

$$\rho v_T^2 = \frac{1}{2}(C_{11} - C_{12}) \tag{9}$$

where ρ is the density, v_L is the LA velocity, and v_T is the TA velocity. The pressure dependence of C_{11} and C_{12} for 20% gelatin is shown in Figure 11. We fit the data with the second-order polynomial function to obtain the pressure derivative of the moduli ($\frac{\partial C_{11}}{\partial P} = 9.3$ and $\frac{\partial C_{12}}{\partial P} = 5.8$ for 20% gelatin). C_{12} is approximately half of C_{11} for gelatin under pressure. C_{11} increases much faster with pressure than C_{44} or C_{12}. Substituted into Equation (9), the ambient pressure transverse TA velocities would be expected to be 1050 m/s for gelatin, corresponding to Brillouin frequency shifts of approximately 2.4 GHz in our measurements. The frequency shifts place the TA peak on the wings of the Rayleigh line, which is, in principle, experimentally resolvable. However, C_{11} and C_{12} approach one another near ambient pressure (Figure 11), such that in the limit of ambient pressure, the TA velocity goes to zero, collapsing into the Rayleigh line. Furthermore, we compared Young, bulk, and shear moduli obtained for 20% gelatin solution to that of VCE polymer in Figure 12.

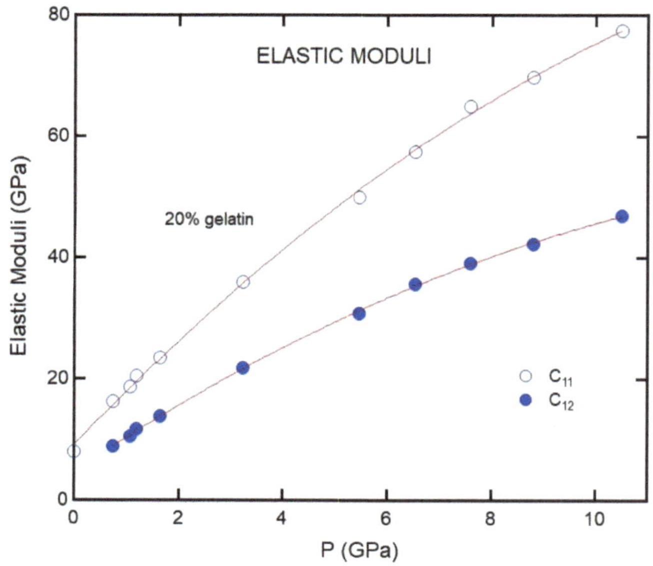

Figure 11. Pressure dependence of the C_{11} and C_{12} elastic constants for 20% gelatin. The curves are second-order polynomial fit to the data.

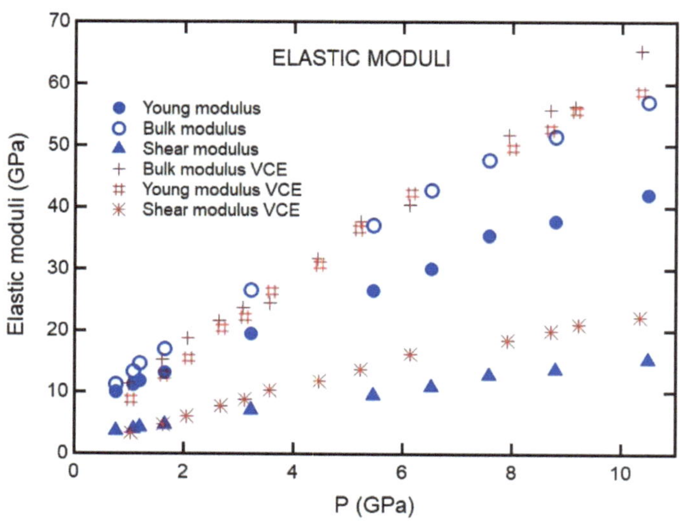

Figure 12. Pressure dependence of bulk, shear, and Young's moduli.

3.5. Poisson's Ratio

Isotropic materials have two independent elastic constants: C_{11} and C_{12}. The pressure dependencies of elastic constants of 20% gelatin are shown in Figure 11. Unfortunately, our lack of knowledge of the initial density of 4% and 10% gelatin and lamb tissue complicates calculating the pressure dependence of density and high-pressure elastic constants. Instead, we can calculate Poisson's ratio:

$$\sigma = \frac{v_L^2 - 2v_T^2}{2(v_L^2 - v_T^2)} \tag{10}$$

The pressure dependencies of Poisson's ratio σ for 10% and 20% gelatin solutions and lamb muscle tissue are shown in Figure 13. The σ value for 20% gelatin is 0.36 at 10 GPa, which is comparable with that of silicate glass at the same pressure [16]. The value for a stable, isotropic, and elastic material cannot be less than -1 nor greater than 0.5 because Young's modulus, the shear modulus, and the bulk modulus must be positive; most materials have a value between 0 and 0.5. Liquid or rubber deformed elastically under small strain has σ near 0.5, whereas for metals, the value is around 0.35. For gelatin solutions, particularly at lower pressures, the TA velocities are zero; thus, Poisson's ratio is 0.5 for that pressure regime. The TA velocities in the Brillouin measurement appear above 0.5 GPa, and σ is 0.36, comparable with that of metals and other materials.

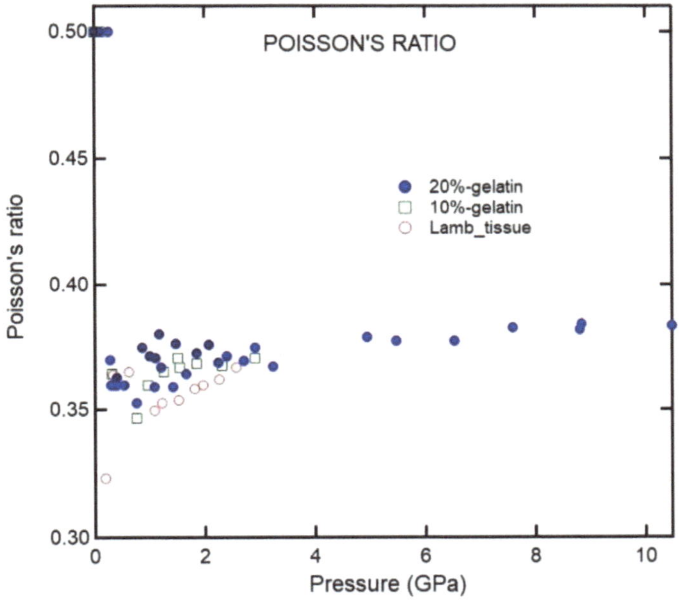

Figure 13. Pressure dependencies of Poisson's ratio at room temperature. The error in pressure is about 0.2 GPa; given the weakness of TA peaks, Poisson's ratio has larger relative errors at lower pressures.

3.6. Cauchy-Like Relations

The Lame coefficient, λ, is given by the following equation:

$$\lambda = K - \frac{2}{3}G \tag{11}$$

where K is the bulk modulus and G is the shear modulus. For Cauchy's identity [29], valid for an isotropic substance composed of particles interacting with two-body central forces, the λ is identical to G, or $K = \frac{5}{3}G$. It is also equal to $M = 3G$, where M is the longitudinal modulus [29]. The M/G ratio is an indicator of the character of the force field: if atoms interact through a central potential, $M/G = 3$. However, most glasses exhibit behavior suggesting principal effective interactions that are far from those of a central potential or ideal Cauchy relation.

Zwanzig and Mountain [30] generalized the Cauchy-like relation to $M_\infty = 3G_\infty + F(T,P)$ for isotropic liquids, where M_∞ is an unrelaxed longitudinal modulus and G_∞ is an unrelaxed shear modulus, and the term $F(T,P)$ depends on both temperature and pressure. Expanding this argument, Yamura et al. [31] and Krüger et al. [32] found the Cauchy-like relation

$$M_\infty = A + BG_\infty \tag{12}$$

to hold across the glass transition for many different liquids at ambient pressure, where A is a system-dependent parameter and B is a parameter found close to 3. Departure from B of 3 indicates deviation from a central potential and possible effects of anharmonicity.

Figure 14 shows the M as a function of G for various materials. Fitting the data with Equation (12) for 20% gelatin gives $A = -0.2$ (± 0.1) and $B = 5.1$ (± 1.3). The B coefficient is larger than the ideal case: $B = 3$, the original definition of Cauchy relation. Many glass-forming liquids follow a Cauchy-like relation with $B = 3$ at ambient pressure. For example, Scarponi et al. [26] obtained $A = 1.4$ and $B = 2.8$ in a fit of $M_\infty = A + BG_\infty$ to data for glycerol across the glass transition temperature at ambient pressure. The closeness of the B parameter to 3 suggests that the relevant interaction for glycerol molecules in this regime can be described by a two-body central potential. However, we have found B values equal or greater than 4 for many polymers under pressure: VCE polymer has $B = 4.2$ for VCE [6], $B = 4.9$ for Kel-F 800 [7,8], and $B = 4.4$ for amber glass [27]. Thus, $B = 5.1$ for 20% gelatin shows the largest deviation from Cauchy-like behavior, suggesting a strong departure from two-body central interactions within this model.

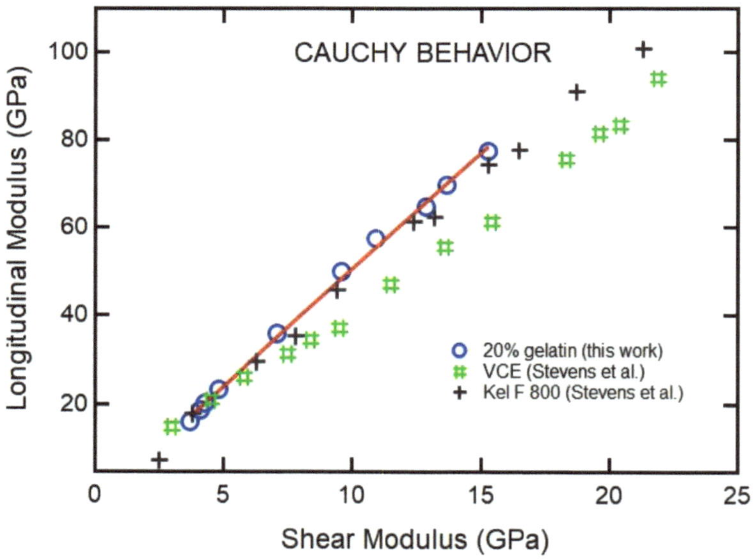

Figure 14. Longitudinal modulus (M) as the function of shear modulus (G). Open circles represent the data obtained from Brillouin scattering experiments. Linear fit shows a Cauchy-like relationship with the coefficients of $A = -0.19 \pm 0.1$ (GPa) and $B = 5.12 \pm 1.3$. For comparison, we also listed the data for VCE polymer [9] with $A = -\pm 1.1$ and $B = 4.2$ and Kel-F 800 polymer [11] with $A = -1.9 \pm 1.1$ and $B = 4.9 \pm 0.1$.

4. Discussion

We first discuss the EOS analysis. As shown in Figure 10 and Table 1, the empirical EOSs we used (Vinet and Tait) do not fit the whole P-V range well. However, the Vinet EOS fit data between 0.7 and 12 GPa well. On the other hand, the Tait and Vinet EOSs fit well to the data below 0.5 GPa and provide similar bulk moduli for 20% gelatin (Table 1). The inability of a single EOS to fit data over a wide range of compression is also evident for phases of materials that exhibit large pressure stability fields (i.e., ice VII in Table 1).

As regards gelatin, we note that the change in the square of inverse bulk velocity with pressure is appreciable (Figure 7). Such a change can be attributed to the collapse of the free volume in polymers on compression. This is consistent with the finding that a single empirical EOS cannot describe the *P-V* relation over a wide pressure range (Figure 10). Our results allow us to compare the pressure-dependent elastic moduli of 20% gelatin solution to those of other selected polymers (Figures 12–14) [6]. The bulk modulus for 20% gelatin is comparable to that of VCE and sylgard polymers [9]. However, the pressure derivatives of these materials vary considerably (Table 1). In addition, the lamb muscle and brain tissues suggest a lack of shear strength at zero frequency and suggest a strong frequency dependence of velocities and moduli.

Table 1. Gelatin and comparison with several polymers and H_2O. The reference ambient pressure density of 20% gelatin solution used in the analysis is 1.007 gm/cm^3.

Material	K_0 (GPa)	K_0'	References
20% gelatin	1.7 (±0.2)		This work (Tait < 0.5 GPa)
20% gelatin	1.6 (±0.2)	53 (±3)	This work (Vinet < 0.5 GPa)
20% gelatin	6.1 (±0.2)	7.5 (±0.2)	This work (Vinet)
VCE	2.05	9.99	[9]
estane	2.84	17.1	[9]
sylgard	1.13	8.95	[9]
Kel-F 800	7.5	10	[10]
n-pentane/isopentane	0.45	10.5	[7]
methanol	0.96		[33]
water	2.2		[34]
ice VII	5	8.1	[35]
	23.7 (±0.9)	4.15 (±0.07)	[36]

The absence of observed shear modes for gelatin at ambient pressure arises from the liquid as opposed to the glassy character of the material. Atomic mobility and even the local structure change across the glass transition, whether determined as a function of temperature (T_g) or pressure (P_g). For example, in PMMA, the pressure dependence of T_g initially shifts by 200°/GPa, asymptotically approaching a limiting high-pressure value around 0.6 GPa [28]. The ambient pressure T_g of −73 °C of the gelatin [29] indicates that it would be difficult to observe the shear mode at room temperature and ambient pressure. The shear mode appears above 1 GPa for 20% gelatin at room temperature. Similar behavior is observed for sylgard, with a T_g of −120 °C and the shear mode observed at 3 GPa by Brillouin scattering. On the other hand, a shear mode is observed for Kel-F 800 polymer at ambient pressure, and its T_g is near room temperature. A lower T_g equates to a more flexible polymer backbone and more "liquid-like" behavior.

We point out that, in optical Brillouin scattering, the measured sound wave is in the GHz range. Because of the dependence of the elastomer mechanical properties on frequency, the Brillouin data appear stiffer than the equilibrium measurements regardless of pressure. Indeed, in such complex polymeric systems, the elastic moduli are expected to have a frequency dependence, especially over the pressures measured. Although assessment of this dispersion of the moduli is beyond the scope of this paper, we can make qualitative comments. In glass-forming materials such as gelatin solutions according to the viscoelastic theory [18], the low-frequency-limit velocity (v_0) approaches the ultrasonic frequencies (Hz–MHz) at high temperature or low pressure with respect to the glass transition temperature T_g or P_g; when the temperature or pressure approaches the glass transition, the v_0 transitions to a high-frequency-limit (GHz) velocity (v_∞). Such a change in velocity has been observed in glass-forming liquids [18] as a function of temperature [37] or pressure [7,8,21].

LA Brillouin peaks measured at high temperatures [37] and pressures [7,8,21] can be broad, indicating coupling of modes and strong relaxation effects. In high-pressure experiments on fluids, the TA mode is observed when material approaches to the glass transition pressure [7,8]. In the 20% gelatin solution, we observed the TA mode above 0.5 GPa, indicating that the glass transition pressure is near. Therefore, the observed velocity at low pressure is slightly higher than v_0. This observation is consistent with the shock experiment and our data (Figure 10). Additional studies are needed to address these questions.

Author Contributions: M.A. and R.J.H. equally contributed to conceptualization, methodology, experiments, analysis, and writing. All authors have read and agreed to the published version of the manuscript.

Funding: US Army Research Laboratory and Army Research Office (W911NF-06-1-0170). Infrastructure and instrumentation were supported by CDAC (DE-NA-0003975).

Institutional Review Board Statement: Not applicable.

Informed Consent Statement: Not applicable.

Data Availability Statement: All original data will be made available on request.

Acknowledgments: We are grateful to Stephen Gramsch for useful discussions.

Conflicts of Interest: The authors declare no conflict of interest.

References

1. Custom Collagen. Available online: http://www.vyse.com/gelatin_for_ballitic_testing.htm (accessed on June 2001).
2. Zhao, P.; Vanderwal, J.J. Brillouin scattering study of gelatin gel. *Polym. Gels Netw.* **1997**, *5*, 23. [CrossRef]
3. Marqusee, J.A.; Deutch, J.M. Brillouin light scattering from polymer gels. *J. Chem. Phys.* **1981**, *75*, 5239. [CrossRef]
4. Chalikian, T.V. Volumetric properties of proteins. *Annu. Rev. Biophys. Biomol. Struct.* **2003**, *32*, 207. [CrossRef] [PubMed]
5. Endo, H. The adiabatic compressibility of nonelectrolyte aqueous solutions in relation to the structure of water and solutions. *Bul. Chem. Soc. Jpn.* **1973**, *46*, 1106. [CrossRef]
6. Naseem, B.; Arif, I.; Jamal, M.A. Kosmotropic and chaotropic behavior of hydrated ions in aqueous solution in term of expansibility and compressibility parameters. *Arab. J. Chem.* **2021**, *14*, 103405. [CrossRef]
7. Oliver, W.F.; Herbst, C.A.; Lindsay, S.M.; Wolf, G.H. High-pressure viscoelastic properties and equation of state of liquids derived from Brillouin data. *Phys. Rev. Lett.* **1991**, *67*, 2795. [CrossRef]
8. Takagi, Y.; Ahart, M.; Yano, T.; Kojima, S. The liquid-glass transition in n-propanol: The pressure dependence of the Brillouin spectra. *J. Phys. Condens. Matter* **1997**, *9*, 6995. [CrossRef]
9. Stevens, L.L.; Orler, E.B.; Dattlbaum, D.M.; Ahart, M.; Hemley, R.J. Brillouin-scattering determination of the acoustic properties and their pressure dependence for three polymeric elastomers. *J. Chem. Phys.* **2007**, *127*, 104906. [CrossRef]
10. Benjamin, A.S.; Ahart, M.; Gramsch, S.A.; Stevens, L.L.; Orler, E.B.; Dattelbaum, D.M.; Hemley, R.J. Acoustic properties of Kel- F 800 copolymer up to 85 GPa. *J. Chem. Phys.* **2012**, *137*, 014514. [CrossRef]
11. Stevens, L.L.; Dattelbaum, D.M.; Ahart, M.; Hemley, R.J. High-pressure elastic properties of a florinated copolymer: Poly(chlorotrifluoroethylene-co-vinylidene fluoride) (Kel-F 800). *J. Appl. Phys.* **2011**, *112*, 023523. [CrossRef]
12. Aihaiti, M.; Hemley, R.J. 2011. Available online: https://apps.dtic.mil/sti/citations/ADA546054 (accessed on June 2011).
13. Bassett, W.A. Diamond anvil cell, 50th birthday. *High Press. Res.* **2009**, *29*, 163. [CrossRef]
14. Dewaele, A.; Torrent, M.; Loubeyre, P.; Mezouar, M. Compression curve of transition metals in the Mbar range: Experiments and projector augmented-wave calculations. *Phys. Rev. B* **2008**, *78*, 104102. [CrossRef]
15. Brillouin, L. Diffusion de la lumiere des rayonnes X per un corp transparent homogene; influence del'agitation thermique. *Ann. Phys.* **1922**, *17*, 88. [CrossRef]
16. Gross, E. Change of wavelength of light due to elastic waves at scattering in liquids. *Nature* **1930**, *126*, 400. [CrossRef]
17. Kabakova, I.; Scarcelli, G.; Yun, S.H. Brillouin light scattering in biological systems. *Semicond. Semimet.* **2022**, *110*, 313.
18. Montrose, C.J.; Solovyev, V.A.; Litovitz, T.A. Brillouin scattering and relaxation in liquids. *J. Acoust. Soc. Am.* **1968**, *43*, 117. [CrossRef]
19. Polian, A.; Grimsditch, M. Brillouin scattering from H_2O: Liquid, ice VI, and ice VII. *Phys. Rev. B* **1983**, *27*, 6409. [CrossRef]
20. Zha, C.S.; Mao, H.K.; Hemley, R.J. Elasticity of MgO and primary pressure scale to 55 GPa. *Proc. Nat. Acad. Sci. USA* **2000**, *97*, 13494. [CrossRef]
21. Loubeyre, L.; Ahart, M.; Gramsch, S.; Hemley, R.J. Density dependence of dynamic heterogeneity in fluid methanol. *J. Chem. Phys.* **2013**, *138*, 174507. [CrossRef]

22. Zha, C.S.; Hemley, R.J.; Mao, H.K.; Duffy, T.S.; Meade, C. Acoustic velocities and refractive index of SiO_2 glass to 57.5 GPa by Brillouin scattering. *Phys. Rev. B* **1994**, *50*, 13105. [CrossRef]
23. Ahart, M.; Yarger, J.L.; Lantzky, K.M.; Nakano, S.; Mao, H.K.; Hemley, R.J. High-pressure Brillouin scattering of amorphous BeH_2. *J. Chem. Phys.* **2006**, *124*, 014502. [CrossRef] [PubMed]
24. Angel, R.J. Equations of state. *Rev. Mineral. Geochem.* **2000**, *41*, 35. [CrossRef]
25. Huang, Y. 2015. Available online: https://apps.dtic.mil/sti/pdfs/ADA621865.pdf (accessed on September 2015).
26. Neece, G.A.; Squire, D.R. On the Tait and related empirical equation of state. *J. Phys. Chem.* **1968**, *72*, 128. [CrossRef]
27. Vinet, P.; Smith, J.R.; Ferrante, J.; Rose, J.H. Temperature effects on the universal equation of state of solids. *Phys. Rev. B* **1987**, *35*, 1945. [CrossRef]
28. Shepherd, C.J.; Appleby-Thomas, G.J.; Hazell, P.J.; Allsop, D.F. The dynamic behavior of ballistic gelatin. *AIP Conf. Proc.* **2010**, *1195*, 1399.
29. Born, M.; Huang, K *Dynamical Theory of Crystal Lattices*; Clarendon Press: Oxford, UK, 1956.
30. Zwangzig, R.; Mountain, R.D. High-frequency elastic moduli of simple fluids. *J. Chem. Phys.* **1965**, *43*, 4464. [CrossRef]
31. Yamura, H.; Matsukawa, M.; Otani, T.; Ohtori, N. Brillouin scattering study on the elastic properties of epoxy adhesive layer. *Jpn. J. Appl. Phys.* **1999**, *38*, 3175. [CrossRef]
32. Kruger, J.K.; Baller, J.; Britz, T.; Le Coutre, A.; Peter, R.; Bactavatchalou, R.; Schreiber, J. Cauchy-like relation between elastic constants in amorphous materials. *Phys. Rev. B* **2002**, *66*, 012206. [CrossRef]
33. Brown, J.M.; Slutsky, L.J.; Nelson, K.A.; Cheng, L.T. Velocity of sound and equation of state for methanol and ethanol in a diamond anvil cell. *Science* **1988**, *241*, 65. [CrossRef]
34. Kell, G.S.; Whalley, E. The PVT properties of water, I. Liquid water in the temperature range 0 to 150 °C and pressure up to 1 kbar. *Philos. Trans.* **1965**, *258*, 565.
35. Ahart, M.; Somayazulu, M.; Gramsch, S.A.; Boehler, R.; Mao, H.K.; Hemley, R.J. Brillouin scattering of H_2O ice to megabar pressures. *J. Chem. Phys.* **2011**, *134*, 124518. [CrossRef] [PubMed]
36. Hemley, R.J.; Jephcoat, A.P.; Mao, H.K.; Zha, C.S.; Finger, L.F.; Cox, D.E. Static compression of H_2O-ice to 128 GPa (1.28 Mbar). *Nature* **1987**, *330*, 737. [CrossRef]
37. Pinnow, A.D.; Candau, S.J.; LaMacchia, J.T.; Litovitz, T.A. Brillouin scattering: Viscoelastic measurements in liquids. *J. Acoust. Soc. Am.* **1968**, *43*, 131. [CrossRef]

materials

Review

Brillouin Light Scattering from Magnetic Excitations

Akira Yoshihara

Faculty of Science and Engineering, Ishinomaki Senshu University, Ishinomaki 986-8580, Japan; yosihara@isenshu-u.ac.jp

Abstract: Brillouin light scattering (BLS) has been established as a standard technique to study thermally excited sound waves with frequencies up to ~100 GHz in transparent materials. In BLS experiments, one usually uses a Fabry–Pérot interferometer (FPI) as a spectrometer. The drastic improvement of the FPI contrast factor over 10^{10} by the development of the multipass type and the tandem multipass type FPIs opened a gateway to investigate low energy excitations ($\hbar\omega \leq 1$ meV) in various research fields of condensed matter physics, including surface acoustic waves and spin waves from opaque surfaces. Over the last four decades, the BLS technique has been successfully applied to study collective spin waves (SWs) in various types of magnetic structures including thin films, ultrathin films, multilayers, superlattices, and artificially arranged dots and wires using high-contrast FPIs. Now, the BLS technique has been fully established as a unique and powerful technique not only for determination of the basic magnetic constants, including the gyromagnetic ratio, the magnetic anisotropy constants, the magnetization, the SW stiffness constant, and other features of various magnetic materials and structures, but also for investigations into coupling phenomena and surface and interface phenomena in artificial magnetic structures. BLS investigations on the Fe/Cr multilayers, which exhibit ferromagnetic-antiferromagnetic arrangements of the adjacent Fe layer's magnetizations depending on the Cr layer's thickness, played an important role to open the new field known as "spintronics" through the discovery of the giant magnetoresistance (GMR) effect. In this review, I briefly surveyed the historical development of SW studies using the BLS technique and theoretical background, and I concentrated our BLS SW studies performed at Tohoku University and Ishinomaki Senshu University over the last thirty five years. In addition to the ferromagnetic SW studies, the BLS technique can be also applied to investigations of high-frequency magnetization dynamics in superparamagnetic (SPM) nanogranular films in the frequency domain above 10 GHz. One can excite dipole-coupled SPM excitations under external magnetic fields and observe them via the BLS technique. The external field strength determines the SPM excitations' frequencies. By performing a numerical analysis of the BLS spectrum as a function of the external magnetic field and temperature, one can investigate the high-frequency magnetization dynamics in the SPM state and determine the magnetization relaxation parameters.

Keywords: Brillouin light scattering; spin wave; magnetic thin film; ultrathin film; magnetic multilayer; magnetic superlattice; magnetic anisotropy; interlayer exchange coupling; nanogranular film; superparamagnetic excitation

Citation: Yoshihara, A. Brillouin Light Scattering from Magnetic Excitations. *Materials* **2023**, *16*, 1038. https://doi.org/10.3390/ma16031038

Academic Editor: Seiji Kojima

Received: 4 December 2022
Revised: 30 December 2022
Accepted: 19 January 2023
Published: 24 January 2023

1. Introduction

Since the early 1960s, Brillouin light scattering (BLS) has been widely applied to study acoustic properties near ferroelectric and ferroelastic phase transitions [1]. Usually, a Fabry–Pérot interferometer (FPI) has been used as a spectrometer for BLS. For a traditional single-pass FPI, the contrast factor C_1, which is defined by the ratio between the maximum transmission and the minimum transmission, was limited to about 10^3 at most. A traditional FPI with a higher contrast factor is a dark FPI with lower transmission efficiency. For successful BLS studies, high-quality transparent samples, which have no inclusions, polished surfaces, and dimensions larger than several mm, are strongly required to reduce the elastic scattering (Rayleigh scattering: RS) intensity. Although the index-matching technique is also helpful to reduce the RS intensity, it is rather difficult to find a suitable index-matching liquid. For these reasons, BLS in 1960–1970's was a much more difficult experiment than Raman scattering. One of the great improvements in Fabry–Pérot interferometry was achieved by the piezoelectrically scanned FPI, which can be electrically stabilized and enable setting of an appropriate spectrum accumulation time, depending on BLS signal intensity [2]. With a piezo-scanning FPI, it is rather easy to assemble a computer-controlled BLS system, including the FPI stabilization, spectrum accumulation on the computer memories, and spectrum analysis [3]. Other great improvements were the development of the multipass-type FPI and the vernier tandem multipass-type FPI by Sandercock and coworkers2) in the early 1980s [4,5]. The contrast factor of a multipass FPI is given by $C_p = (C_1)^p$ (p is the total number of passes through an FPI or FPIs). The drastic improvement of the contrast factor of the multipass FPI without losing transmission efficiency has made BLS studies much easier than the BLS studies reported in 1960 through the 1970s. The vernier tandem multipass FPI can eliminate the overlapping effect with the adjacent interference-order spectrum of a traditional FPI (a multipass FPI still retains the overlapping effect), which has enabled the monitoring of overlap-free BLS spectra up to a few hundred GHz from collective excitations in various research fields of condensed matter physics [6].

2. Historical Survey of Brillouin Light Scattering from Spin Waves

2.1. Experiment

Reviews on the early stage of BLS from SWs have been already given by Borovik-Romanov and Kreines [7], Patton [8], Sandercock [9], and Grünberg [10] by the middle 1980's. Hillebrands gives a list of publications on SW BLS up to 1999 [11].

Since the mid-1970s, the BLS technique has been intensively applied to study spin waves (SWs) from opaque surfaces. The first observation of BLS from SWs was reported by Grüenberg and Mitawe from ferromagnetic semiconductor EuO (T_C = 69 K) [12]. Sandercock and Wettling reported SW BLSs from Fe and Ni at room temperature [13]. Many SW BLS results have been subsequently reported. Readers can find them in the list of references [11]. Thanks to the developments of high-quality thin film preparation techniques, such as the sputtering technique, the MBE technique, and so on, the BLS technique has been successfully applied to study collective SWs in various types of magnetic structures (thickness of L), including thin films, ultrathin films, multilayers, superlattices, and artificially arranged dots and wires [14–21]. For BLS from metallic surfaces, it is important to recognize that there is an essential difference between BLS phenomena from transparent materials and from opaque surfaces. Visible laser light penetrates at most a few hundreds of angstroms from the illuminated surface due to the skin effect (in other words, the absorption effect) [22]. The skin effect strongly violates the momentum conservation law for light scattering. For transparent materials, the momentum conservation law is fully conserved during scattering process. For description of the optical property of metals (for convenience's sake, this example is of an isotropic metal), one should introduce a complex refractive index (n, κ). Here, n is the real part and κ is the imaginary part of the refractive index. The imaginary part κ is usually larger than the real part n for visible light in metals. Then, one should take into account the large uncertainty of $\Delta q_\perp/q_\perp$~2 κ/n in the momentum conservation law for the surface normal (perpendicular) component $q_\perp$ of the light momentum. The in-plane (surface parallel) component $Q_{//}$ of the wave vector is defined as:

$$Q_{//} = \frac{2\pi}{\lambda}(\sin\vartheta_{in} + \sin\vartheta_s) \quad (1)$$

Here, λ is the vacuum wavelength of the laser light, and ϑ_{in} and ϑ_s are the incident and scattering angles measured from the surface normal. Usually, the standard backscattering geometry is employed, in which one sets $\vartheta_{in} = \vartheta_s = \vartheta$ as shown in Figure 1.

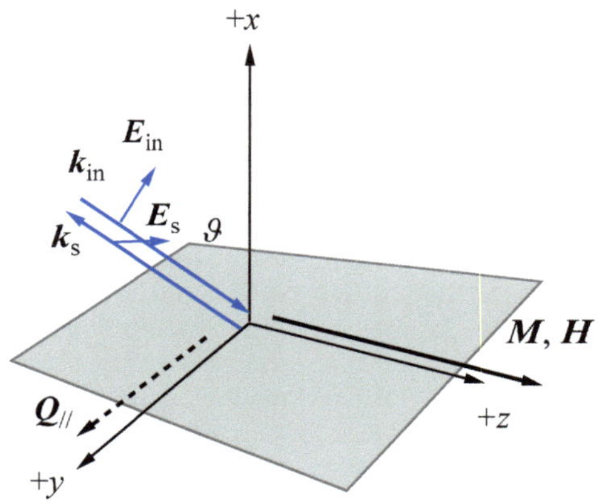

Figure 1. Schematic illustration of the present scattering geometry and the coordinate system for BLS spectrum calculation. The *s*-polarized component of the scattered beam is selected by using a polarizing beam splitter placed in front of the tandem FPI.

In contrast to the momentum conservation law for the perpendicular component $q_\perp$, the momentum conservation law for $Q_{//}$ is always conserved, just the same as transparent materials. It is convenient for later discussions to define the surface dispersion parameter, $Q_{//}L$.

The magnetic structures of thickness L are deposited on appropriate substrate. Within the skin depth of a magnetic structure, both of the SWs and surface acoustic waves (Rayleigh and the Sezawa waves) coexist and can be simultaneously observed in a BLS spectrum. For a thin magnetic structure which satisfies a condition of $Q_{//}L < 1$ (though in this example, this condition is satisfied for L by less than a few tens of nanometers for λ = 5320 Å (1 Å = 10^{-8} cm = 0.1 nm) laser light and $\vartheta = 45°$), the SAW frequencies are merely controlled by the elastic properties of the substrate and independent of the external magnetic fields [23]. Hence, the SAWs are not the subjects of present interest. By examining the external magnetic field's dependence on the observed peaks, one can readily identify the SAW peaks. As discussed later, the selection rule for SW scattering helps eliminate the SAW contributions in a BLS spectrum. Because the SWs interact with laser photons within the skin depth from the laser-illuminated surface, BLS can give us information on both bulk and surface-localized SWs. Note that the tail portion of the bulk SW within the skin depth reflects the surface's pinning conditions for the SWs. The BLS technique can directly determine the pinning states. This is one of the most important reasons for the effectiveness of SW BLS in thin film magnetism studies. As already mentioned, the in-plane momentum conservation law (and also the energy conservation law) is satisfied, and one can expect sharp peaks for the surface-localized SWs in a BLS spectrum. On the other hand, one can expect broad bulk SW peaks due to the large uncertainty of $\Delta q_\perp$ and possibly the bulk SW dispersion from the exchange coupling.

2.2. Theory

Theoretical developments on SWs and SW BLSs from magnetic films were another motive force. Damon and Eshbach have already discussed magnetostatic SWs in a ferromagnetic slab and discussed the surface-localized SW now known as the Damon–Eshbach (DE) mode by employing standard magnetic boundary conditions [24]. Beyond the magnetostatic framework of the film SW theory, the dipole-exchange framework was developed [25]. In this approach, one must introduce additional boundary conditions. These are known as the Rado–Weertman boundary conditions [26] and the Hoffman boundary conditions [27]. These are related to the SW pinning effects at the interfaces.

In the late 1970s, many theoretical efforts were devoted to calculating a SW BLS spectrum from an opaque surface. Cottam developed a BLS theory for a finite-thickness ferromagnetic slab in terms of the response functions within the magnetostatic framework [28]. Another description of the BLS spectrum calculation from opaque semi-infinite ferromagnetic surfaces was published by Camley and Mills (CM) in 1978 [29]. Readers can refer to reviews by Cottam [30] and Mills [31]. Camley, Rahman, and Mills successively developed a quantitative theory for SW BLS from a ferromagnetic thin film, taking into account both the exchange coupling and the surface pinning conditions [32]. In spite of excellent agreement between the observed and calculated standing spin-wave (SSW) BLS spectra, their theory was too complicated. Another effort to calculate the SSW BLS spectrum was proposed by Cochran and Dutcher [33]. With these theoretical efforts, a quantitative comparison between the observed and calculated SW BLS spectra became possible.

Other efforts were devoted to calculating SW frequencies in thin magnetic film beyond magnetostatic approximation and in layered magnetic structures. Rado and Hicken calculated the SW frequencies from an epitaxial Fe thin film on W substrate taking into account the exchange coupling, MAE, and the surface pinning energies [34]. Grünberg discussed SWs in a trilayer, in which two magnetic layers sandwich a nonmagnetic spacer layer, by adapting the magnetostatic boundary conditions at the top and bottom surfaces and each interface [35,36]. Grünberg and Mika extended the trilayer approach to more stacked multilayer films [37]. Their approach was quite intuitive, but it requires handling a large boundary condition determinant (BCD) as the stacked multilayer increases. The transfer matrix method was developed by Barnas and found to be effective in treating the SWs in magnetic superlattices [38]. On the other hand, Camley, Rahman, and Mills developed a theory of SWs in a superlattice consisting of ferromagnetic and nonmagnetic layers within the magnetostatic framework [39]. Although the theory by Camley, Rahman, and Mills was clear in the thread of the argument and much easier to handle than the theory by Grünberg and Mika, it seems to be difficult to extend beyond the magnetostatic framework. Vohl, Barnas, and Grünberg developed a SW theory based the dipole-exchange model in which the interlayer exchange coupling between ferromagnetic layers across the nonmagnetic spacer layer was taken into account [40]. I will give brief outlines of these theories as I discuss each subject.

For SWs in magnonic crystals, which consist of artificial periodic structures instead of mathematical descriptions, micromagnetic calculations have been widely utilized [41].

3. Spin Wave Light Scattering as Dynamic Magneto-Optic Effects

The most dominant interaction between laser photon and spin waves is not the Zeeman interaction but the electro-dipole interaction, which is given by [42] as below:

$$H_{\text{int}} = -\sum_{i,j} (E_s)_i \delta\varepsilon(M)_{ij} (E_{in})_j \tag{2}$$

in which $\delta\varepsilon(M)$ is a magnetization-dependent dielectric constant. A phenomenological description of SW scattering with the magnetization-dependent dielectric constant was developed by Wettling, Cottam, and Sandercock [43]. For brevity's sake, we assume there is a transparent magnet which belongs to cubic (O_h) symmetry and has a dielectric constant ε_0. Spontaneous magnetization M_z is directed along the z-axis. Because the spontaneous magnetization appears as a result of breaking the time-reversal symmetry, not due to the symmetry lowering in the case of ferroelectrics, the dielectric tensor $\delta\varepsilon(M)$ should be invariant under the O_h symmetry operations. For example, we can apply the $C_{4z.}$ operation, which is $\pi/2$ rotation around the z axis, as below:

$$\delta\varepsilon(M_z) = \begin{pmatrix} \delta\varepsilon_{xx} & \delta\varepsilon_{xy} & \delta\varepsilon_{xz} \\ \delta\varepsilon_{yx} & \delta\varepsilon_{yy} & \delta\varepsilon_{yz} \\ \delta\varepsilon_{zx} & \delta\varepsilon_{zy} & \delta\varepsilon_{zz} \end{pmatrix} = \begin{pmatrix} \delta\varepsilon_{yy} & -\delta\varepsilon_{yx} & -\delta\varepsilon_{yz} \\ -\delta\varepsilon_{xy} & \delta\varepsilon_{xx} & \delta\varepsilon_{xz} \\ -\delta\varepsilon_{zy} & \delta\varepsilon_{zx} & \delta\varepsilon_{zz} \end{pmatrix} = C_{4z}^{-1} \delta\varepsilon(M_z) C_{4z}. \tag{3}$$

By comparing the matrix elements before and after the operation, we can readily obtain

$$\delta\varepsilon_{xx}(M_z) = \delta\varepsilon_{yy}(M_z) \neq \delta\varepsilon_{zz}(M_z),\ \delta\varepsilon_{xy}(M_z) = -\delta\varepsilon_{yx}(M_z) \tag{4}$$

and

$$\delta\varepsilon_{xz}(M_z) = \delta\varepsilon_{zx}(M_z) = \delta\varepsilon_{yz}(M_z) = \delta\varepsilon_{zy}(M_z) = 0 \tag{5}$$

We can expand $\delta\varepsilon(M_z)$ into a power series of the magnetization M_z with complex coefficients K and G as follows:

$$\delta\varepsilon(M_z)_{\alpha\beta} \cong \sum_{\gamma}(K'_{\alpha\beta z} + iK''_{\alpha\beta z})M_z + \sum_{\gamma,\delta}(G'_{\alpha\beta zz} + iG''_{\alpha\beta zz})M_z^2\ (\alpha, \beta, \ldots = x, y, z) \tag{6}$$

The dielectric constant should obey the Onsager's reciprocal theorem [43] given by, $\delta\varepsilon(M_z)_{\alpha\beta} = \delta\varepsilon(-M_z)_{\beta\alpha}$, and we obtain

$$K_{\alpha\beta z} = -K_{\beta\alpha z} \text{ and } G_{\alpha\beta zz} = G_{\beta\alpha zz} \tag{7}$$

It is obvious that the diagonal elements of the dielectric matrix should be even functions of M_z, and that the expansion coefficient K should satisfy the following relations:

$$K_{\alpha\alpha\gamma} = K_{\alpha\gamma\alpha} = K_{\gamma\alpha\alpha} = 0 \text{ and } K_{xy,z} = -K_{yx,z} \tag{8}$$

It is known that a second-order tensor can be decomposed into the Hermitian part ($\varepsilon^H_{\alpha\beta} = \varepsilon^{H*}_{\beta\alpha}$) and the anti-Hermitian part ($\varepsilon^A_{\alpha\beta} = -\varepsilon^{A*}_{\beta\alpha}$) as follows:

$$\varepsilon_{\alpha\beta} = \frac{\varepsilon_{\alpha\beta} + \varepsilon^*_{\beta\alpha}}{2} + \frac{\varepsilon_{\alpha\beta} - \varepsilon^*_{\beta\alpha}}{2} = \varepsilon^H_{\alpha\beta} + \varepsilon^A_{\alpha\beta} \tag{9}$$

Here, the asterisk means the complex conjugate. Combining with the Onsager's theorem [44], one can readily obtain

$$\delta\varepsilon^H_{\alpha\beta}(M_z) = \delta\varepsilon^H_{\beta\alpha}(M_z)^* = \delta\varepsilon^H_{\alpha\beta}(-M_z)^* \tag{10}$$

and

$$\delta\varepsilon^A_{\alpha\beta}(M_z) = -\delta\varepsilon^A_{\beta\alpha}(M_z)^* = -\delta\varepsilon^A_{\alpha\beta}(-M_z)^* \tag{11}$$

From Equations (6), (7), (10), and (11), one obtains

$$K^{\mathrm{H}}_{\alpha\beta,\gamma} = K''_{\alpha\beta,\gamma},\ G^{\mathrm{H}}_{\alpha\beta,\gamma\delta} = G'_{\alpha\beta,\gamma\delta},\ K^{\mathrm{A}}_{\alpha\beta,\gamma} = K'_{\alpha\beta,\gamma}, \text{ and } G^{\mathrm{A}}_{\alpha\beta,\gamma\delta} = G''_{\alpha\beta,\gamma\delta}. \tag{12}$$

Note that the real and the imaginary parts of the expansion coefficients K and G are fully separated from each other. Because we are now considering a transparent magnet, there is no optical absorption; the dielectric matrix should contain only the Hermitian components. The dielectric matrix is given as the following:

$$\varepsilon(M_z) = \varepsilon_0\mathbf{1} + \delta\varepsilon(M_z) = \varepsilon_0\mathbf{1} + \begin{pmatrix} G'_{12}M_z^2 & iK''_{63}M_z & 0 \\ -iK''_{63}M_z & G'_{12}M_z^2 & 0 \\ 0 & 0 & G'_{11}M_z^2 \end{pmatrix}. \tag{13}$$

Here, we have introduced $G'_{11} = G'_{zz,zz}$, $G'_{12} = G'_{xx,zz} = G'_{yy,zz}$, and $K''_{63} = K''_{xy,z}$ in accordance with the conventional tensor index assignment. Although Equation (13) is the fundamental equation to discuss the magneto-optic effects, it depends only on static magnetization. For discussions on SW scattering from opaque magnets, we should take into account both the contributions from the small amplitude SW variables m_x and m_y and from the anti-Hermitian components of the dielectric matrix as well as the Hermitian components. We can replace M_z in Equation (6) by the magnetization vector M and expand Equation (6) up to the first-order terms of m_x and m_y. In accordance with the angular momentum operators in quantum mechanics [45], we can introduce the ladder operators $m_\pm$ as follows: $m_\pm = m_x \pm i m_y$. Then, the m_- operator describes the SW creation (Stokes) process, and the m_+ operator describes the SW annihilation (anti-Stokes) process. Of course, one can introduce the magnon creation and annihilation operators through the Holstein–Primakoff representation [46]. Finally, we obtain the dielectric constant matrix, which describes SW scattering in terms of the SW operators, $m_\pm$, as follows:

$$\delta\varepsilon_{SW}^{(\pm)} = \frac{1}{2}\begin{pmatrix} 0 & 0 & \zeta_{13}^{(\pm)} \\ 0 & 0 & \zeta_{23}^{(\pm)} \\ \zeta_{31}^{(\pm)} & \zeta_{32}^{(\pm)} & 0 \end{pmatrix} m_\pm, \tag{14}$$

in which we can define:

$$\zeta_{13}^{(\pm)} = \left(G'_{44}M_z \mp K''_{63}\right) + i\left(G''_{44}M_z \pm K'_{63}\right) \tag{15}$$

$$\zeta_{23}^{(\pm)} = \left(\pm G''_{44}M_z + K'_{63}\right) \mp i\left(G'_{44}M_z \mp K''_{63}\right) \tag{16}$$

$$\zeta_{31}^{(\pm)} = \left(G'_{44}M_z \pm K''_{63}\right) + i\left(G''_{44}M_z \mp K'_{63}\right) \tag{17}$$

$$\zeta_{32}^{(\pm)} = \pm\left(G''_{44}M_z \mp K'_{63}\right) \mp i\left(G''_{44}M_z \pm K''_{63}\right) \tag{18}$$

Here we used the tensor index assignment of $G_{44} = G_{xz,xz} = G_{yz,yz}$. Based on Equation (14), we can summarize characteristic features of light scattering from SWs as follows:

1. The polarization of the SW scattered light should be cross-polarized from the polarization of the incident light. For example, we can consider the p-polarized incident beam in Figure 1 with the polarization vector $e_p = \begin{pmatrix}\sin\vartheta & -\cos\vartheta & 0\end{pmatrix}$. Then, the scattered beam should be s-polarized with the polarization vector $e_s = \begin{pmatrix}0 & 0 & 1\end{pmatrix}$ and vice versa. Because the SAW scattering is observed in the p-p or s-s scattering geometry, we can eliminate the SAW structure from a SW BLS spectrum by inserting an analyzer in front of FPI.
2. The Stokes and anti-Stokes scattering intensities are generally different. In the above example, this can be expressed as follows:

$$\begin{aligned}\left|\zeta_{13}^{(+)}\right|^2 &= \left|G'_{44}M_z - K''_{63}\right|^2 + \left|G''_{44}M_z + K'_{63}\right|^2 \\ &\neq \left|\zeta_{13}^{(-)}\right|^2 = \left|G'_{44}M_z + K''_{63}\right|^2 + \left|G''_{44}M_z - K'_{63}\right|^2\end{aligned} \tag{19}$$

Therefore, we can observe an asymmetrical SW spectrum around the elastic Rayleigh peak. This is in contrast with the phonon BLS spectrum, which is symmetrical around the Rayleigh peak. Furthermore, when we reverse the spontaneous magnetization M_z to $-M_z$ by changing the polarity of the magnetic field, the Stokes and anti-Stokes spectra are interexchanged. Figure 2 gives a schematic illustration of the reason why the Stokes and anti-Stokes intensities are different.

When the dynamical magnetization $m(t)$ rotates around the static magnetization M_z, the Faraday geometry $k//m(t)$, and the Voigt geometry $k\perp m(t)$ that coexist during one cycle. Therefore, in SW scattering, two different magneto-optic effects simultaneously contribute to and interfere with each other. For a magnet with the real refractive index $\overline{n}$ and negligibly weak optical absorption, the real and imaginary parts of the expansion coefficients in Equation (6) can be related to the magneto-optic coefficients as follows [43]:

$$\Phi_{MCB} = \frac{\pi}{\lambda\overline{n}}\Delta\varepsilon''_H = \frac{\pi}{\lambda\overline{n}}K''_{63}M_z : \text{Magneticcircularbirefringence}, \tag{20}$$

$$\Phi_{MCD} = \frac{4\pi}{\lambda\overline{n}}\Delta\varepsilon'_A = \frac{4\pi}{\lambda\overline{n}}K'_{63}M_{zz} : \text{Magneticcirculardichroism}, \tag{21}$$

$$\Psi_{MLB} = \frac{2\pi}{\lambda\overline{n}}\Delta\varepsilon'_H = \frac{2\pi}{\lambda\overline{n}}G'_{12}M_z^2 : \text{Magneticlinearbirefringence}, \quad (22)$$

$$\Psi_{MLD} = \frac{4\pi}{\lambda\overline{n}}\Delta\varepsilon''_A = \frac{4\pi}{\lambda\overline{n}}G''_{12}M_z^2 : \text{Magneticlineardichroism}. \quad (23)$$

3. Because the matrix components $\zeta_{13}^{(\pm)}$ and $\zeta_{23}^{(\pm)}$ are not equal to the $\zeta_{31}^{(\pm)}$ and $\zeta_{32}^{(\pm)}$ components, the $p \rightarrow s$ scattering and the $s \rightarrow p$ scattering intensities will be different in general. Because of the Brewster angle, the *p*-polarized incident arrangement is preferable to the *s*-polarized incident arrangement in Figure 1.

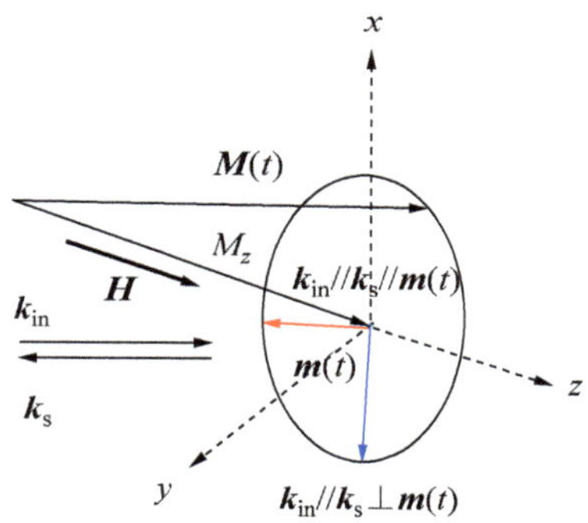

Figure 2. Schematic illustration of the precession motion of the magnetic moment and scattering geometry. It shows that two magneto-optical geometries take place during a cycle of precession motion.

Quantum mechanical description of the coupling coefficient *K* was given by Fleury and Loudon [46]. Hereafter, we use the $\hbar = 1$ unit for convenience's sake. In accordance with their theory, let us consider a very simple example of a 3d electron magnetic crystal placed in the electromagnetic radiation specified by the vector potential *A*. We assume that the ferromagnetic ground state is specified by $L = 0$ and $S = 1/2$, and we also assume the $L = 1$ and $S = 1/2$ intermediate states. We took into account the orbital quenching effect at the ground state. The intermediate states split into a $J = 3/2$ quartet and a $J = 1/2$ doublet with an energy separation of $3\zeta/2$ due to the spin-orbit coupling $\zeta L{\cdot}S$. The Hamiltonian interaction between an electron at *r* within a magnetic ion and the uniform or long wavelength radiation in second quantized notation is given as follows:

$$H' = -er \cdot E = er \cdot \frac{\partial A}{\partial t} \propto \sum_{\lambda}(r \cdot \varepsilon_\lambda)\left[a_\lambda^+ \exp(i\omega_\lambda t) - a_\lambda \exp(-i\omega_\lambda t)\right] \quad (24)$$

where a and a^+ are the photon annihilation and creation operators, ε_λ is the polarization vector of the radiation, and we omitted the coefficient of the vector potential, which is not important for our discussion. We use the $|L, L_z\rangle|S, S_z\rangle$ notation for electron wave functions. The wave functions for the ground state and for the spin excited state are given by $|G\rangle = |0,0\rangle|1/2,1/2\rangle$ and $|G^*\rangle = |0,0\rangle|1/2,-1/2\rangle$, respectively. The intermediate wave functions responsible for the dipole transition are given by [45] as follows:

$$\Phi_{3/2\ 1/2} = \sqrt{\frac{2}{3}}|1,0\rangle|1/2,1/2\rangle + \sqrt{\frac{1}{3}}|1,1\rangle|1/2,-1/2\rangle \quad (25)$$

and

$$\Phi_{3/2\ -1/2} = \sqrt{\frac{1}{3}}|1,-1\rangle|1/2,\ 1/2\rangle + \sqrt{\frac{2}{3}}|1,0\rangle|1/2\,,-1/2\rangle \quad (26)$$

for the quartet, and

$$\Phi_{1/2\,1/2} = -\sqrt{\frac{1}{3}}|1,0\rangle|1/2\,,1/2\rangle + \sqrt{\frac{2}{3}}|1,1\rangle|1/2,-1/2\rangle \tag{27}$$

and

$$\Phi_{1/2\,-1/2} = -\sqrt{\frac{2}{3}}|1,-1\rangle|1/2,1/2\rangle + \sqrt{\frac{1}{3}}|1,0\rangle|1/2,-1/2\rangle \tag{28}$$

for the doublet, respectively. The photon state changes from the initial state given by $|n_1\rangle|n_2\rangle$ to the final state given by $|n_1-1\rangle|n_2+1\rangle$. Here, subscripts 1 and 2 refer to the incident and scattered field quantities. The transition from the $|G\rangle$ state to the $|G^*\rangle$ state is interpreted as the SW creation process (the Stokes process). We can then perform perturbation calculations on the SW creation process with these wave functions and the Hamiltonian interaction, and obtain the final result given by [22,42,47] as below:

$$\frac{e^2\zeta}{2\sqrt{2}}\langle 00|z|\,10\rangle\langle 1-1|x-iy|00\rangle[\varepsilon_{1z}\varepsilon_{2+}-\varepsilon_{1+}\varepsilon_{2z}]\left[\frac{1}{(\varepsilon_{CF}-\omega_1)^2}-\frac{1}{(\varepsilon_{CF}+\omega_2)^2}\right] \tag{29}$$

where ε_{CF} is the crystal field splitting energy given by $\varepsilon_{J=1/2}-\varepsilon_G$. It is important to recognize that the spin-orbit coupling constant ζ is the key parameter to determine the scattering efficiency from a 3d electron magnet.

4. Experimental

The experimental setup for SW BLS is a rather standard one for conventional BLS studies with a 3 + 3 pass tandem FPI, except for the backscattering geometry. Figure 1 shows a standard scattering geometry for SW BLS and the coordination system employed in our SW calculations. The magnetic field H is applied along the z-direction within the plane of the sample and perpendicular to the x-z scattering plane defined by the incident and scattered light beams. We always measure the SWs propagating perpendicular to the magnetic field. We can employ three different scattering geometries: (A) magnetic fields applied in the in-pane easy direction (shown in Figure 1), (B) magnetic fields applied in the in-plane hard axis, and (C) a constant magnetic field rotated from the easy direction to the hard direction. Note that the magnetization $\boldsymbol{M}$ and the external magnetic field $\boldsymbol{H}$ are not collinear with each other in the (B) and (C) geometries. The incident angle ϑ is measured from the surface normal along the x direction and is chosen to be the same as the scattered angle. The incident angle ϑ can be arbitrarily changed between 25° and 65°. The scattering geometries (B) and (C) can be employed for the in-plane MAE studies. In general, scattering intensity from the surface SW mode increases for larger incident angles, and in contrast, the bulk SW mode intensities increase for smaller angles.

At the early stage of our SW BLS studies, we used a laboratory-constructed Sandercock-type 3-pass (or 5-pass) FPI as an interferometer, depending on the surface quality of the prepared films. The FPI was assembled at the machine shop of the Research Institute for Scientific Measurements (RISM), Tohoku University [48,49]. For some sputtered films, we observed SW spectra by using the 3-pass FPI [50–52]. The spectra were excited by the 5145 Å or 4880 Å line of an argon ion laser in single-mode operation and detected by a thermoelectrically cooled photomultiplier tube (PMT) for a dark count of less than 1 cps. Later, we constructed a Sandercock-type 3 + 3 pass vernier-tandem FPI in 1994 at the machine shop of RISM, Tohoku University [53], and the multipass type FPI was replaced by the tandem FPI. The 5145 Å or 4880 Å line of argon ion laser can be replaced by the 5320 Å or 4730 Å line of diode-pumped solid-state (DPSS) single-mode laser. Various types of DPSS laser are now commercially available. The DPSS laser is much easier to use and also more economical compared with a water-cooled argon ion laser. The PMT can be also replaced by an avalanche photo-diode (APD) detector, which possesses higher quantum efficiency than the PMT detectors.

In our system, a laser beam was introduced by a 45° right-angle prism with 3 × 3 mm^2 input-face into an optical axis of the FPI. A camera lens (50 mm *f*1.2) focused the beam on the sample and also collected the backscattered light. We used a spatial filter that consisted of two camera lenses (135 mm *f*2.8 and 50 mm *f*1.2) and a 200 μm pinhole. To eliminate the scattering from the SAWs and most of the intense Rayleigh peaks and FPI ghost peaks, a cross-polarizing beam splitter (extinction ratio = 1/200) was inserted in front of the FPI. We found that the elastically scattered light was still intense enough in many cases even after the polarization selection. In order to protect the highly sensitive PMT or APD from optical damage due to elastically scattered light and ghosts, we introduced a tandem acousto-optic modulator (AOM), which was activated around the Rayleigh and ghost peaks as an intensity attenuator. We also added a mechanical shutter, which was activated only when the peak intensity was getting higher than a preset level.

In some cases, we experienced that a sample exposed to an intense laser beam (even less than 50 mW) in air was easily damaged by the local heating and oxidizing effects. Therefore, we found that we should place the sample inside a vacuum chamber or under an appropriate atmosphere during BLS measurements. In order to make possible the degree of atmosphere control and also the low-temperature studies under magnetic fields, we prepared a liquid He cryostat which could be used as a vacuum chamber. Furthermore, a closed-cycle refrigerator was used to generate low temperatures down to 15 K. In order to perform variable-temperature and magnetic field studies, we assembled a refrigerator tip suitable for BLS study under magnetic fields of up to 4.5 kOe at the RISM machine shop. During a spectral accumulation time over several hours, the lowest temperature of 15 K could be fully stabilized within ±0.5 K.

5. BLS Results

5.1. Semi-Infinite Magnet

Figure 3 shows typical BLS spectra obtained from nanogranular Co-Al-O films of 1~2 μm thickness prepared by means of radio frequency-reactive magnetron sputtering onto glass substrates at the Research Institute of Electric and Magnetic Materials (RIEMM) [54].

These spectra were excited by the *p*-polarized 4880 Å line from an Ar$^+$ laser operated in a single-cavity mode with the output power below 30 mW to protect films from local heating by the laser beam. Typical spectrum accumulation time was about 4 h. An external magnetic field of H = 2.0 kOe was applied parallel to the film plane and perpendicular to the scattering plane (x-z plane). The incident angle was chosen to be the same as the scattered angle (ϑ = 45°) in these measurements. We observed very similar BLS spectra from a sputtered Fe film and from Fe-Al-O nanogranular films deposited on oxidized Si(001) substrates at the Institute of Multidisciplinary Research for Advanced Materials (IMRAM), Tohoku University. These spectra were excited by the *p*-polarized 5320 Å line from a DPSS laser [55] The TM-Al-O nanogranular films (TM = Fe, Co), which consist of crystalline TM particles with several tens of angstrom in diameter, were surrounded by the Al-O grain boundary. The Al-O grain boundary was of ~10 Å in thickness. The BLS technique observes SWs with an in-plane wavelength $\lambda_{//}$ defined by $2\pi/Q_{//}$. The in-plane wavelength $\lambda_{//}$ (typically ~3500 Å) is much longer than the characteristic lengths of granules but much shorter than the in-plane lengths of magnetic structures. For long-wavelength SWs observed with the BLS technique, the real magnetic structure may not be important, and the magnetic properties averaged over in-plane wavelength $\lambda_{//}$ within the laser-illuminated area determine the SW response in BLS spectra. The peak assignment of these spectra was quite obvious as will be discussed soon. The labels DE and B refer to the Damon-Eshbach (DE) surface wave and the bulk SWs, respectively. Note that the DE peak appears only on the anti-Stokes side in these spectra in contrast to the bulk SW peaks. When we changed the polarity of the external magnetic field, the DE peak appeared on the Stokes side of a spectrum. Another interesting observation is the line shape of the bulk peaks. The bulk peaks are asymmetric with tails to higher-frequency sides. This is due to the relaxation of the momentum conservation law as already mentioned and the SW dispersion (energy as a function of the wave vector) [54].

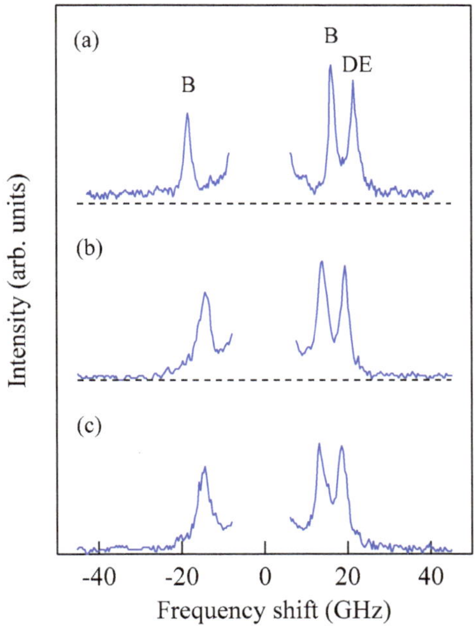

Figure 3. BLS spectra observed from ferromagnetic Co-Al-O nanogranular films in a magnetic field of H = 2.0 kOe. (**a**) $Co_{74}Al_{18}O_{18}$, (**b**) $Co_{66}Al_{11}O_{23}$, and (**c**) $Co_{63}Al_{3}O_{243}$. The broken lines show the baselines of each spectrum [54].

In order to understand the characteristic feature of the SW spectra and the magnetic field dependence of the SW frequencies for qualitative discussions and determination of the magnetic constants, I will describe a standard magnetostatic theory rather in-detail for readers not familiar in SW BLS [22,56]. Because we are interested in the magnetization dynamics below 10^{11} Hz, which is well below the optical frequency of ~6×10^{14} Hz, we ignore the time-dependent terms in Maxwell's equations. We consider a magnetic film of the magnetization M and thickness L prepared on a nonmagnetic substrate. For the sake of convenience, we can ignore the exchange coupling and the magnetic anisotropy energy (MAE) at this stage, and then set the surfaces as x = 0 and $-L$. Let us introduce the SW variables m as the small amplitude precession motion around the static magnetization $\boldsymbol{M}$ and the demagnetization field $\boldsymbol{h}$. The Landau–Lifshitz (LL) equation of motion on $M(t) = M + m(t)$ is given as follows:

$$\frac{1}{\gamma}\frac{dM}{dt} = M \times H_{eff} \tag{30}$$

where γ is the gyromagnetic ratio ($\gamma/2\pi$ = 1.4g GHz/kOe, where g is the Lande's g factor). The effective magnetic field H_{eff} consists of the external magnetic field H and the demagnetization field $\boldsymbol{h}$ as below:

$$H_{eff} = H + h \tag{31}$$

The demagnetization field $\boldsymbol{h}$ should satisfy Maxwell's magnetostatic equations, given below:

$$\nabla \times h = 0 \text{ and } \nabla \cdot (h + 4\pi m) = 0. \tag{32}$$

The first equation of Equation (32) guarantees the introduction of a magnetic scalar potential φ satisfying the below condition:

$$h = -\nabla\varphi \tag{33}$$

We consider SWs propagating along the $(0,\ \cos\theta, -\sin\theta)$ direction measured from the y-axis to be within the film plane, and we assume plane-wave-type space-time dependence for the dynamical variables:

$$\begin{pmatrix} m(r,t) \\ \varphi(r,t) \end{pmatrix} = \begin{pmatrix} m \\ \varphi \end{pmatrix} \exp[i\omega t - iq_\perp x - i(Q_{//}\cos\theta)y + i(Q_{//}\sin\theta)z] \tag{34}$$

Furthermore, the continuity conditions of the variables φ and $b_x = h_x + 4\pi m_x$ at $x = 0$ and $-L$ should be satisfied. Outside of the magnet, the scalar potentials are given

$$\varphi_{\text{out}}(r,t) = \varphi_{\mp}\exp(\mp Q_{//}x)\exp[i\omega t - i(Q_{//}\cos\phi)y + i(Q_{//}\sin\phi)z] \tag{35}$$

Here, we should choose the $\exp(-Q_{//}x)$ term for $x \geq 0$ and the $\exp(Q_{//}x)$ term for $x \leq -L$. Inside the magnet, we have

$$\begin{aligned}\varphi_{\text{in}}(r,t) = {} & [\varphi(+)\exp(-iq_\perp x) + \varphi(-)\exp(iq_\perp x)] \\ & \times \exp[i\omega t - i(Q_{//}\cos\theta)y + i(Q_{//}\sin\theta)z]\end{aligned} \tag{36}$$

We can solve the LL equation in terms of the susceptibilities:

$$m_x(\omega) = \chi_{xx}(\omega)h_x + \chi_{xy}(\omega)h_y = \frac{-MH}{H^2 - (\omega/\gamma)^2}\left(\frac{\partial\varphi_{\text{in}}}{\partial x}\right) + \frac{iM(\omega/\gamma)}{H^2 - (\omega/\gamma)^2}\left(\frac{\partial\varphi_{\text{in}}}{\partial y}\right) \tag{37}$$

and

$$m_y(\omega) = \chi_{yx}(\omega)h_x + \chi_{yy}(\omega)h_y = \frac{-iM(\omega/\gamma)}{H^2 - (\omega/\gamma)^2}\left(\frac{\partial\varphi_{\text{in}}}{\partial x}\right) + \frac{-MH}{H^2 - (\omega/\gamma)^2}\left(\frac{\partial\varphi_{\text{in}}}{\partial y}\right). \tag{38}$$

Combining Equations (32) and (36)–(38), we obtain

$$q_\perp^2 + Q_{//}^2 + 4\pi\left(Q_{//}^2\cos^2\theta + q_\perp^2\right)\chi_{xx} = 0 \tag{39}$$

The perpendicular wave vector $q_\perp$ is found to be

$$q_\perp^2 = -\frac{1 + 4\pi\chi_{xx}\cos^2\theta}{1 + 4\pi\chi_{xx}}Q_{//}^2 = \frac{(\omega/\gamma)^2 - H(H + 4\pi M\cos^2\theta)}{H(H + 4\pi M) - (\omega/\gamma)^2}Q_{//}^2 \tag{40}$$

Because $q_\perp$ should be real for the bulk mode, we obtain a SW band given by

$$H\left(H + 4\pi M\cos^2\theta\right) \leq \left(\frac{\omega}{\gamma}\right)^2 \leq H(H + 4\pi M) \tag{41}$$

The SW band became a single level at $\theta = 0$ (along the y-axis) and gradually spread wider as θ approached $\pi/2$ (along the z-axis) as shown in Figure 4a. It is important to note that at the upper-bound SW frequency, the perpendicular component $q_\perp$ became much larger than the in-plane component $Q_{/\!/}$. On the contrary, we have $q_\perp = 0$ for the lower-bound frequency.

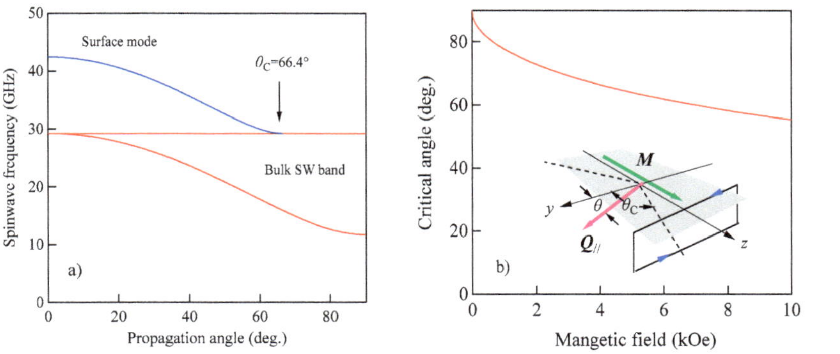

Figure 4. (**a**) Propagation angle development of the bulk SW band (—) and the DE mode frequency (—). We used the following set of magnetic constants: g = 2.09, H = 4.0 kOe, and $4\pi M$ = 21.0 kG. Here, the upper and lower bounds of the bulk SW band are displayed for clarity's sake. (**b**) The magnetic field dependence of the critical angle θ_C. The inset shows a schematic illustration of the nonreciprocal propagation characteristics of the DE mode. The critical angle is indicated by the broken lines. The allowed propagation direction of the DE mode localized on the top and bottom magnetic surfaces is indicated by the arrows on the rectangle.

There is another type of SW solution which satisfies the magnetic continuity conditions at the boundaries. This solution is the surface-localized SW and is known as the Damon–Eshbach (DE) mode. For discussions of the surface mode, it is convenient to rewrite Equations (34) and (40) as follows:

$$\begin{pmatrix} m(r,t) \\ \varphi(r,t) \end{pmatrix} = \begin{pmatrix} m \\ \varphi \end{pmatrix} \exp[i\omega t + q_\perp x - i(Q_{//}\cos\theta)\, y + i(Q_{//}\sin\theta)\, z] \tag{42}$$

and

$$q_\perp = \left[\frac{1+4\pi\chi_{xx}\cos^2\theta}{1+4\pi\chi_{xx}}\right]^{1/2} Q_{//} = \alpha Q_{//} \tag{43}$$

By eliminating the potential amplitudes $\varphi_\mp$ outside of the magnet using the boundary conditions, we obtain a set of homogeneous equations on $\varphi(\pm)$ to determine the surface-localized SW frequency. Per the requirement for nontrivial solutions of the homogeneous equations, we obtain:

$$\begin{gathered} -\beta\left[\alpha\left(1+\frac{4\pi MH}{H^2-(\omega/\gamma)^2}\right)-1\right]^2+\beta\left[\frac{4\pi M(\omega/\gamma)}{H^2-(\omega/\gamma)^2}\cos\theta\right]^2 \\ +\left[\alpha\left(1+\frac{4\pi MH}{H^2-(\omega/\gamma)^2}\right)+1\right]^2-\left[\frac{4\pi M(\omega/\gamma)}{H^2-(\omega/\gamma)^2}\cos\theta\right]^2=0. \end{gathered} \tag{44}$$

Here, we define $\beta = \exp(-2Q_{//}L)$. To obtain the DE mode frequency, we must solve numerically Equation (44). Fortunately, we can obtain an analytical expression of the DE mode frequency in following two cases.

Case 1: $\theta = 0$ ($\alpha = 1$)

$$\left(\frac{\omega}{\gamma}\right)^2 = (H+2\pi M)^2 - (2\pi M)^2\exp(-2Q_{//}L) \tag{45}$$

For a film with a thickness larger than $L \cong \lambda/2$, the exponential term in Equation (45) can be safely neglected. It means that a film thicker than $L \cong \lambda/2$ can be treated as a semi-infinite magnet.

Case 2: $\beta = 0$

$$\alpha\left(1 + \frac{4\pi MH}{H^2 - (\omega/\gamma)^2}\right) + 1 \pm \frac{4\pi M(\omega/\gamma)}{H^2 - (\omega/\gamma)^2}\cos\theta = 0 \tag{46}$$

After some calculations, we obtain the below:

$$2\pi\Delta\nu_{\mathrm{DE}} = \pm\frac{\gamma}{2}\left[\frac{H}{\cos\theta} + (H + 4\pi M)\cos\theta\right] \tag{47}$$

Equation (47) at $\theta = 0$ gives the frequency which is exactly the same as the frequencies given by Equation (45) because of $\beta = 0$. However, we should check whether these modes are truly eligible for the surface mode or not. From the boundary conditions, we obtain

$$q_\perp = -Q_{//}\frac{4\pi(-i\chi_{xy})\cos\theta + 1}{1 + 4\pi\chi_{xx}} \tag{48}$$

Because the semi-infinite magnet occupies the space below $x \leq 0$, the perpendicular component $q_\perp$ should be positive for the eligible surface mode. By substituting the frequencies in Equation (47) into Equation (48), we find that the positive frequency in Equation (47) always gives $q_\perp < 0$ and fails to satisfy the localization condition. We should abandon the positive frequency solution. Meanwhile, the negative frequency gives

$$q_\perp = -Q_{//}\frac{\left(4\pi M\cos^2\theta + H\sin^2\theta\right)}{\left[H - (H + 4\pi M)\cos^2\theta\right]^2}\left[H - (H + 4\pi M)\cos^2\theta\right] \tag{49}$$

Therefore, the negative frequency mode propagating within the critical angle given by Equation (50) can be the surface-localized DE mode [24]:

$$\cos\theta \geq \left(\frac{H}{H + 4\pi M}\right)^{1/2} = \cos\theta_C \tag{50}$$

Figure 4a shows the propagation angle θ development of the bulk SW band given by Equation (41) and the DE mode frequency given by Equation (47). At the critical angle θ_C, the DE mode frequency coincided with the upper bound of the bulk SW band. In this calculation, we used a set of the magnetic constants suitable for Fe: $g = 2.09$, $H = 4.0$ kOe, and $4\pi M = 21.0$ kG. Figure 4b shows the critical angle θ_C as a function of the external magnetic field for the same set of the magnetic constants. The critical angle was gradually squeezed with the increasing magnetic field. The inset displays a schematic illustration of the nonreciprocal propagation characteristics of the DE mode. The DE mode always propagates from the left to the right across the magnetization as indicated by the arrows. At an angle θ beyond the critical angle θ_C, the attenuation factor for the DE mode $q_\perp$ in Equation (49) becomes negative, and the DE mode is no longer allowed above θ_C. The existence of the critical angle is the reason why the DE peak appears on only one side of a SW BLS spectrum. The nonreciprocal propagation characteristics of the DE mode are schematically illustrated in Figure 4c. Note that the counterpart of the DE mode propagating along the opposite direction is located on the bottom surface of the magnet. The incident laser photon can never interact with the counterpart DE mode because of the absorption effect of visible light, as we have already mentioned. The DE mode frequency was always above the bulk SW frequency band. At the critical angle, the DE mode frequency was just on the upper bound of the bulk SW band (see Equation (41)). Sandercock and Wettling nicely presented how the DE mode behaves as the propagation angle approaches the critical angle [13]. Their results clearly show that the DE mode decays into the bulk SW band, and no surface mode is allowed beyond the critical angle.

Figure 5 shows the SW frequencies as a function of the magnetic field for $Fe_{64}Al_{19}O_{17}$ nanogranular film [54]. The inset shows a BLS spectrum observed at $H = 0.5$ kOe. Because we have not performed the polarization selection for the scattered beam, SAW peaks appear as a pair of small peaks just below the B-peaks. I will soon explain the solid lines, the broken line, and the dots and dashes.

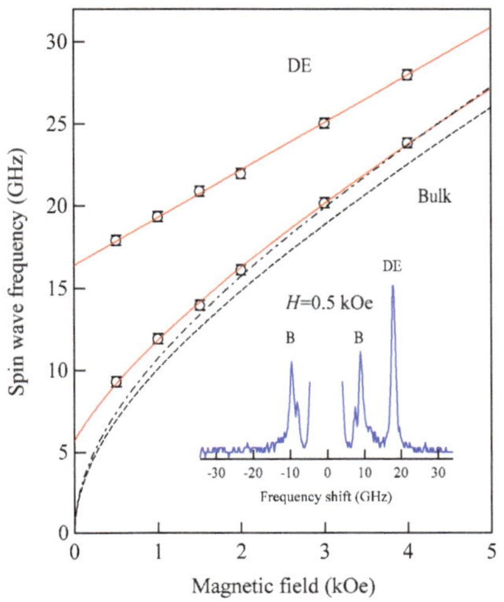

Figure 5. SW frequencies of $Fe_{64}Al_{19}O_{17}$ film as a function of the magnetic field. A BLS spectrum at $H = 0.5$ kOe is also shown. The solid lines are the calculated SW frequencies using Equation (33) with $g = 2.07$, $4\pi M_B = 11.1$ kG, $4\pi M_{DE} = 11.3$ kG, and $H_{ex} = 0.34$ kOe. The broken line ($4\pi M_B = 11.1$ kG) and dots and dashes ($4\pi M_B = 12.7$ kG) are the calculated frequencies without the H_{ex} term [54].

In these measurements, both the bulk and DE modes were propagating perpendicular to the magnetization, and their frequencies are given by

$$\Delta\nu_B = \frac{\gamma}{2\pi} H^{1/2}(H + 4\pi M)^{1/2} \quad (51)$$

and

$$\Delta\nu_{DE} = \frac{\gamma}{2\pi}(H + 2\pi M) \quad (52)$$

Because the frequency shifts $\Delta\nu_B$ and $\Delta\nu_{DE}$ were directly obtained from the BLS spectrum, both frequencies should be reproduced by the same magnetic constants as a function of the magnetic field. However, sometimes we encountered somewhat different $4\pi M$ values for the bulk and DE modes. The broken line and dots and dashes in Figure 5 are the calculated bulk SW frequencies using Equation (51) by changing the $4\pi M$ value. It is clear that Equation (51) fails to reproduce the observed magnetic field dependence of the bulk SW frequency. Because our model is oversimplified in the first attempt, we will try to include the exchange coupling term into Equation (51). For long wavelength SWs, the exchange coupling can be represented by the differential operator $H_{ex} = -D\nabla^2$. Because the external magnetic field and the magnetization are collinear, we can replace the external magnetic field H with $H + DQ^2$. Here, Q is the wave vector of the SW, and D is the SW stiffness constant and related to the exchange stiffness constant A through the relation of $D = 2A/M$. With the exchange field term, Equation (51) is replaced by

$$\Delta\nu_B = \frac{\gamma}{2\pi}\left(H + DQ^2\right)^{1/2}\left(H + DQ^2 + 4\pi M\right)^{1/2} \quad (53)$$

The solid lines in Figure 5 are calculated bulk SW frequencies with the exchange field term ($DQ^2 = 0.34$ kOe) in Equation (53). Although the exchange field value is usually much smaller than the other fitting parameters, typically less than 0.5 kOe, agreements between the observed and calculated bulk SW frequencies are excellent. In spite of the importance of the exchange field term for qualitative fitting of the bulk SW frequencies, as shown in Figure 5, we cannot determine the D constant from the fitting because we have no information on the SW wave vector Q in Equation (53) due to the relaxation of the momentum conservation law. The DE mode frequency is rather insensitive to the exchange term because of the linear dependence of the frequency on the external magnetic field. Furthermore, the existence of the DE mode was derived from the boundary conditions, and the negligibly small $DQ_{//}^2$ term was completely masked by the other quantities in Equation (52).

Up to this stage, we considered soft ferromagnetic materials with negligibly small MAE. For such small-MAE cases, it may be an easy- or hard- axis type, and we can readily align the magnetization along the external magnetic field. Let us consider the uniaxial in-plane MAE given by

$$E_K = -K_{//}\left(\frac{M_z}{M}\right)^2 \tag{54}$$

Here, $K_{//}$ is the in-plane MAE constant. When we apply the magnetic field along the easy direction and examine SWs propagating perpendicular to the magnetization (θ = 0), the SW frequencies are given by

$$\Delta\nu_B = \frac{\gamma}{2\pi}\left(H + H_{K//} + DQ^2\right)^{1/2}\left(H + H_{K//} + DQ^2 + 4\pi M\right)^{1/2} \tag{55}$$

and

$$\Delta\nu_{DE} = \frac{\gamma}{2\pi}(H + H_{K//} + 2\pi M) \tag{56}$$

Here, we defined the in-plane anisotropy field $H_{K//} = 2K_{//}/M$. Meanwhile, for the magnetic field along the hard direction, we have

$$\Delta\nu_B = \frac{\gamma}{2\pi}\left[\left(H + DQ^2\right)\left(H - H_{K//} + DQ^2 + 4\pi M\right) - 4\pi M H_{K//}\left(\frac{Q_\perp}{Q}\right)^2\right]^{1/2} \tag{57}$$

and

$$\Delta\nu_{DE} = \frac{\gamma}{2\pi}\left(H + 2\pi M - \frac{H_{K//}}{2}\right) \tag{58}$$

The bulk SW in an isotropic magnet forms the bulk SW band given by Equation (41). The bandwidth depends on the in-plane propagation direction θ. On the other hand, when the external magnetic field is applied along the hard direction, the MAE introduces the SW band for the bulk SW propagating even for the θ = 0 direction. The bandwidth given by $4\pi M H_{K//}$ depends on both the strength of the MAE and the perpendicular component of the SW wave vector. The main contribution for the SW BLS is from $Q_\perp/Q < 0.5$, and the bandwidth $4\pi M H_{K//}$ is usually smaller than the bulk SW's peak width. Hence, it is impractically difficult to determine the MAE parameters from BLS measurement by itself.

Next, we consider the out-of-plane type MAE given by

$$E_K = -K_\perp\left(\frac{M_x}{M}\right)^2 \tag{59}$$

We can also define the out-of-plane anisotropy field $H_{K\perp} = 2K_\perp/M$. For a weak anisotropy field, which satisfies the in-plane magnetization condition given by $4\pi M - H_{K\perp} \geq 0$, the magnetization is confined within the film plane and aligned colinear to the external magnetic field. In this case, the upper and lower bounds of the SW band are given by

$$\left(\frac{\omega}{\gamma}\right)^2 = (H + DQ^2)(H + DQ^2 + 4\pi M - H_{K\perp}) - 4\pi M H_{K\perp}\left(\frac{Q_{//}}{Q}\right)^2 = \begin{cases} (H + DQ^2)(H + DQ^2 + 4\pi M - H_{K\perp})(Q_{//}/Q = 0) \\ (H + DQ^2 - H_{K\perp})(H + DQ^2 + 4\pi M)(Q_{//}/Q = 1) \end{cases} \quad (60)$$

and the DE mode frequency is given by

$$\frac{\omega}{\gamma} = H + 2\pi M - \frac{H_{K\perp}}{2} \quad (61)$$

In this case, the main contribution for the bandwidth is from $Q_{\perp}/Q > 0.5$. For weak magnetic fields which satisfy $H + DQ^2 - H_{K\perp} \leq 0$, the lower bound of the SW band should be set to zero. On the contrary, the out-of-plane MAE is large enough to overcome the in-plane magnetization condition, and the perpendicular magnetization state is the ground state under zero magnetic field. We will discuss this case later.

5.2. Thin Films

As already mentioned, a magnetic film thicker than ~$\lambda/2$ can be treated as a semi-infinite magnet. Now, what happens for thinner films? For thinner films with thicknesses less than ~1000 Å, new aspects of SWs, known as the standing SWs (SSWs), appear in a BLS spectrum. The first BLS observation of the SSWs was reported by Grimsditch and Malozemoff on metallic amorphous $Fe_{80}B_{20}$ films. They determined the SW stiffness constant of D_{BLS} = (1.4 ± 0.2) × 10^{-9} Oe·cm^2 [57]. Successively, BLS from the SSWs has been reported on various ferromagnetic thin films [11]. Neutron scattering is the best technique to investigate SW dynamics in the entire Brillouin zone and can be used to determine the SW stiffness constant D_{NS}. However, neutron scattering requires a reactor and eventually becomes a huge project. In this section, in order to distinguish the SW stiffness constant obtained from BLS and neutron scattering, we add the subscript BLS and NS to the D constant. Otherwise, we simply use the symbol D for the BLS SW stiffness constant. By observing the SSW structure, we can precisely determine the SW stiffness constant D_{BLS} even in a small optical laboratory. This is one of the virtues of the SW BLS technique. The research group in Brookhaven National Laboratory (NBL) extensively investigated SWs in Fe, Ni, and Co in the 1960s [58]. Note that the BLS technique gives information near the Brillouin zone center thanks to visible laser light as an excitation source. We can change the D_{BLS} value in the 10^{-9} Oe·cm^2 unit to the D_{NS} value in the meV·Å^2 unit used in neutron scattering and magnetization studies by a formula of D_{BLS} = 1.728 × $10^{-2}D_{NS}/g$. Neutron scattering and BLS give D_{NS}~280 meV·Å^2 for Fe, and these results are in good agreement [58,59].

Figure 6 shows an example of SSW spectrum observed from a 450 ± 10 Å thick epitaxial $(10\bar{1}0)$ Co film deposited on a 500 Å thick Cr (211) buffer layer prepared on MgO (110) substrate at IMRAM, Tohoku University [60,61].

This spectrum was excited by the p-polarized 5320 Å line from a DPSS laser. Because we have not performed the polarization selection for the scattered light in this measurement, the SAW structures, indicated by phonons, were also observed. The peaks indicated by labels 1 and 2 are the first and second SSW peaks. The peak indicated by 1 + DE on the anti-Stokes side consists of the first SSW and the DE peaks. Note that the DE mode in thinner films also retained the nonreciprocal propagating character. When we define the critical angle θ_C, at which the DE mode frequency given by Equation (45) is equal to the upper bulk band frequency, we obtain Equation (50) again by using Equation (44). For brevity's sake we ignore the exchange term for the upper bulk band. For thinner magnetic films, the perpendicular component $q_{\perp}$ of the SW wave vector was quantized into $q_{\perp}(n) = n\pi/L$ (n = 1, 2, ...). In this case, the perpendicular components $q_{\perp}(n)$ were well-defined, and the momentum conservation law during the scattering process recovered. This is the reason why we can observe sharp SSW peaks in our spectrum. On our BLS results from this epitaxial $(10\bar{1}0)$ Co film, we will discuss details later.

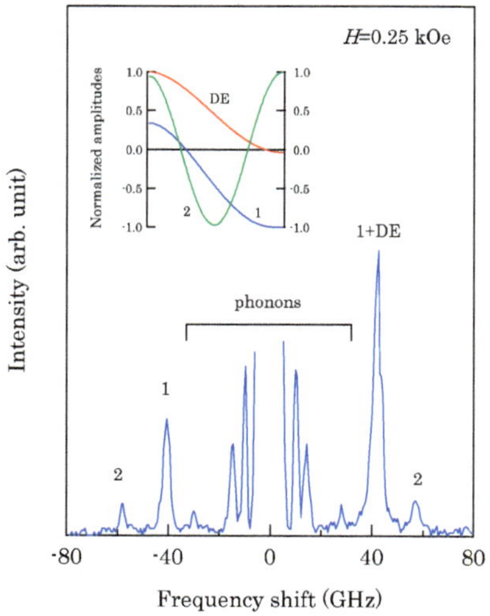

Figure 6. BLS spectrum observed from an epitaxial $(10\bar{1}0)$ Co thin film 45 nm thick at room temperature. Because no polarization selection has been done, scattering from SWs and SAWs was observed. The structure labeled "phonons" within ±30 GHz around the elastic RS peak was due to SAWs. These were assigned to the Rayleigh wave and the first and second order Sezawa waves with increasing frequency. SW peaks up to the second order SSW appeared above ±40 GHz. Note that the phonon peak intensities are symmetrical for the Stokes and anti-Stokes peaks. On the other hand, the SW peaks were asymmetric. The insert shows calculated n = 1 (—) and 2 (—) SSW and DE mode (—) profiles [60]. For clarity's sake, I display the profiles of the DE mode localized on the top surface and the SSW modes for the bottom surface.

Figure 7 shows another example of SSWs observed from 1000 ± 50 Å-thick sputtered $Co_{85}Nb_{12}Zr_3$ film on a glass substrate prepared at RISM, Tohoku University [62].

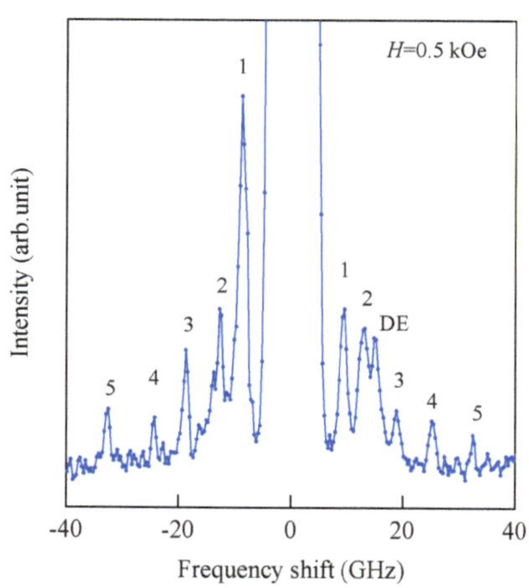

Figure 7. BLS spectrum observed from a 1000 ± 50 Å thick $Co_{85}Nb_{12}Zr_3$ film prepared by means of rf-magnetron sputtering from alloy target onto glass substrate. Labels 1–5: SSW mode number; DE: DE surface mode. (Copyright (1994) The Japan Society of Applied Physics) [62].

We set $\vartheta = 15^\circ$ and $H = 0.5$ kOe. We could observe the SSW peaks up to the fifth order in this spectrum. Note that the peak intensities are highly asymmetric between the Stokes and anti-Stokes sides. This is a characteristic feature of SW BLS, as I have already mentioned, as the interference effects it. The DE peak appears only on the anti-Stokes side. The DE peak intensity is not high compared to the SSW peaks because of the small incident angle. In fact, when we increase the incident angle ϑ, the DE peak intensity gradually increases. Figure 8 shows the SW frequencies as a function of the magnetic field.

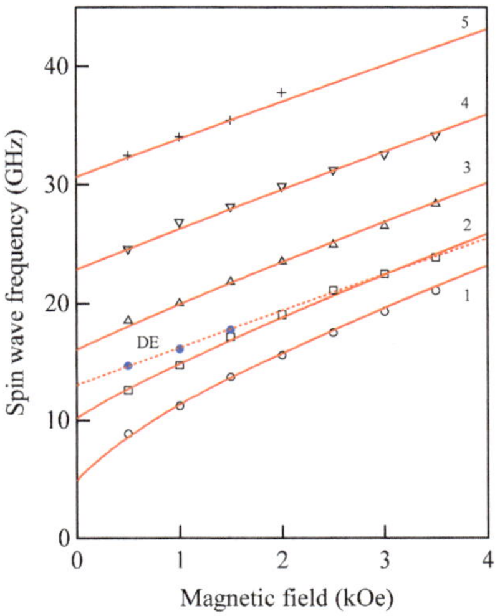

Figure 8. Magnetic field dependence of the SSW and DE mode frequencies. The open symbols and + symbol are for the SSW modes, and (•) is for the DE mode. Here, the labels 1 to 5 stand for the SSW mode number. The solid and broken lines were calculated from Equations (59) and (60) with the magnetic constants given in the text (Copyright (1994) The Japan Society of Applied Physics) [62].

The open symbols stand for the SSWs, and the filled circles stand for the DE mode. Above $H = 1.0$ kOe, we could not fully resolve the DE mode from the second-lowest-order SSW peak. In order to determine the magnetic constants of the $Co_{85}Nb_{12}Zr_3$ film while taking into account the quantization effect on the $q_\perp$ components, we employed a conventional formula given below:

$$\Delta\nu(n) = \frac{\gamma}{2\pi}\left[H + D\left\{Q_{//}^2 + \left(\frac{n\pi}{L}\right)^2\right\}\right]^{1/2}\left[H + D\left\{Q_{//}^2 + \left(\frac{n\pi}{L}\right)^2\right\} + 4\pi M\right]^{1/2} (n = 1, 2, \ldots) \tag{62}$$

and

$$\Delta\nu_{DE} = \frac{\gamma}{2\pi}\left[(H + 2\pi M)^2 - (2\pi M)^2 \exp(-2Q_{//}L)\right] \tag{63}$$

It can be readily recognized from Equation (62) that the D constant governs the splitting between the SSW frequencies. The solid lines in Figure 8 are the calculated SSW frequencies from Equation (62), and the broken line is the calculated DE mode frequency from Equation (63) with the magnetic constants listed in Table 1.

We obtained $4\pi M_{VSM} = 10.1 \pm 0.2$ (kG) from a VSM measurement and evaluated the exchange stiffness constant $A = 0.98 \pm 0.14$ ($\times 10^{-6}$ erg/cm) using the magnetic constants. With these constants, an excellent agreement between the calculation and observation was obtained. This agreement is not only for this case. Usually, Equations (62) and (63) gave good agreement between calculations and observations.

Table 1. Magnetic constants of $Co_{85}Nb_{12}Zr_3$, bcc Fe, and hcp Co films used for the SW frequency calculations.

Magnetic Constants	$Co_{85}Nb_{12}Zr_3$	bcc Fe	hcp Co
$\gamma/2\pi$ (GHz/kOe)	2.98	2.09	2.17
$4\pi M_B$ (kG)	10.0	21.5	17.8
$4\pi M_S$ (kG)	10.0	21.5	17.8
D (10^{-9} Oe·cm^2)	2.47	2.34	3.39
K_1 (10^6 erg/cm^3)	–	0.45	3.45
K_2 (10^6 erg/cm^3)	–	–	0.9

Figure 9 shows the SW stiffness constant D in $Co_{100-x}Cr_x$ binary alloy of 300~500 Å in thickness as a function of the Cr at % [63,64].

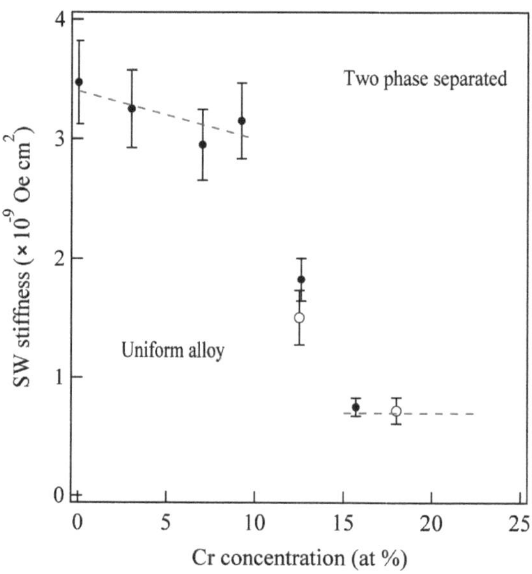

Figure 9. A summary of the SW stiffness constant D of the $Co_{1-x}Cr_x$ binary system (x in at % unit) as a function of the Cr content x. We added the D_{BLS} values for $x = 12.5$ from [65] and $x = 18$ from [66]. The colored area shows the transition region between the uniform alloy state and the two-phase separated state. The broken lines give $D(x) = (3.41 - 0.043x) \times 10^{-9}$ Oe·cm^2 and 0.7×10^{-9} Oe·cm^2, respectively [63,64].

These alloys were prepared at RISM, Tohoku University. For the CoCr binary alloy system, phase separation occurred from the Co-rich uniform state below x~10 at % to the phase-separated state. In the phase-separated state, the Co-rich ferromagnetic regions were surrounded by the nonmagnetic Cr-rich grain boundaries above x~12 at %. Because the exchange coupling is purely a quantum mechanical effect due to the electron itineracy, overlapping of the electron wave functions, or both, we can expect that the exchange coupling strength is very sensitive to the microscopic atomic structure inside a film. Figure 9 clearly shows that a drastic change of the exchange coupling scheme from the direct coupling in the Co-rich uniform state to a weak indirect coupling via Cr-rich regions took place around x = 10~15 at %. In this way, the BLS technique can provide quantitative information on these magnetic interactions.

Because we have phenomenologically introduced Equation (62), we must take into account the exchange coupling and the MAE and derive more rigorous descriptions of the SW frequencies. We applied the dipole-exchange model with continuum approximation to discuss the SWs. The exchange coupling can be calculated with the second derivative operator given by $H_{ex} = -D\nabla^2$ in this model. Recent developments in film preparation techniques allowed us to examine various types of epitaxial structures. For materials with hexagonal or tetragonal structures, the MAE played more important roles than for the cubic structures. Thinking of the epitaxial Co $(10\bar{1}0)$ films, we considered the uniaxial in-plane MAE up to the fourth order as below [60]:

$$E_K = K_1 \sin^2 \phi + K_2 \sin^4 \phi \tag{64}$$

Figure 10 shows the coordinate systems and scattering geometry used in our discussions. The crystallographic coordinates are shown by (x_c, y_c, z_c).

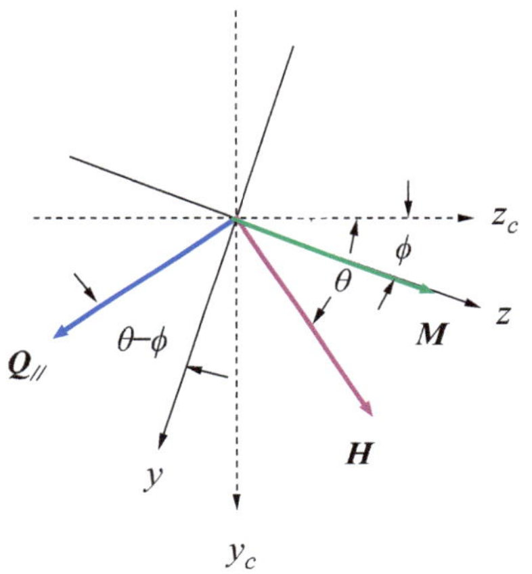

Figure 10. Schematic illustration on the magnetization rotation by the external magnetic field applied in the y–z plane. The magnetic field was applied along the θ direction measured from the crystal easy axis z_C. The rotation angle of the magnetization measured from the easy axis ϕ can be determined by Equation (65). BLS observed the SWs propagating along the y direction, which is always perpendicular to the external magnetic field.

Here, the x_c axis is along the surface normal direction, and the film surfaces are located at $x = \pm L/2$. The easy axis is along the z_c direction. We applied a magnetic field H to make an angle θ between the magnetic field H and the z_c axis, and we always measured SWs propagating perpendicular to H, as shown in Figure 10. Then, the magnetization M rotates from the easy axis. However, M will be not collinear with H because of the MAE. For convenience's sake, we introduced the magnetization coordinates (x,y,z) by rotating the crystallographic coordinates around the x_c axis by an angle ϕ. The z direction is along the magnetization direction. The rotation angle ϕ can be determined by the competition between the external magnetic field and the MAE as given by

$$H \sin(\theta - \phi) = \sin 2\phi \left(\frac{K_1}{M} + \frac{2K_2}{M} \sin^2 \phi \right) \tag{65}$$

We introduced the SW variable m as the small amplitude precession motion around the static magnetization M. The LL equation of motion on $M(t) = M + m(t)$ is already given in Equation (30). The effective magnetic field H_{eff} consists of the external magnetic field H, the uniaxial magnetic anisotropy field $H_K = -\nabla_M E_K$, the exchange field $H_{ex} = (D/M)\nabla^2 M$, and the demagnetization field $\boldsymbol{h}$ as follows:

$$H_{eff} = H + H_K + H_{ex} + h \tag{66}$$

When H and M are not collinear with each other, we must include the longitudinal components m_z and h_z in addition to the their transverse components. Because the MAE in Equation (39) is defined in the crystallographic coordinates, we must rewrite it in the magnetization coordinate variables in order to evaluate the anisotropy field. It is given by

$$\begin{aligned} E_K \cong & \sin 2\phi\left(K_1+2K_2\sin^2\phi\right)\left(\tfrac{m_y}{M}\right) \\ & +\left\{K_1\cos 2\phi+2K_2\left(\sin^2 2\phi-\sin^2\phi\right)\right\}\left(\tfrac{m_y}{M}\right)^2 \\ & +\cos^2\phi\left(K_1+2K_2\sin^2\phi\right)\left(\tfrac{m_x}{M}\right)^2. \end{aligned} \tag{67}$$

The linearized LL equation into a compact form given by

$$\frac{1}{\gamma}\frac{d}{dt}m_x=\left(H_1-D\nabla^2\right)m_y-Mh_y, \tag{68}$$

$$\frac{1}{\gamma}\frac{d}{dt}m_y=-\left(H_2-D\nabla^2\right)m_x+Mh_x, \tag{69}$$

and

$$\frac{1}{\gamma}\frac{d}{dt}m_z=\left(H\sin\theta\cos\phi-H_a\sin\phi\right)m_x=0 \tag{70}$$

Here, we define H_a, H_1, and H_2 as follows:

$$H_a=H\cos\theta+\left(\frac{2K_1}{M}+\frac{4K_2}{M}\sin^2\phi\right)\cos\phi, \tag{71}$$

$$H_1=H\cos(\theta-\phi)+\frac{2K_1}{M}\cos 2\phi+\frac{4K_2}{M}\left(\sin^2 2\phi-\sin^2\phi\right), \tag{72}$$

and

$$H_2=H\cos(\theta-\phi)+\frac{2K_1}{M}\cos^2\phi+\frac{K_2}{M}\sin^2 2\phi. \tag{73}$$

Equation (70) confirms that the longitudinal component m_z does not contribute to the SWs. We assume plane-wave-type space-time dependence for the dynamical variables:

$$\begin{pmatrix} m(r,t) \\ h_d(r,t) \end{pmatrix}=\begin{pmatrix} m \\ h_d \end{pmatrix}\exp\left[i\omega t-iq_\perp x-iQ_{//}\cos(\theta-\phi)y+iQ_{//}\sin(\theta-\phi)z\right] \tag{74}$$

The first equation of Equation (32) gives only two independent equations—for example,

$$(\nabla\times h)_y=Q_{//}\sin(\theta-\phi)h_x+q_\perp h_z=0 \tag{75}$$

and

$$(\nabla\times h_d)_z=q_\perp h_{dy}-Q_{//}\cos(\theta-\phi)h_{dx}=0 \tag{76}$$

Combining these equations, we obtain the $(\nabla\times h)_x$ component. The second equation of Equation (32) can be regarded as an additional equation of motion to the LL equation.

$$q_\perp(h_x+4\pi m_x)+Q_{//}\cos(\theta-\phi)\left(h_y+4\pi m_y\right)-Q_{//}\sin(\theta-\phi)h_z=0 \tag{77}$$

A set of five homogeneous equations for the five variables $\left(m_x,m_y,h_x,h_y,h_z\right)$ gives an equation for the nontrivial solutions:

$$\begin{aligned} & \left(P^2\right)^3+\tfrac{(H_1+H_2+4\pi M)}{DQ_{//}^2}\left(P^2\right)^2 \\ & +\tfrac{\left\{H_1(H_2+4\pi M)-4\pi MDQ_{//}^2\sin^2(\theta-\phi)-(\omega/\gamma)^2\right\}}{\left(DQ_{//}^2\right)^2}\left(P^2\right) \\ & +\tfrac{4\pi M\left\{H_2\cos^2(\theta-\phi)-H_1\right\}}{\left(DQ_{//}^2\right)^2}=0. \end{aligned} \tag{78}$$

Here, we define $P^2 = \left(q_\perp^2 + Q_{//}^2\right)/Q_{//}^2$ and adapt the partial wave technique in which SWs are constructed by a sum of bulk partial waves. By adjusting coefficients for the partial waves, we made the constructed SWs satisfy the proper boundary conditions. Equations (75) and (76) give six $q_\perp$ solutions allowed for the bulk partial waves [34].

Combining the linearized LL equation with the effective magnetic field given by Equation (66) and Maxwell's equations, as seen in Equation (32), we obtain in terms of the partial waves

$$m_x = \sum_{j=1}^{6} M \frac{\left(H_1 + DQ_j^2\right)h_{jx} - i\widetilde{\omega}h_{jy}}{\left(H_1 + DQ_j^2\right)\left(H_2 + DQ_j^2\right) - \widetilde{\omega}^2} = \sum_{j=1}^{6} M \frac{\left(H_1 + DQ_j^2\right) - i\widetilde{\omega}\frac{Q_{//}}{q_{\perp j}}\cos(\theta - \phi)}{\left(H_1 + DQ_j^2\right)\left(H_2 + DQ_j^2\right) - \widetilde{\omega}^2} h_{jx} \quad (79)$$

and

$$m_{\Gamma} = \sum_{j=1}^{6} M \frac{i\widetilde{\omega}h_{jx} + \left(H_2 + DQ_j^2\right)h_{jy}}{\left(H_1 + DQ_j^2\right)\left(H_2 + DQ_j^2\right) - \widetilde{\omega}^2} = \sum_{j=1}^{6} M \frac{i\widetilde{\omega} + \left(H_2 + DQ_j^2\right)\frac{Q_{//}}{q_{\perp j}}\cos(\theta - \phi)}{\left(H_1 + DQ_j^2\right)\left(H_2 + DQ_j^2\right) - \widetilde{\omega}^2} h_{jx} \quad (80)$$

Here, we introduced $\widetilde{\omega} = \omega/\gamma$ for convenience and used the demagnetization fields instead of the magnetic potential φ. By adapting the magnetic boundary conditions at $x = \pm L/2$, we can eliminate the demagnetization fields outside of the magnet and obtain a set of two equations given by

$$\sum_{j=1}^{6}\left[1 + 4\pi M \frac{\left(H_1 + DQ_j^2\right) - i\widetilde{\omega}\frac{Q_{//}}{q_{\perp j}}\cos(\theta - \phi)}{\left(H_1 + DQ_j^2\right)\left(H_2 + DQ_j^2\right) - \widetilde{\omega}^2} \pm i\frac{Q_{//}}{q_{\perp j}}\right]\exp(\mp i q_{\perp j}L/2)\, h_{xj} = 0 \quad (81)$$

The upper compound symbols in Equation (81) are for the top surface at $x = L/2$ and the lower symbols for the bottom surface at $x = -L/2$, respectively. Because we have six unknown variables h_{xj}, we need four more boundary conditions. Rado and Weertman derived generalized boundary conditions, which is known as the Rado–Weertman surface pinning conditions [26], based on the LL equation. It is given by

$$M \times \left(\frac{D}{M}\frac{\partial}{\partial n}m - \nabla_M E_{surf}\right)\bigg|_{x=\pm L/2} = 0 \quad (82)$$

where E_{surf} is the surface magnetic anisotropy (SMA) energy and we assume a SMA energy with the in-plane and out-of-plane terms as given by

$$E_{\text{surf}} = -k_\perp\left(\frac{m_x}{M}\right)^2 - k_{//}\left(\frac{m_y}{M}\right)^2 \quad (83)$$

Now, we have a set of six homogeneous equations for six unknown variables. The SW frequencies can be obtained by solving numerically a 6×6 boundary condition determinant (BCD) equation. When we have no SMEs (no surface pinning) and the limiting case of $Q_{//} \approx 0$, we can obtain asymptotic expression for the SW frequencies given by

$$\Delta\nu(n) = \frac{\gamma}{2\pi}\left[H + D\left(\frac{n\pi}{L}\right)^2\right]^{1/2}\left[H + D\left(\frac{n\pi}{L}\right)^2 + 4\pi M\right]^{1/2} \quad (n = 1, 2, \ldots) \quad (84)$$

and

$$\Delta\nu_{DE} = \frac{\gamma}{2\pi}\left[(H + 2\pi M)^2 - (2\pi M)^2\exp(-2Q_{//}L)\right] \quad (85)$$

These are exactly the same as Equations (62) and (63). As already we have seen, these expressions can describe well the magnetic field dependence of the SW frequencies.

Figure 11 shows an example of the SW frequencies of bcc Fe propagating with $Q_{//}$ = 1.85×10^5 cm^{-1} along the [110] direction on the (001) surface at H = 3.0 kOe as a function of film thickness down to 10 Å.

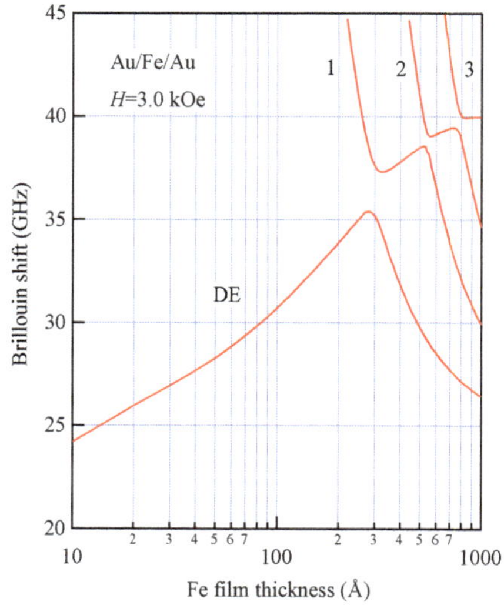

Figure 11. Calculated SW frequencies (—) propagating along the [110] direction on the epitaxial (001) Fe film as a function of the film thickness. The magnetic field was applied along the [1–10] direction, and the cubic MAE and the effective demagnetization factor $D_{\perp}$ were included in this calculation. The broken lines are just for eye.

The SW frequencies were obtained by solving the 6 × 6 BCD equation. We used a set of the magnetic parameters listed Table 1 in this calculation. It is known that the demagnetization factor 4π for thick films should be replaced by the effective demagnetization factor $4\pi f_{\perp}$. The $f_{\perp}$ factor was given by $1-0.4245/n$ for the bcc (001) structure and $1-0.2338/n$ for the fcc (001) structure [67]. Here, n is the number of the atomic layers stacked on the film. We have included the $4\pi D_{\perp}$ term in our calculations. In Figure 11, the label DE stands for the DE mode frequency, and the labels 1, 2, and 3 stand for the first, second, and third SSW frequencies. For Fe films with thicknesses less than 100 Å, only the DE mode appears in a BLS spectrum, and the SSW peaks appear well above 50 GHz. Another interesting observation is the anticrossing effect between the DE mode and each SSW mode. Therefore, it is clear that Equations (84) and (85) give a good description of the DE and SSW frequencies when these frequencies are well separated. For thinner films with thicknesses less than ~300 Å, the DE mode localized on the opposite surface may appear in a spectrum. The amplitude of the opposite DE mode was roughly estimated to be $\exp(-Q_{//}L)$~0.57 on the laser-illuminated surface. Of course, the spectrum exhibited quite asymmetric peak intensities.

Various structures of thin Co films, including polycrystalline films [63,68], bcc films [69–71], and fcc, bcc, and hcp films [72], have been extensively investigated by BLS since the early stage of the BLS SW studies. One of the most pronounced magnetic properties of Co in the hcp structure is the large uniaxial MAE, which is about one order of magnitude larger than the MAE of cubic Fe and Ni. The MAE for hcp Co is given by Equation (64). The K_1 and K_2 constants take positive values of ~10^6 erg/cm^3 at room temperature. Because of the large MAE, hcp Co-based alloys have been widely applied for many industrial applications. As I have already mentioned, it is possible to grow epitaxial $(10\bar{1}0)$ Co films which possess both the easy direction (hcp [001] axis) and the hard direction (hcp [010] axis) within a film plane. A BLS analysis from the epitaxial $(10\bar{1}0)$ Co thin films was performed by Grimsditch, Fullerton, and Stamps [73]. The SW stiffness constant of hcp Co has been a subject of controversy. The D_{NS} values of 490–510 meV·Å^2 for hcp Co crystals by the BNL group [58] were obviously larger than the recent BLS values of D_{BLS} = 430–470 meV·Å^2 [63,68,71,73].

We performed BLS measurements on the epitaxial hcp $(10\bar{1}0)$ Co thin films [73,74]. When we adapt the scattering geometry (A) in which the magnetic fields were applied along the easy direction, only the K_1 term in Equation (61) contributed to the SW frequencies because $\theta = \phi = 0°$ in Equations (72) and (73). An example of a BLS spectrum observed from the scattering geometry (A) is shown in Figure 6. The first and second order SSWs and the DE mode can be seen. The inset shows the mode profiles of these SSWs. In order to determine the K_2 constant, we must adapt the scattering geometries (B) and (C). The magnetic field dependence of the SW frequencies obtained from the scattering geometry (B) is shown in Figure 12.

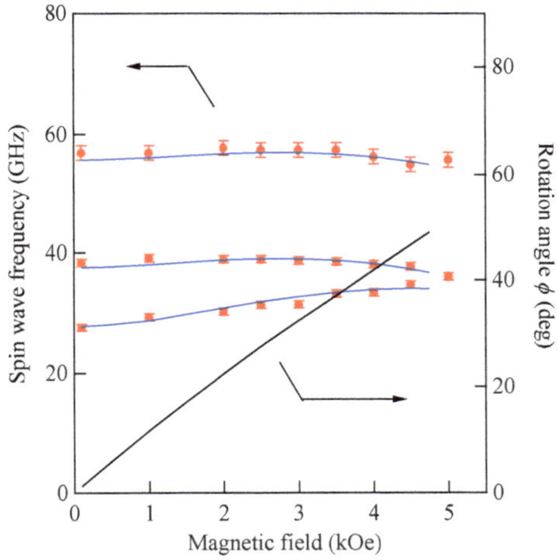

Figure 12. Magnetic field dependence of the SW frequencies (•) in the scattering geometry (B). The solid lines were calculated for the SW frequencies (–) and the rotation angle ϕ of the magnetization (–) using the full magnetic constants given in the text [74].

In this case, the magnetic field was always applied along the y_c-direction in Figure 10. The calculated SW frequencies and the rotation angle are shown by the solid lines. We summarized the parameters used in our calculations in Table 1. These MAE constants, K_1 and K_2, and the SW stiffness constant are in good agreement with the ones reported by Grimsditch et al. [73], The present SW stiffness constant of $D_{BLS} = 3.39 \times 10^{-9}$ Oe·cm^2 is equivalent to D_{BLS} = 427 meV·Å^2 and in good agreement with the previous values. There are independent NS results on hcp Co reported by using the Kraków neutron spectrometer [75]. Both NS groups employed a two-parameter model for SW dispersion given by $\varepsilon_Q = D_{NS}Q^2(1 - \beta Q^2)$. The Kraków group obtained $D_{NS} = 437 \pm 20$ meV·Å^2 and $\beta = 0.345$ Å^2, whereas the BNL group obtained a set of D_{NS} = 510 meV·Å^2, β = 1.8 Å^2, D_{NS} = 490 meV·Å^2, and β = 3.3 Å^2. The calculated dispersion curves with these three parameter sets are very close in the limited range of Q between 0.08Å and 0.25 Å. More extensive studies on bulk Co are strongly recommended.

The mode profile calculation revealed complicated mode conversion schemes as a function of the magnetic field. The lowest mode possessed a uniform amplitude across the film (n = 0 SSW mode) under zero magnetic field and gradually changed into the DE-like mode as field strength increased, finally changing into the n = 1 SSW mode at well above H = 5.0 kOe. The second-lowest mode retained the n = 1 SSW character up to H~5.0 kOe and finally changed into the DE mode at well above H = 5.0 kOe. Figure 13 shows the SW frequencies obtained from the scattering geometry (C) as a function of the field direction at $H = 3.0$ kOe.

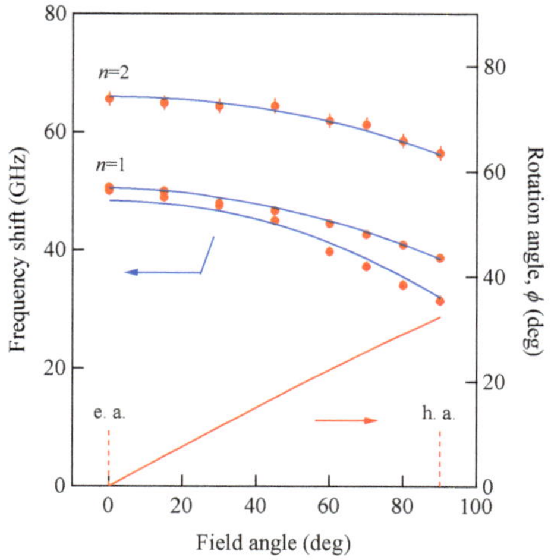

Figure 13. Field angle dependence of the SW frequencies (•) in the scattering geometry (C). The solid lines were calculated for the SW frequencies (−) and the rotation angle ϕ of the magnetization (−) using the full magnetic constants given in the text [74]. Here, e.a. and h.a. mean the easy axis (y_C-direction) and the hard axis (z_C-direction) in Figure 10.

The solid lines give the calculated SW frequencies and the rotation angle ϕ of the magnetization. We also performed the mode profile calculation in this geometry. The lowest-frequency mode at $\theta = 0°$ was the $n = 1$ SSW, which gradually changed into the 0th SSW mode at $\theta = 90°$. The second-lowest mode changed the character from the generalized DE mode at $\theta = 0°$ into the $n = 1$ SSW mode at $\theta = 90°$. As shown in Figure 6, the $n = 1$ SSW mode and the De mode frequencies were very close. In this interpretation, we took account of the anti-crossing effect shown in Figure 11.

The SW stiffness constants of the Heusler compounds and alloys, Co_2MnSi, $Co_2MnAl_xSi_{1-x}$, Co_2FeAl, and $Co_2Cr_{0.6}Fe_{0.4}Al$ were intensively investigated with BLS [76–78].

5.3. Ultrathin Films

Let us define magnetic films with thicknesses less than the exchange length $\ell_{ex} = (D/4\pi M)^{1/2}$ [67], which is estimated to be ~30 Å for Fe as an ultrathin film. It is possible to prepare epitaxial ultrathin films by means of the molecular beam epitaxy (MBE) technique. Magnetic properties of such ultrathin films will be strongly affected by the SMA [67]. Here we have a question: what does BLS observe from such ultrathin films? When the film thickness L goes to zero, the DE mode frequency is well-separated from the bulk SSW frequencies as shown in Figure 11. Because the surface dispersion parameter $Q_{//}L$ is quite small, the DE mode amplitude is almost uniform across the film. It seems to be preferable to name this the uniform DE (UDE) mode or simply as the uniform mode. I will use the term "UDE mode" in this section. The DE mode was originally discussed for an isotropic slab taking account of the magnetic boundary conditions. The UDE mode is quite different from the slab DE mode because of the SMAs. Then, how can we take into account the SMAs in our discussions? If it is possible, we can positively apply the BLS technique to investigate the SMAs of ultrathin films.

We consider an ultrathin film with surfaces at $x = 0$ and $-d$, and the LL equation given by

$$\frac{1}{\gamma}\frac{dM}{dt} = M \times \left(H + H_K + h_d + \frac{D}{M}\frac{d^2M}{dx^2} \right) \tag{86}$$

Here we ignored the in-plane exchange term. For an ultrathin film, the UDE mode profile is regarded as uniform across the film. Then, we can readily integrate the LL equation across the film and obtain

$$\frac{1}{\gamma}\frac{dM}{dt} = M \times (H + H_K + h_d) + \frac{1}{d} \cdot \frac{D}{M} M \times \left. \frac{dm}{dx} \right|_{x=-d}^{x=0} \tag{87}$$

When we notice $\partial/\partial n = -d/dx$ in this case, the last term in the right-hand side can be replaced by the Rado-Weertman pinning boundary condition given by Equation (82), thus producing

$$\frac{1}{\gamma}\frac{dM}{dt} = M \times \left(H + H_{K,eff} + h \right) \tag{88}$$

and

$$H_{K,eff} = -\nabla_M E_K - \frac{2}{d}\nabla_M E_{surf} \tag{89}$$

where $H_{K,eff}$ is the effective magnetic anisotropy field consisting of the MAE and IMA SMA terms. We assumed the same SMA for both surfaces. When we adapt the bulk MAE given by Equations (54) and (59) and the SMA given by Equation (83), we obtain

$$\frac{1}{\gamma}\frac{dm_x}{dt} - \left[H - \frac{2}{M}\left(K_{//} + \frac{2k_{//}}{d} \right) \right] m_y + Mh_{dy} = 0 \tag{90}$$

and

$$\frac{1}{\gamma}\frac{dm_y}{dt} + \left[H - \frac{2}{M}\left(K_{\perp} + \frac{2k_{\perp}}{d} \right) \right] m_x - Mh_{dx} = 0 \tag{91}$$

With these results, we obtain the UDE frequency as follows:

$$\left(\frac{\omega}{\gamma}\right)^2 = \left[H - \frac{2}{M}\left(K_{//} + \frac{2k_{//}}{d} \right) \right] \left[H - \frac{2}{M}\left(K_{\perp} + \frac{2k_{\perp}}{d} \right) + 4\pi M \right] \tag{92}$$

Here, we define the effective saturation magnetization as

$$4\pi M_{\text{eff}} = 4\pi M - \frac{2}{M}\left(K_{\perp} + \frac{2k_{\perp}}{d} \right) \tag{93}$$

The $1/d$ factor in Equation (93) gives a multiplication factor of $10^8/d$ in the angstrom unit, and the SMA term dominates the out-of-plane magnetic anisotropy field for the UDE mode. As already discussed, the $4\pi M_{\text{eff}}$ variable is the essential parameter to distinguish the magnetization state under zero magnetic field. For a positive $4\pi M_{\text{eff}}$, the film is in the in-plane magnetization state under zero magnetic field, and it is in the perpendicular magnetized state for a negative $4\pi M_{\text{eff}}$.

We performed a BLS study on an ultrathin epitaxial Fe wedge with Fe layer thicknesses up to 8.9 Å under the magnetic fields of up to 4.5 kOe [79]. The wedge was prepared at the Electrotechnical Laboratory (ETL), Tsukuba using the MBA technique. Figure 14 shows a schematic illustration of the structure of the MBE-prepared wedge with the crystallographic coordinate systems showing the epitaxial relations.

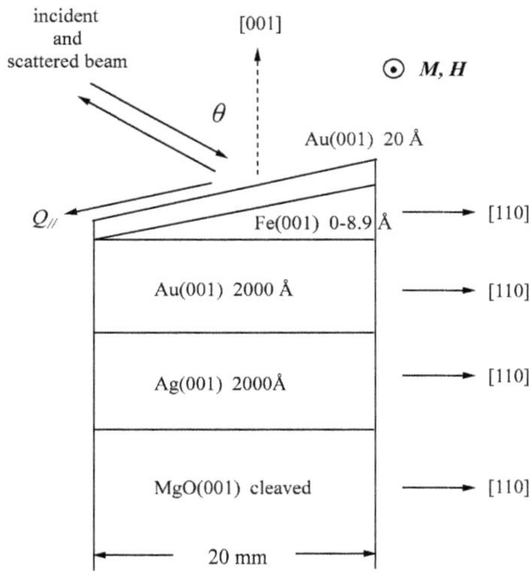

Figure 14. Schematic illustration of the crystallographic wedge structure and scattering geometry [79].

A 20 Å-thick Au (001) cap layer was deposited to protect the wedge from crucial surface deterioration. We applied the magnetic field along the crystallographic [1–10] direction, and the SWs propagating along the [110] direction were measured. For a cubic symmetry crystal, the MAE is given by

$$E_K = \frac{K_1}{M^4}\left(M_x^2 M_y^2 + M_y^2 M_z^2 + M_z^2 M_x^2\right) \tag{94}$$

When the magnetization is directed along the [1–10] direction, in terms of the SW variables, the MAE can be written as [35]

$$E_K \cong \text{const.} - \frac{K_1}{2M^2}\left(2m_y^2 - m_x^2\right) \tag{95}$$

Taking account of the possible tetragonal distortion along the surface normal direction, the SMA energy can be reduced into a simple form given by

$$\begin{aligned} E_{surf} &\cong -k_{//}^{(s)}\left(\frac{m_y}{M}\right)^2 - \left(k_u^{(s)} - \frac{k_{//}^{(s)}}{2}\right)\left(\frac{m_x}{M}\right)^2 \\ &= -k_{//}^{(s)}\left(\frac{m_y}{M}\right)^2 - k_{\perp}^{(s)}\left(\frac{m_x}{M}\right)^2, \end{aligned} \tag{96}$$

in which $k_u^{(s)}$ is the uniaxial out-of-plane SMA constant due to tetragonal distortion.

Figure 15 shows the thickness development of the BLS spectra of the wedge at $H = 3.0$ kOe. The thickness was indicated on each spectrum.

These spectra were excited by the *p*-polarized 4880 Å line of an Ar^+ ion laser in a single-cavity mode with a power of 80 mW directed at the wedge. The incident angle ϑ was fixed to 45° ($Q_{//} = 1.82 \times 10^5$ cm^{-1}). The Fe layer thickness was thin enough to observe scattering from the UDE modes existing on both surfaces. Typical accumulation time for a spectrum was only less than 1 h. Because we had not performed polarization selection, both the USW and SAW peaks were observed in a spectrum. The SAW peaks were masked by the Rayleigh peak in Figure 15. The intensity asymmetry between the Stokes side and the anti-Stokes side is one of the characteristic features of SW scattering, as already discussed. The UDE frequency rapidly decreased from about 20 GHz to 12 GHz with decreasing Fe thickness. Figure 16 shows the external field development of BLS spectra observed at an Fe thickness of d = 2.6 Å.

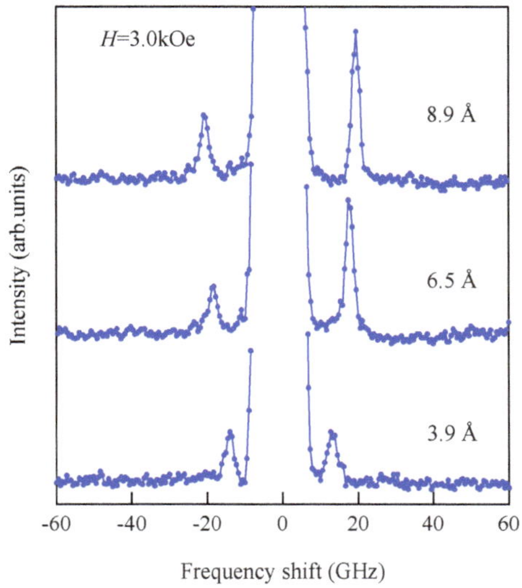

Figure 15. Thickness development of BLS spectra observed at H = 3.0 kOe. The thickness was indicated on each spectrum [79].

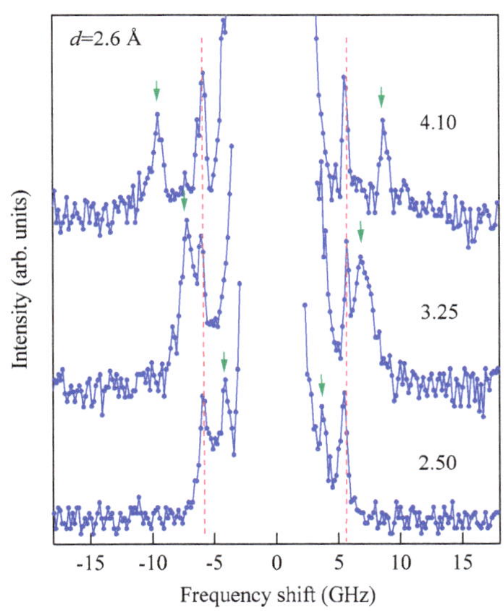

Figure 16. Magnetic field development of BLS spectra observed at d = 2.6 Å. The applied field is indicated on each spectrum. SW peaks are indicated by arrows. The peaks on the broken lines are due to the Rayleigh surface acoustic waves, which mainly reflect the elastic properties of the Au/Ag/MgO structure [79].

The applied magnetic field was indicated on each spectrum. The UDE peaks are indicated by the arrows, and the SAW peaks are indicated by the broken lines. The sharp UDE peaks shown in Figures 15 and 16 indicate that a well-defined ferromagnetic order was realized in the wedge at room temperature even at 2.6 Å, which corresponds to a ~1.8 atomic layer. Taking into account the extremely short photon–UDE interaction length, although multiple reflections would take place, the large BLS efficiency seems to have been closely related to the enhancement of the magneto-optical Kerr rotation and ellipticity observed around 2.5 eV of incident photon energy due to the plasma-edge effect of the Au layers [80]. This energy is close to our 4880 Å laser photon energy (2.54 eV). Figure 17 shows the intensity ratio between the Stokes peak and the anti-Stokes peak.

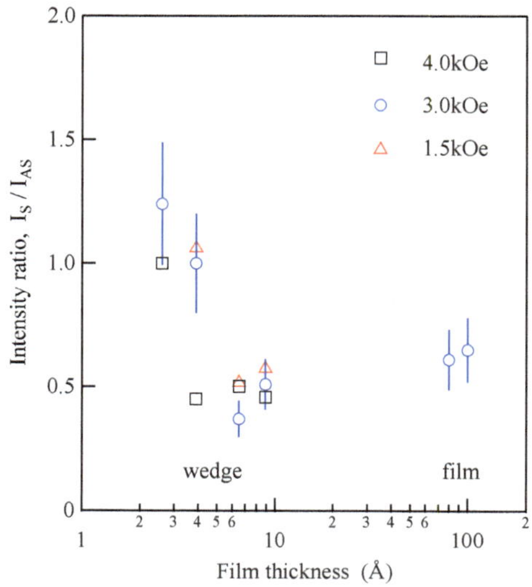

Figure 17. Film thickness and magnetic field dependences of the intensity ratio between the Stokes peak and the anti-Stokes peak of the Au/Fe/Au films and the wedge [61].

Above $d > 40$ Å, the intensity ratio was about 0.5 in a wide range of Fe thicknesses, whereas it became larger than 1 below $d < 40$ Å. The intensity ratio for the 45° incident *p*-polarized geometry can be written as follows in terms of the matrix components defined in Equations (15)–(18):

$$\frac{I_S}{I_{aS}} = \frac{\left|\bar{\varsigma}_{31}^{-} - \varsigma_{32}^{-}\right|^2}{\left|\bar{\varsigma}_{31}^{+} - \varsigma_{32}^{+}\right|^2} \tag{97}$$

As shown in Equation (29), the *K* coefficient and possibly also the *G* coefficient are sensitive to the spin-orbit coupling constant and the wave functions for the ground state and intermediate state. Because the ultrathin Fe layer was sandwiched by the thicker Au layers, the Fe wave functions were strongly modified by mixing between the Au wave functions at $d < 40$ Å. This situation can be regarded as a quantum well effect.

Figure 18 shows the thickness development of the SW frequencies of the Au/Fe/Au films, including the wedge and the epitaxial films, with thicknesses between 80 and 1000 Å for $H = 3.0$ kOe.

These thin Au/Fe/Au epitaxial films were also prepared at ETL, Tsukuba using the MBE technique. The solid circles are the DE and UDE mode frequencies, and the open circles are the 1-st SSW frequencies. The solid lines are the calculated SW frequencies with the same magnetic constants used in Figure 11. It is obvious that the observed UDE frequencies decreased more rapidly than the calculated frequency as the Fe thickness decreased below 100 Å. In contrast to the present Au/Fe/Au films, for the ultrathin Fe (110) film deposited on the W (110) substrate, the UDE frequency increased with decreasing Fe thickness because of the negative $k_{\perp}^{(s)}$ constant (this means that the surface normal is the hard direction) [34,81]. Because we have not included the SMA terms in Equations (92) and (93) in our calculations, our calculations gave higher SW frequencies below 100 Å.

Figure 19 shows the UDE frequencies as a function of the external magnetic field for four Fe thicknesses: 8.9, 6.5, 3.9, and 2.6 Å.

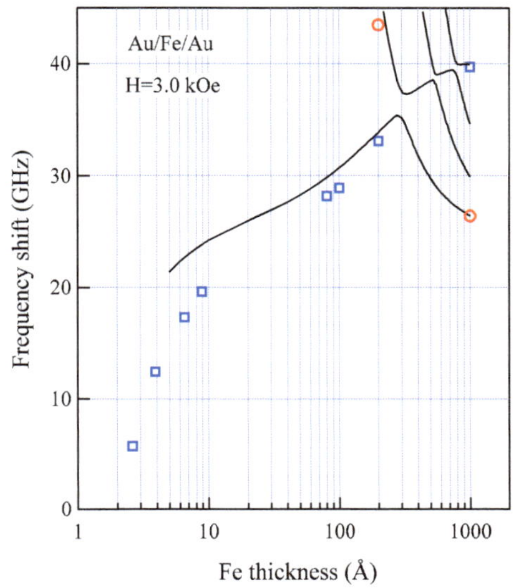

Figure 18. SW frequencies (□ for the DE mode, ○ for the $n = 1$ SSW mode) observed from the SWs propagating along the [110] direction on the epitaxial (001) thin Fe films and Fe wedge as a function of the Fe thickness. The solid lines are the calculated SW frequencies, and the results above $d \geq 1$ nm are already shown in Figure 11. Although the MAE and the effective demagnetization factor were included in the calculation, the surface perpendicular anisotropy term was not included [61]. The broken lines are just for eye.

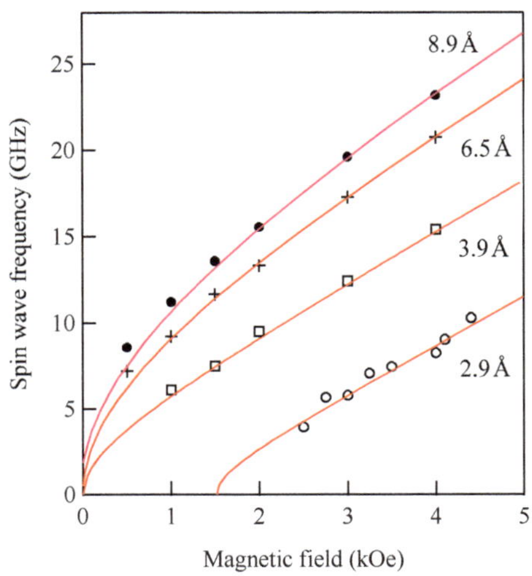

Figure 19. SW frequencies for four Fe thicknesses, 8.9, 6.5, 3.9, and 2.6 Å, as a function of the magnetic field. The solid lines are the SW frequencies calculated by solving numerically the BCD equation [79].

In order to analyze these results, we solved numerically the 6×6 BCD equation with the SMA terms given by Equation (95). The solid lines in Figure 19 are the calculated UDE frequencies by solving the BCD equation as a function of the magnetic field. Because we found that the bulk MAE was negligibly small, we retained only the SMA terms and a bulk spin wave stiffness constant of $D = 2.34 \times 10^{-9}$ Oe·cm^2 throughout the present BCD analyses, which can also be applied for the in-plane magnetized films. For d = 2.6 Å, we can solve the BCD equation above H = 1.57 kOe. Figure 20 shows the in-plane and out-of-plane SMA constants and the $4\pi M_{\text{eff}}$ defined in Equation (93) as a function of Fe thickness. Because the in-plane SMA constant $k_{//}^{(s)}$ was negligibly small, as shown in Figure 20, the out-of-plane SMA constant $k_{\perp}^{(s)}$ stems from the uniaxial SMA constant $k_u^{(s)}$, which reflects the tetragonal distortion of the bcc Fe layer. Although $k_{\perp}^{(s)}$ gradually decreased with decreasing Fe thickness, the d^{-1} factor in Equation (93) increased much faster than the decrease of $k_{\perp}^{(s)}$. Therefore, $4\pi M_{\text{eff}}$ changes its sign from positive to negative at around d~3 Å, and the stable direction of the magnetization under zero magnetic field turns over from the in-plane direction to the surface normal direction. The in-plane to out-of-plane transition of the stable magnetization configuration under zero magnetic field has been observed in several ultrathin films [81–84].

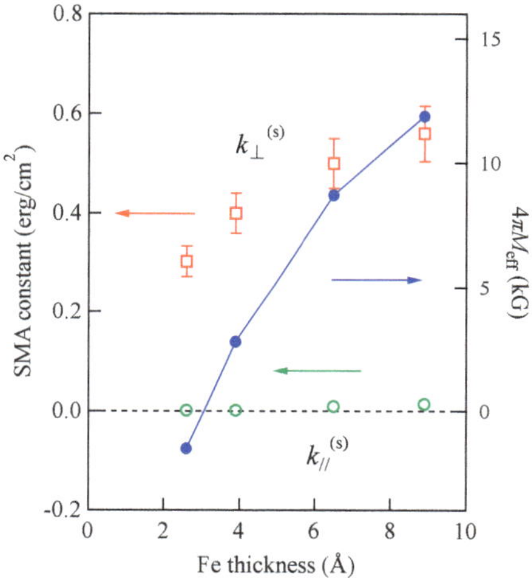

Figure 20. SMA constants for $k_{\perp}^{(s)}$ (□) and $k_{//}^{(s)}$ (○) and $4\pi M_{\text{eff}}$ (•) for four Fe thicknesses. The solid lines are just for eye [79].

Apart from the ultrathin films, let us consider SWs in magnetic films with the out-of-plane MAE as given by Equation (59). Dutcher et al. [82] and Rahman and Mills [85] treated this problem using the magnetostatic framework. Here, the $K_{\perp}$ coefficient can be regarded as an effective constant including both the bulk and surface terms, which depend on the crystallographic structure of the magnet and the film thickness. The magnet occupies the x–z plane between x = 0 and $-L$, and the applied magnetic field is always fixed to the z-axis. The magnetization always lies in the x–z plane with the equilibrium angle ϕ measured from the x-axis, as shown in Figure 21.

We adapted the LL equation of motion with the effective fields given by Equation (66). Because the magnetization $\boldsymbol{M}$ is tilted in the x–z plane, we should add the demagnetization field H_{d} term given by

$$H_d = -4\pi(M\cos\phi + m_x,\ 0,\ 0) \tag{98}$$

Then, we obtain a set of linearized LL equations:

$$\frac{1}{\gamma}\frac{dm_x}{dt} = \left(H - D\sin\phi\nabla^2\right)m_y - M\sin\phi\, h_y \tag{99}$$

$$\frac{1}{\gamma}\frac{dm_y}{dt} = M\cos\phi[H_C\sin\phi - H] + \cos\phi\left(H_C - D\nabla^2\right)m_z + D\sin\phi\nabla^2 m_x + M(\sin\phi\, h_x - \cos\phi\, h_z), \tag{100}$$

and

$$\frac{1}{\gamma}\frac{dm_z}{dt} = -\cos\phi\left(H_C - D\nabla^2\right)m_y + M\cos\phi\, h_y \tag{101}$$

Here, we define the critical field H_C as

$$H_C = \frac{2K_\perp}{M} - 4\pi M = H_{K\perp} - 4\pi M \tag{102}$$

The equilibrium angle ϕ is determined from the torque-free condition around the y axis given by

$$\cos\phi[H_C\sin\phi - H] = 0 \tag{103}$$

We obtain three angles:

(1) $\phi = \pi/2$ for $4\pi M > H_{K\perp}$,

(2) $\phi = \sin^{-1}(H/H_C)$ for $H_C \geq H > 0$,

and

(3) $\phi = \pi/2$ for $H > H_C > 0$.

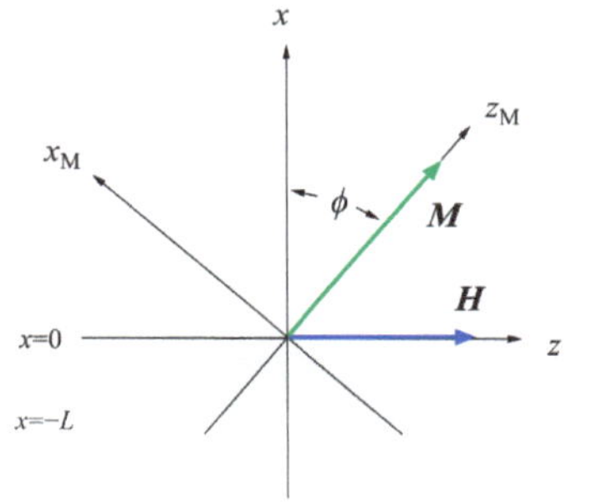

Figure 21. Schematic illustration on the magnetization rotation by the external magnetic field H applied the in-plane z direction ($H \leq H_C$). The equilibrium angle ϕ can be determined by Equation (103). By counterclockwise $\pi/2$-ϕ rotation around the y-axis, we can introduce the magnetization-based coordinates (x_M, y_M, z_M). In this coordinate choice, we can decouple the LL equation of motion on the z_M component from the set of equations on the x_M and y_M components.

We introduce a 3 × 3 susceptibility matrix χ by $m = \chi \cdot h$ through the LL equation. For the present scattering geometry shown in Figure 21, among the nine components of χ, four components, χ_{xx}, χ_{xy}, χ_{yx}, and χ_{yy}, are relevant, and their explicit expressions are given as

$$\chi_{xx}(\omega) = M\frac{\left(H + DQ^2\sin\phi\right)\sin\phi}{(\Omega/\gamma)^2 - (\omega/\gamma)^2}, \tag{104}$$

$$\chi_{yy}(\omega) = M\frac{\left(H_K - 4\pi M + DQ^2\right)\cos^2\phi + \left(H - \left(H_K - DQ^2\right)\sin\phi\right)\sin\phi}{(\Omega/\gamma)^2 - (\omega/\gamma)^2}, \tag{105}$$

and

$$\chi_{yx} = -\chi_{xy} = M\frac{i(\omega/\gamma)\sin\phi}{(\Omega/\gamma)^2 - (\omega/\gamma)^2} \tag{106}$$

We also define

$$(\Omega/\gamma)^2 = \left[H - \left(H_K - DQ^2\right)\sin\phi\right]\left(H + DQ^2\sin\phi\right) + \cos^2\phi\left(H_K - 4\pi M + DQ^2\right)^2 \tag{107}$$

We have obtained the relevant susceptibilities, and we can follow our calculations performed for the in-plane thin films. Therefore, we will not repeat them here. When we ignore the exchange terms, we can summarize our results on the bulk SW as follows:

For $4\pi M > H_{K\perp}$

$$\left(\frac{\omega}{\gamma}\right)^2 = H(H + 4\pi M - H_K) - 4\pi M H_K \frac{Q_{//}^2}{Q^2} \tag{108}$$

The upper and lower bound frequencies are given by

$$H(H + 4\pi M - H_{K\perp}) \qquad :Q_{//}^2/Q^2 = 0 \tag{109}$$

and

$$(H - H_{K\perp})(H + 4\pi M) \qquad :Q_{//}^2/Q^2 = 1 \tag{110}$$

For $H > H_C > 0$

$$\left(\frac{\omega}{\gamma}\right)^2 = H(H - H_C) - 4\pi M H_K \frac{Q_{//}^2}{Q^2} \tag{111}$$

The upper and lower bound frequencies are given by

$$H(H - H_C) \qquad :Q_{//}^2/Q^2 = 0 \tag{112}$$

and

$$(H - H_{K\perp})(H + 4\pi M) \qquad :Q_{//}^2/Q^2 = 1 \tag{113}$$

and for $H_C \geq H > 0$

$$\left(\frac{\omega}{\gamma}\right)^2 = {H_C}^2 - H^2 + \frac{Q_{//}^2}{Q^2}\left[H_C(H_K - H_C) - \frac{H_K^2}{H_C^2}H^2 + H^2\right] \tag{114}$$

We have two boundary frequencies given by

$$H_C^2 - H^2 \qquad :Q_{//}^2/Q^2 = 0 \tag{115}$$

and

$$H_C H_K - \frac{H_K^2}{H_C^2}H^2 \qquad :Q_{//}^2/Q^2 = 1 \tag{116}$$

Note that these boundaries are crossed at $H = H_C[H_C/(H_C + H_{K\perp})]^{1/2}$ in Equation (114). We also performed numerical fits using the upper bounds in Equations (108) and (111) with our BLS results shown in Figure 19, and we obtained reasonable agreement between the calculated and observed SW frequencies.

The discussions on the DE and UDE modes are a little more complicated. The surface mode frequency is determined by a BCD equation given as

$$\begin{vmatrix} \alpha(1 + 4\pi\chi_{xx}) - (1 - i4\pi\chi_{xy}) & \alpha(1 + 4\pi\chi_{xx}) + (1 - i4\pi\chi_{xy}) \\ \alpha(1 + 4\pi\chi_{xx}) + (1 + i4\pi\chi_{xy}) & [\alpha(1 + 4\pi\chi_{xx}) - (1 + i4\pi\chi_{xy})]\beta^2 \end{vmatrix} = 0 \tag{117}$$

Here we define $q_\perp = \alpha Q_{//} = [(1 + 4\pi\chi_{yy})/(1 + 4\pi\chi_{xx})]^{1/2} Q_{//}$ and $\beta^2 = \exp(-2\alpha Q_{//}L)$. We rewrite Equation (117) as

$$\tanh(\alpha Q_{//}L) = -\frac{2\alpha(1 + 4\pi\chi_{xx})}{\alpha^2(1 + 4\pi\chi_{xx})^2 + (1 - i4\pi\chi_{xy})(1 + i4\pi\chi_{xy})} \tag{118}$$

Note that the case of $\tanh(\alpha Q_{//}L) = 0$ corresponds to an ultrathin film, and the case of $\tanh(\alpha Q_{//}L) = 1$ corresponds to a semi-infinite magnet. It is not difficult to show that the DE and UDE modes are always allowed for in-plane films ($\phi = \pi/2$) with the frequency given by Equation (62). On the other hand, because of negative α, these surface modes are forbidden for the out-of-plane films above $H > H_C$.

Next let us consider an out-of-plane film under a magnetic field of $H \leq H_C$. For a semi-infinite magnet, Equation (115) becomes a simple form given by

$$\left[\alpha(1+4\pi\chi_{xx})+\left(1+i4\pi\chi_{xy}\right)\right]\left[\alpha(1+4\pi\chi_{xx})+\left(1-i4\pi\chi_{xy}\right)\right]=0 \qquad (119)$$

These equations can be analytically solved and give solutions with opposite signs. We solved the first one here and obtained a surface mode solution with a frequency given by $\omega/\gamma = H_C^2/2H$ in a limited range of the external field $\sqrt{H_C/2H_K} < H/H_C < 1/\sqrt{2}$. Figure 22 shows a simulation of the SW frequencies as a function of the magnetic field for an out-of-plane, magnetic, semi-infinite slab calculated using the magnetic constants obtained from the Fe wedge:

$$\gamma/2\pi = 2.8\text{ GHz/kOe},\ H_C = 1.57\text{ kOe},\ 4\pi M = 18.6\text{ kG, and } H_k = 20.17\text{ kOe}.$$

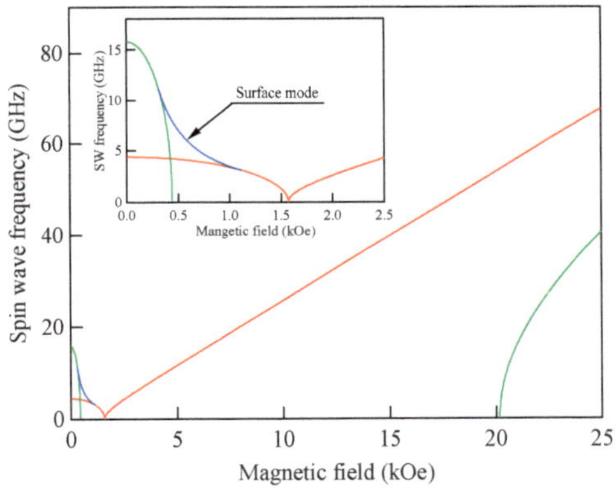

Figure 22. A simulation of the SW frequencies for an out-of-plane magnetized ferromagnetic slab as a function of the magnetic field. Here, the red lines and the green lines show the upper and lower bound of the bulk SW band. We used the magnetic constants obtained from the Fe wedge given in the text. The inset shows the lower field result below 2.5 kOe. Note the boundary crossing below H~0.4 kOe.

For an ultrathin film, Equation (118) merely gives the bulk upper and lower bounds, and we have no UDE solution, as already discussed. For a finite thickness film, we must solve Equation (118) numerically, but we will discuss no more on this case.

5.4. Multilayers and Superlattices

Figure 23 shows an example of a trilayer structure consisting of a nonmagnetic layer sandwiched between magnetic layers [35].

These magnetic layers need not be the same with respect to their thicknesses and materials [36]. The thicknesses of the magnetic layers are usually set to be less than 100 Å in order to fully separate the bulk SSWs and the DE mode as shown in Figure 11. In our following discussions based on [36], we consider the SWs in the trilayer by combining the DE modes in each magnetic layer. We have two choices on the origin of the x-coordinate. In this example, we use the first setting shown on the left-hand side of Figure 23.

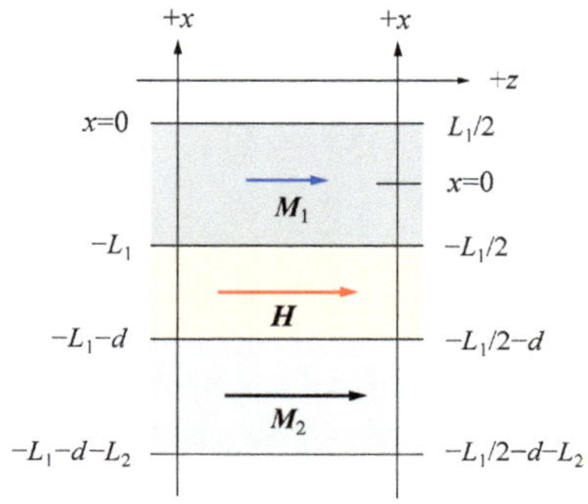

Figure 23. Schematic illustration of a (M1/spacer/M2) trilayer structure and the x-axis setting. We set $x = 0$ on top of the M1 layer in the first setting and in the center of the M1 layer in the second setting.

At first, we consider the SWs propagating along the y-axis in Figure 23 and solve the LL equation for each magnetic layer in terms of the susceptibilities and the magnetic potentials as follows:

$$m_x^{(j)} = \chi_{xx}^{(j)} h_x^{(j)} + \chi_{xy}^{(j)} h_y^{(j)} = -\left(\chi_{xx}^{(j)} \frac{\partial}{\partial x} + \chi_{xy}^{(j)} \frac{\partial}{\partial y}\right) \varphi^{(j)} \tag{120}$$

and

$$m_y^{(j)} = \chi_{yx}^{(j)} h_x^{(j)} + \chi_{yy}^{(j)} h_y^{(j)} = -\left(\chi_{yx}^{(j)} \frac{\partial}{\partial x} + \chi_{yy}^{(j)} \frac{\partial}{\partial y}\right) \varphi^{(j)} \tag{121}$$

Here, the superscript j (=1, 2) specifies the magnetic layer. The magnetic potentials outside and inside of the trilayer are given as follows:

$$\varphi = \varphi_> \exp(i\omega t - Q_{//}x \pm iQ_{//}y) \tag{122}$$

$$\varphi_{M1} = (A\exp(-Q_{//}x) + B\exp(Q_{//}x))\exp(i\omega t \pm iQ_{//}y) \tag{123}$$

$$\varphi_{NM} = (C\exp(-Q_{//}x) + D\exp(Q_{//}x))\exp(i\omega t \pm iQ_{//}y) \tag{124}$$

$$\varphi_{M2} = (E\exp(-Q_{//}x) + F\exp(Q_{//}x))\exp(i\omega t \pm iQ_{//}y) \tag{125}$$

and

$$\varphi = \varphi_< \exp(i\omega t + Q_{//}x \pm iQ_{//}y) \tag{126}$$

The magnetic boundary conditions at each surface and interface give us a set of six homogeneous equations for the unknown variables A to F. To obtain the non-trivial solutions of the set of the homogeneous equations, the BCD should vanish at the SW frequencies.

$$\begin{vmatrix} \Lambda_1^{(+)} - 1 & -\left(\Lambda_1^{(-)} + 1\right) & 0 & 0 & 0 & 0 \\ 1 & \alpha & -1 & -\alpha & 0 & 0 \\ \Lambda_1^{(+)} & -\alpha\Lambda_1^{(-)} & -1 & \alpha & 0 & 0 \\ 0 & 0 & 1 & \beta & -1 & -\beta \\ 0 & 0 & 1 & -\beta & -\Lambda_2^{(+)} & \beta\Lambda_2^{(-)} \\ 0 & 0 & 0 & 0 & \Lambda_2^{(+)} + 1 & -\gamma\left(\Lambda_2^{(-)} - 1\right) \end{vmatrix} = 0 \tag{127}$$

in which we defined

$$\alpha = \exp(-2Q_{//}L_1) \tag{128}$$

$$\beta = \exp(-2Q_{//}(L_1 + d)) \tag{129}$$

$$\gamma = \exp(-2Q_{//}(L_1 + d + L_2)) \tag{130}$$

and

$$\Lambda_\pm^{(j)} = 1 + 4\pi\chi_{xx}^{(j)} \pm i4\pi\chi_{xy}^{(j)} \tag{131}$$

For isotropic (no MAE) magnetic layers, the SW frequencies are given by the solution of Equation (127) as below:

$$\begin{aligned}
&\exp(-2Q_{//}d)[1-\exp(-2Q_{//}L_1)]\,[1-\exp(-2Q_{//}L_2)](2\pi M_1)(2\pi M_2) \\
&\times\left\{\left(H+2\pi M_1 \pm \tfrac{\omega}{\gamma_1}\right)\left(H+2\pi M_2 \mp \tfrac{\omega}{\gamma_2}\right)\right\} \\
&=\left\{(H+2\pi M_1)^2-(2\pi M_1)^2\exp(-2Q_{//}L_1)-\left(\tfrac{\omega}{\gamma_1}\right)^2\right\} \\
&\times\left\{(H+2\pi M_2)^2-(2\pi M_2)^2\exp(-2Q_{//}L_2)-\left(\tfrac{\omega}{\gamma_2}\right)^2\right\}.
\end{aligned} \tag{132}$$

The composite symbols in Equations (131) and (132) correspond to the SW propagation directions given by $\exp(\pm iQ_{//}y)$. It is clear that for identical magnetic layers, Equation (132) gives two SW frequencies which are independent of the propagation direction. When the surface dispersion parameter $Q_{//}d$ is zero, the SW frequencies are given by the DE mode frequency for a $2L$ thick film (set $L \rightarrow 2L$ in Equation (45)), and the bulk SW frequency is given by Equation (51). For the large $Q_{//}d$ values, the SW frequencies approach to the DE mode frequency is given by Equation (45) for two different isolated magnetic layers. Now, we consider parallel and anti-parallel arrangements of two different isotropic magnetic layers [35]. Figure 24a shows the SW frequencies as a function of the $Q_{//}d$ parameter for the parallel arrangement under zero magnetic field.

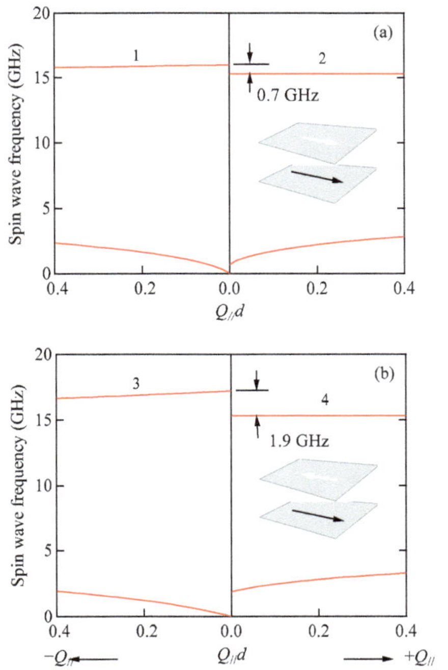

Figure 24. The solid lines are calculated SW frequencies in the (M1/spacer/M2) trilayer at zero magnetic field as a function of $Q_{//}d$ in which d is the spacer thickness. (**a**) Parallel arrangement and (**b**) anti-parallel arrangement. The magnetic and layer parameters used in the calculations are given in the text.

In the calculations, we set the surface wave vector $Q_{//}$ to 2.0 × 10^{-7} m^{-1} and used $4\pi M_1$ = 10.6 kG, g = 2.13, and L_1 = 50 Å for layer 1 (CoNbZr) and $4\pi M_2$ = 19.8 kG, g = 2.17, and L_2 = 150 Å for layer 2 (Co). Figure 24b shows the SW frequencies as a function of the $Q_{//}d$ parameter for the anti-parallel arrangement under zero magnetic field. It is important to note that the anti-parallel arrangement of the magnetizations is unstable even for weak external magnetic fields. We have various spin valve devices [86]. The spin valve structure consists of the pinned layer, in which the magnetization is pinned to prevent free motion against the external magnetic field, and the free layer, in which the magnetization can be freely aligned along the external field even under a weak magnetic field. Because the trilayer consists of two different magnets, the BLS spectrum is eventually asymmetric between the Stokes and anti-Stokes sides for the $Q_{//}d$ parameters below ~1. The frequency difference between levels 3 and 4 in the anti-parallel arrangement was 1.9 GHz, whereas the difference between levels 1 and 2 was 0.7 GHz in the parallel arrangement. Note that the frequency difference of 0.7 GHz is rather difficult to detect by means of the BLS technique; of course, it thus depends on the free spectral range setting. The lower frequency peaks were be masked by the intense Rayleigh peak. When the magnetization arrangement was changed from the parallel to anti-parallel arrangements, the frequency difference between the levels 3 and 1 was 2 GHz, and it was easily detected by the BLS technique.

For thinner spacer layers with thicknesses below the exchange length, we must take into account the interlayer exchange coupling (IEC) between the magnetic layers across the spacer layer [40]. The simplest form of IEC is given by the Heisenberg-type coupling given as

$$H_{\text{IEC}} = -\frac{A_{12}}{M_1 M_2}(M_1 \cdot M_2) \tag{133}$$

in which the IEC constant A_{12} depends on the spacer layer thickness d and changes its sign similar to the Ruderman–Kittel–Kasuya–Yoshida (RKKY) coupling [87]. For a positive A_{12}, the parallel arrangement of M_1 and M_2 is preferable, and the anti-parallel arrangement is preferable for negative A_{12}. The equations of motion for M_1 and M_2 are given as

$$\frac{1}{\gamma}\frac{dM_1}{dt} = \frac{A_{12}}{M_1 M_2}(M_1 \times M_2) \tag{134}$$

and

$$\frac{1}{\gamma}\frac{dM_2}{dt} = \frac{A_{12}}{M_1 M_2}(M_2 \times M_1) \tag{135}$$

The ICE acts as torque for the magnetization on each layer. We can take into account the torque through the Hoffmann boundary conditions [27] at the magnet–spacer interfaces. The Hoffmann boundary conditions are given by

$$M_1 \times \left[\frac{2A_1}{M_1^2}\frac{\partial M_1}{\partial n} - \nabla_M E_{\text{int}}^{(1)}\right] + \frac{2A_{12}}{M_1 M_2} M_1 \times M_2 = 0 \tag{136}$$

and

$$M_2 \times \left[\frac{2A_2}{M_2^2}\frac{\partial M_2}{\partial n} - \nabla_M E_{\text{int}}^{(2)}\right] + \frac{2A_{12}}{M_1 M_2} M_2 \times M_1 = 0 \tag{137}$$

Here, A_j (=$D_j M_j$/2) is the intra-layer exchange stiffness constant in the j-th magnetic layer, and $\partial/\partial n$ is always directed to the inside of the magnets. For brevity's sake, we adapted the same energy form for both of the interfacial magnetic anisotropy energy (IME) and the SME.

$$E_{\text{int}}^{(j)} = -\lambda_{\perp}^{(j)}\left(\frac{m_x^{(j)}}{M_j}\right)^2 - \lambda_{//}^{(j)}\left(\frac{m_y^{(j)}}{M_j}\right)^2 \quad (j = 1, 2) \tag{138}$$

After some calculations, we obtained the explicit expressions of the linearized Hoffmann boundary conditions as follows [40]:

For the magnetic layer 1:

$$\left(\frac{A_1}{M_1}\frac{\partial m_y^{(1)}}{\partial x}+\frac{\lambda_{//}^{(1)}}{M_1}m_y^{(1)}-\frac{A_{12}}{M_1}m_y^{(1)}\right)\Bigg|_{x=-L_1}+\frac{A_{12}}{M_2}m_y^{(2)}\Big|_{x=-L_1-d}=0 \tag{139}$$

and

$$\left(\frac{A_1}{M_1}\frac{\partial m_x^{(1)}}{\partial x}+\frac{\lambda_{\perp}^{(1)}}{M_1}m_x^{(1)}-\frac{A_{12}}{M_1}m_x^{(1)}\right)\Bigg|_{x=-L_1}+\frac{A_{12}}{M_2}m_x^{(2)}\Big|_{x=-L_1-d}=0 \tag{140}$$

For the magnetic layer 2:

$$\left(\frac{A_2}{M_2}\frac{\partial m_y^{(2)}}{\partial x}-\frac{\lambda_{//}^{(2)}}{M_2}m_y^{(2)}+\frac{A_{12}}{M_2}m_y^{(2)}\right)\Bigg|_{x=-L_1-d}-\frac{A_{12}}{M_1}m_y^{(1)}\Big|_{x=-L_1}=0 \tag{141}$$

and

$$\left(\frac{A_2}{M_2}\frac{\partial m_x^{(2)}}{\partial x}-\frac{\lambda_{\perp}^{(2)}}{2M_2}m_x^{(2)}+\frac{A_{12}}{M_2}m_x^{(2)}\right)\Bigg|_{x=-L_1-d}-\frac{A_{12}}{M_1}m_x^{(1)}\Big|_{x=-L_1}=0 \tag{142}$$

Because the Hoffmann boundary conditions were adapted at the magnet–spacer interfaces and the Rado–Weertman boundary conditions at the top and bottom magnetic surfaces, we must use the dipole-exchange framework which has been already discussed for thin films and results in a 6×6 BCD equation to determine the SW frequencies for a single magnetic film. In the present case, we have two mutually coupled magnetic layers and must solve a 12×12 BCD equation to obtain the SW frequencies. We have already given the explicit expressions of m_x and m_y in Equations (37) and (38). Grünberg and his coworkers successfully showed the oscillatory behavior of the A_{12} constant as a function of the spacer layer thickness in Fe/Cr/Fe trilayers using the BLS technique [40]. They found the giant magnetoresistance (GMR) effect in the anti-parallel state of the layer magnetizations [88]. In the anti-parallel magnetization state, the spin-flop phenomena can be induced by the external magnetic field [89,90].

In magnetic multilayers (MMLs) and superlattices (MSLs), a fundamental structure, for example magnetic layer 1 on magnetic layer 2, is stacked an arbitrary number of times. SWs, as a whole of the multilayers and superlattices, are constructed from the DE or UDE mode in each magnetic layer. As an example, we consider a $(M1/M2)_3$ multilayer. Here, the $(M1/M2)_3$ symbol means that the fundamental unit consisting of magnetic layers M1 and M2 is repeated three times. We can successively apply the appropriate boundary conditions at each interface and solve a BCD equation to obtain the MML and MSL SW frequencies [37]. In this example, we employed the second setting for the origin of the x-axis shown in the right-hand side of Figure 23. We set $x = 0$ at the center of the top M1 layer. When we set susceptibilities to zero in the M2 layers, the M2 layers can be treated as nonmagnetic spacer layers. In contrast, when a nonmagnetic layer undergoes a magnetic phase transition, for example Fe/Gd multilayers, we can examine SW dynamics near the phase transition. In order to make our discussions clear and easy, we assumed no MAE and IEC for the magnetic layers. Because the calculations are rather straightforward but tedious even under these simplifications, I only show the final result. It is a 12 × 12 BCD equation given by Equation (143).

$$
\begin{vmatrix}
m_a & m_b & & & & & & & & & & \\
m_1 & m_2 & m_3 & m_4 & & & & & & & & \\
m_5 & m_6 & m_7 & m_8 & & & & & & & & \\
 & & m_9 & m_{10} & m_{11} & m_{12} & & & & & & \\
 & & m_{13} & m_{14} & m_{15} & m_{16} & & & & & & \\
 & & & & m_1 & m_2 & m_3 & m_4 & & & & \\
 & & & & m_5 & m_6 & m_7 & m_8 & & & & \\
 & & & & & & m_9 & m_{10} & m_{11} & m_{12} & & \\
 & & & & & & m_{13} & m_{14} & m_{15} & m_{16} & & \\
 & & & & & & & & m_1 & m_2 & m_3 & m_4 \\
 & & & & & & & & m_5 & m_6 & m_7 & m_8 \\
 & & & & & & & & & & m_c & m_d
\end{vmatrix} = 0 \tag{143}
$$

Here, we give m_a to m_d elements in a matrix form for convenience's sake.

$$
\begin{pmatrix}
m_a & m_b & & & & \\
m_1 & m_2 & m_3 & m_4 & & \\
m_5 & m_6 & m_7 & m_8 & & \\
 & & m_9 & m_{10} & m_{11} & m_{12} \\
 & & m_{13} & m_{14} & m_{16} & m_{16} \\
 & & & & m_c & m_d
\end{pmatrix}
= \begin{pmatrix}
1-\Lambda_+^{(1)} & \alpha^{-1}\left(1+\Lambda_-^{(1)}\right) & & & & \\
1 & \alpha & -1 & -\alpha & & \\
\Lambda_+^{(1)} & -\alpha\Lambda_-^{(1)} & -\Lambda_+^{(2)} & \alpha\Lambda_-^{(2)} & & \\
 & & \gamma^{-1} & \beta & -1 & -\alpha^{-1} \\
 & & \gamma^{-1}\Lambda_+^{(2)} & -\beta\Lambda_-^{(2)} & -\Lambda_+^{(1)} & \alpha^{-1}\Lambda_-^{(1)} \\
 & & & & \gamma\left(1+\Lambda_+^{(2)}\right) & \beta\left(\Lambda_-^{(2)}-1\right)
\end{pmatrix}, \tag{144}
$$

in which $\alpha = \exp(-Q_{//}L_1)$, $\beta = \exp(-Q_{//}L_2)$, $\gamma = \exp(-Q_{//}\Lambda)$ ($\Lambda = L_1 + L_2$), and $\Lambda_\pm^{(j)}$ (j = 1, 2) have been defined in Equation (131). The rest of the determinant and matrix elements are all zero. It is clear that the elements m_1 to m_{16} appear as a set in Equation (143). This set is the algebraic description of the M1/M2 structure. When we insert an additional set into Equation (143), we have SWs in the $(M1/M2)_4$ multilayer. In this way, we can generate an arbitrary number of (M1/M2) stacking MMLs. This approach is quite intuitive and, of course, it is possible to take into account the MAEs and IECs into the above framework of the SW frequency calculations; to do so, we must solve a huge and complicated BCD equation.

Another approach to obtain the MSL SWs was developed by Camley, Rahman, and Mills [39]. For an infinite stack of periodic structures with the periodic length Λ, the property of translational invariance gives us Bloch's theorem. In the present case, Bloch's theorem requires the magnetic potential $\varphi(x)$ to satisfy

$$
\varphi(x+\Lambda) = \exp(-iQ_\perp\Lambda)\varphi(x) \tag{145}
$$

and

$$
\varphi(x) = \exp(-iQ_\perp x)u(x) \tag{146}
$$

where $u(x)$ is a periodic function which satisfies $u(x + \Lambda) = u(x)$, and $Q_\perp$ is a wave vector confined to the first Brillouin zone, $0 \leq Q_\perp \leq \pi/\Lambda$. For nth layer, we can write the magnetic potential in the nth layer as

$$u^{(M1)}(x) = \exp[iQ_\perp(x+n\Lambda)][A\exp\{-Q_{//}(x+n\Lambda)\} + B\exp\{Q_{//}(x+n\Lambda)\}] \quad (147)$$

for $0 \leq x + n\Lambda \leq -L_1$ and

$$u^{(M2)}(x) = \exp[iQ_\perp(x+n\Lambda)][C\exp\{-Q_{//}(x+n\Lambda)\} + D\exp\{Q_{//}(x+n\Lambda)\}] \quad (148)$$

for $-L_1 \leq x + n\Lambda \leq -\Lambda$, respectively. By virtue of Bloch's theorem, four amplitude variables, A to D, are enough for our discussions. We consider MSL SWs propagating along the y-direction and apply the magnetic boundary conditions at $x = -\Lambda$ and $-\Lambda - L_1$. Finally, we obtain a 4 × 4 BCD equation to determine the bulk SL SW frequencies.

$$\begin{vmatrix} \alpha\delta & \delta & -\beta^{-1} & -\gamma \\ 1 & \alpha & -1 & -\alpha \\ \alpha\delta\Lambda_+^{(1)} & -\delta\Lambda_-^{(1)} & -\beta^{-1}\Lambda_+^{(2)} & \gamma\Lambda_-^{(2)} \\ \Lambda_+^{(1)} & -\alpha\Lambda_-^{(1)} & -\Lambda_+^{(2)} & \alpha\Lambda_-^{(2)} \end{vmatrix} = 0 \quad (149)$$

in which $\alpha = \exp(-Q_{//}L_1), \beta = \exp(-Q_{//}L_2), \gamma = \exp(-Q_{//}\Lambda)$, and $\delta = \exp(iQ_\perp\Lambda)$. When we replace magnetic layer 2 with a nonmagnetic layer, $\Lambda_\pm^{(2)} = 1$, Equation (149) can be reduced into

$$[1+\Lambda_+\Lambda_-]\sinh Q_{//}L_1 \sinh Q_{//}L_2 + (\cosh Q_{//}L_1 \cos Q_{//}L_2 - \cos Q_\perp\Lambda)[\Lambda_+ + \Lambda_-] = 0 \quad (150)$$

We write $\Lambda_\pm^{(1)}$ as $\Lambda_\pm$ for convenience's sake. The bulk MSL SW frequency is found to be

$$\left(\frac{\omega}{\gamma}\right)^2 = \frac{H(H+4\pi M)}{1+\Delta} + \frac{\Delta}{1+\Delta}\left[H^2 + (H+4\pi M)^2\right] \quad (151)$$

Here, the MSL band factor $\Delta(Q_{//}, Q_\perp)$ is defined by

$$\Delta(Q_{//}, Q_\perp) = \frac{\sinh Q_{//}L_1 \sinh Q_{//}L_2}{\cosh Q_{//}L_1 \cos Q_{//}L_2 - \cos Q_\perp\Lambda} \quad (152)$$

For a semi-infinite MSL, we cannot apply Bloch's theorem because of the MSL surface, which violates the translational invariance. Taking into account the infinite MSL discussions, let us assume the magnetic potentials at the nth period are as follows:

$$\varphi_n^{(M1)}(x) = \exp(-\varepsilon n\Lambda)[A\exp\{-Q_{//}(x+n\Lambda)\} + B\exp\{Q_{//}(x+n\Lambda)\}] \quad (153)$$

for $0 \leq x + n\Lambda \leq -L_1$ and

$$\varphi_n^{(M2)}(x) = \exp(-\varepsilon n\Lambda)[C\exp\{-Q_{//}(x+n\Lambda)\} + D\exp\{Q_{//}(x+n\Lambda)\}] \quad (154)$$

for $-L_1 \leq x + n\Lambda \leq -\Lambda$. Here, ε is a positive attenuation parameter of MSL surface mode. We can eliminate C and D from the boundary conditions and obtain another homogeneous equation on A and B given as

$$\begin{aligned} &\left(\Lambda_+^{(2)} - \Lambda_+^{(1)}\right)[1 - \exp(-Q_{//}L_1)\exp(-\varepsilon\Lambda)\exp(Q_{//}L_2)]A \\ &\quad + \left(\Lambda_+^{(2)} + \Lambda_-^{(1)}\right)[\exp(-Q_{//}L_1) - \exp(-\varepsilon\Lambda)\exp(Q_{//}L_2)]B = 0, \end{aligned} \quad (155)$$

and

$$\left(\Lambda_+^{(1)} + \Lambda_-^{(2)}\right)[\exp(-Q_{//}L_1)\exp(-\varepsilon\Lambda) - \exp(Q_{//}L_2)]A + \left(\Lambda_-^{(2)} - \Lambda_-^{(1)}\right)[\exp(-\varepsilon\Lambda) - \exp(Q_{//}L_2)\exp(-Q_{//}L_1)]B = 0. \quad (156)$$

From Equations (155) and (156), the attenuation parameter ε is required to satisfy

$$2\left[\Lambda_+^{(2)}\Lambda_-^{(2)} + \Lambda_+^{(1)}\Lambda_-^{(1)}\right]\sinh(Q_{//}L_2)\sinh(Q_{//}L_1) + \left(\Lambda_+^{(2)}\Lambda_-^{(1)} + \Lambda_+^{(1)}\Lambda_-^{(2)}\right)[\cosh\{Q_{//}(L_2 - L_1)\} - \cosh(\varepsilon\Lambda)] + \left(\Lambda_+^{(1)}\Lambda_+^{(2)} + \Lambda_-^{(1)}\Lambda_-^{(2)}\right)[\cosh\{Q_{//}(L_2 + L_1)\} - \cosh(\varepsilon\Lambda)] = 0. \quad (157)$$

Finally, we must include the boundary conditions at the surface. By eliminating the magnetic potential outside MSL, we obtain

$$\alpha\left[\Lambda_+^{(1)} - 1\right]A - \left[\Lambda_-^{(1)} + 1\right]B = 0 \quad (158)$$

The dispersion relation of the MSL surface SWs is given as

$$\begin{vmatrix} \left(\Lambda_+^{(2)} - \Lambda_+^{(1)}\right)(1 - \alpha\beta^{-1}\kappa) & \left(\Lambda_-^{(1)} + \Lambda_+^{(2)}\right)(\alpha - \beta^{-1}\kappa) \\ \alpha\left(\Lambda_+^{(1)} - 1\right) & -\left(\Lambda_-^{(1)} + 1\right) \end{vmatrix} = 0. \quad (159)$$

Here, we defined $\kappa = \exp(-\varepsilon\Lambda)$. Equation (159) gives the dispersion relation of the surface modes.

$$\left(\Lambda_-^{(1)} + 1\right)\left(\Lambda_+^{(2)} - \Lambda_+^{(1)}\right)\left(1 - \alpha\beta^{-1}\kappa\right) + \left(\Lambda_+^{(1)} - 1\right)\left(\Lambda_-^{(1)} + \Lambda_+^{(2)}\right)\left(\alpha^2 - \alpha\beta^{-1}\kappa\right) = 0 \quad (160)$$

Now, we replace magnetic layer 2 by a nonmagnetic layer again and write $\Lambda_\pm^{(1)}$ as $\Lambda_\pm$ for convenience's sake. Equation (160) gives

$$\begin{gathered}(\Lambda_- + 1)(1 - \Lambda_+)(1 - \alpha\beta^{-1}\kappa) + (\Lambda_+ - 1)(\Lambda_- + 1)(\alpha^2 - \alpha\beta^{-1}\kappa) = 0 \\ (\Lambda_+ - 1)(\Lambda_- + 1)\sinh Q_{//}L_1 = 0\end{gathered} \quad (161)$$

Equation (161) gives us three solutions: $\Lambda_+ - 1 = 0$, $\Lambda_- + 1 = 0$, and $\sinh Q_{//}L_1 = 0$. However, all three of these solutions cannot be the eligible surface SWs. Readers should note that $\Lambda_\pm$ in this review corresponds to $\Lambda_\mp$ in the original paper. For example, we obtained $A \neq 0$ and $B = 0$ for the $\Lambda_+ - 1 = 0$ solution from Equation (158). Equation (155) is automatically satisfied, and Equation (156) gives $\exp(-Q_{//}L_1)\exp(-\varepsilon\Lambda) - \exp(Q_{//}L_2) = 0$. Because this equation gives a negative ε, the $\Lambda_+ - 1 = 0$ solution cannot be an eligible surface mode. On the other hand, the $\Lambda_- + 1 = 0$ solution gives a positive ε for $L_1 > L_2$ and thus can be an eligible surface mode. In fact, Grimsditch et al. observed BLS from the surface mode only for an MSL with a magnetic layer thicker than the nonmagnetic spacer layer [91]. The surface mode frequency was found to be [39]

$$\frac{\omega}{\gamma} = H + 2\pi M \quad (162)$$

This frequency is exactly the same as the DE mode frequency given by Equation (52) for a semi-infinite magnet. We can readily reject the $\sinh Q_{//}L_1 = 0$ solution for SWs propagating along the y direction because of the real quantity $Q_{//}L_1 \neq 0$. This mode can exist within the bulk SW band given by Equation (41).

Because a large number of reports on BLS from MMLs and MSLs have been already published, I have not enough space to mention them. I recommend that the reader refer to the references in this review and the most recent publications.

We examined the SWs in [Fe (30 Å)/Cr(x Å)] (x = 8–60 Å) MMLs prepared at IMR by rf-sputtering on quartz substrate [92]. The total thickness of each MML was fixed at ~1000 Å. BLS spectra were excited by the 4880 Å line from an Ar^+ laser with a power of ~80 mW with $\vartheta = 15^\circ$ ($Q_{//} = 0.67 \times 10^5\ \mathrm{cm}^{-1}$). Figure 25a shows the SW frequencies as a function of the magnetic field for Fe (30 Å)/Cr(21 Å) MML. The Fe (30 Å)/Cr(21 Å) MML shows a typical ferromagnetic loop.

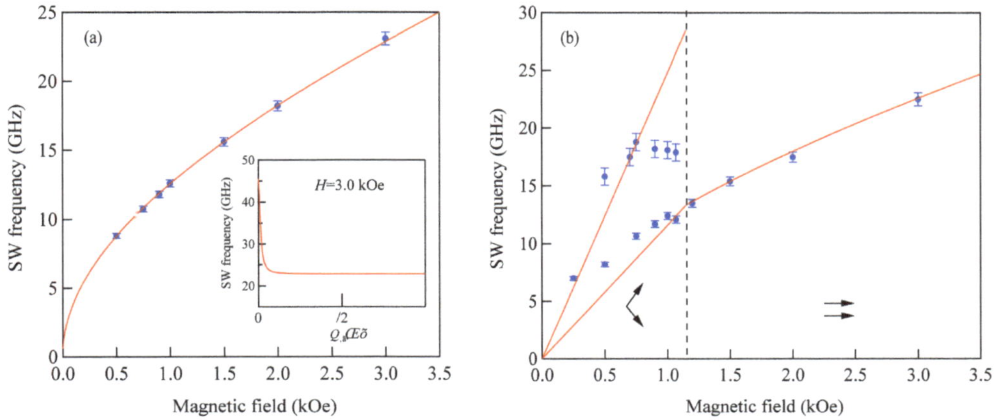

Figure 25. (**a**) SW frequencies as a function of the magnetic field for the [Fe(30 Å)/Cr(21 Å)] MML. The solid lines show the calculated SW frequencies by using Equation (151) with $Q_{//} = 0.67 \times 10^5\ \mathrm{cm}^{-1}$, $\gamma/2\pi = 2.88$ GHz/kOe, and $4\pi M = 18.0$ kG. The inset shows the SW frequencies as a function of the $Q_{//}\Lambda$ parameter. (**b**) SW frequencies as a function of the magnetic field for [Fe(30 Å)/Cr(13 Å)] MML. The solid lines were calculated by using Equations (162)–(164) below $H = 1.15$ kOe and by Equation (151) above $H = 1.15$ kOe [92].

The solid line in Figure 25a was calculated by using Equations (151) and (152) with $g = 2.06$ ($\gamma/2\pi = 2.88$ GHz/kOe), $4\pi M = 18.0$ kG, $L_1 = 30$ Å, and $L_2 = 21$ Å. The calculated line shows excellent agreement with the observed frequencies. The inset shows the SW frequencies at $H = 3.0$ kOe as a function of the $Q_\perp\Lambda$ parameter in Equation (152). It is clear that the SW band frequency became independent of the $Q_\perp\Lambda$ parameter above $Q_\perp\Lambda$~$\pi/10$. It means that BLS observed the lower bound of the bulk SW band. Because the magnetic layer thickness of the tested MML was thicker than the nonmagnetic Cr spacer layer, we expected to observe scattering from the surface SW peak at the frequency given by Equation (162). The surface SW frequency was expected to be ~27 GHz at $H = 0.5$ kOe and ~35 GHz at $H = 3.0$ kOe. Because our FPI mirror spacing was set to 5 mm in this study, the surface SW peak was unfortunately masked by the intense ghost peak from the adjacent interference order. Figure 25b shows the SW frequencies as a function of the magnetic field for Fe (30 Å)/Cr(13 Å) MML. I performed an additional measurement at $H = 0.9$ kOe and added the result in Figure 25b. The in-plane hysteresis loop from the Fe (30 Å)/Cr(13 Å) MML indicates the antiferromagnetic structure of the magnetizations between adjacent Fe layers under zero magnetic field. The loop indicates the in-plane canted structure below ~1.2 kOe and the ferromagnetic aligned structure above ~1.2 kOe. We can define the transition of the magnetic field H_C from the canted state to the ferromagnetic state. We observed an SW doublet on both frequency sides below $H = 1.1$ kOe. To the contrary, we observed an SW singlet above 1.2 kOe. When we take account the hysteresis loop result, the SW doublet is clearly related to the canted magnetization state. Generally speaking, the magnetic unit cell in the canted state is essentially the same as in the antiferromagnetic state and is double the unit cell in the ferromagnetic state. Therefore, we can expect a doublet of SW peaks in the canted state and a singlet bulk SW peak in the ferromagnetic state. Nörtemann et al. calculated SW frequencies of the dipole modes in an exchange-coupled MML with a canted ground state in terms of the effective-medium theory [90]. The most striking feature of their results is the mode crossing between the upper bulk SW band and the surface mode around ~$H_C/2$. For a semi-infinite canted stack without the MAE and the SWs propagating perpendicular to the net magnetization, the SW frequencies, except the mode crossing region around ~$H_C/2$, in the canted state are given as

$$\omega_1/\gamma = (H)^{1/2}(H + 4\pi M_{SL}\cos\alpha)^{1/2} \tag{163}$$

and

$$\omega_2/\gamma = H + 2\pi M_{SL}\cos\alpha \tag{164}$$

The equilibrium canting angle α (note that 2α is the true canting angle between the adjacent magnetizations) is given as

$$\cos\alpha = \frac{H}{2(H_E/N)} = \frac{H}{H_C} \tag{165}$$

Here, H_E is the interlayer exchange field and N is the number of atomic layers in the MML. The solid lines in Figure 25b were calculated by Equations (163)–(165) below H_C and by Equation (151) above H_C with the magnetic constants g = 2.06 ($\gamma/2\pi$ = 2.88 GHz/kOe), $4\pi M$ = 17.5 kG, H_C =1.15 kOe, L_1 = 30 Å, and L_2 = 13 Å. Agreement in the canted state below H_C was not satisfactory. For a possible reason, we consider that our MML consists of only 11~12 canted units, which is given by 2[Fe (30 Å)/Cr(13 Å)]. On the other hand, the effective-medium theory assumes a semi-infinite or a large number of canted stacks with ideal sharp interfaces.

So far, our discussion on IEC has been based on the assumption that the asymptotic limit is applicable. The term "asymptotic limit" means that the coupling is independent of the ferromagnetic layer thickness, and that the interlayer thickness is large as compared to the Fermi wavelength $\lambda_F = 2\pi/k_F$, which is typically either ~5 Å or a few monolayers (MLs) in many metals [93]. For thicknesses below λ_F, IEC cannot be described by our previous theoretical framework, and we should apply more a fundamental numerical method, for example, ab initio, by the self-consistent full-potential linearized augmented-plane-wave (FLAPW) method. Fine-layered [Fe (n ML)/Au (n ML)]$_m$ SLs with n = 1 to 5, for which we use the abbreviation $(n)_m$, were prepared by means of the MBE on MgO(001) substrates at IMR, Tohoku University. The total numbers of Fe and Au atomic planes were kept constant. We use the term "fine-layered" SLs (FLSLs) for SLs with layer thicknesses comparable or smaller than the Fermi wavelength. The Fe(1 ML)/Au(1 ML) FLSL corresponds to the ordered alloys with the tetragonal $L1_0$ structure, which exists in the equilibrium phase diagram for Fe_1Pt_1 alloy but not for the Fe_1Au_1 alloy.

BLS spectra at 300 K were excited by the p-polarized 5320 Å/150 mW line from a DPSS laser, and a cross-polarized analyzer was inserted in front of a tandem FPI to eliminate scattering from SAWs [94,95]. Magnetic fields of up to 7 kOe were applied along the crystallographic $(1\bar{1}0)$ direction, and the SWs propagating along the (110) direction were examined. Typical spectrum accumulation time was around 2h. Figure 26 displays BLS spectra observed from $(n)_m$ FLSLs in an external magnetic field of 3.0 kOe.

Here, the symbol of $n = 2 \pm \delta$ indicates an average ML, and δ = 0.25 and 0.5. The total thickness L_{SL} of the $(1.5)_{30}$ FLSL is 145 Å, and it is 269 Å for the $(2.5)_{30}$ FLSL. The SSW structure can be clearly seen, and the labels 1 to 4 on each spectrum stand for the corresponding SSW mode number. Figure 27 shows a comparison of the BLS spectra observed from the integer-type FLSLs.

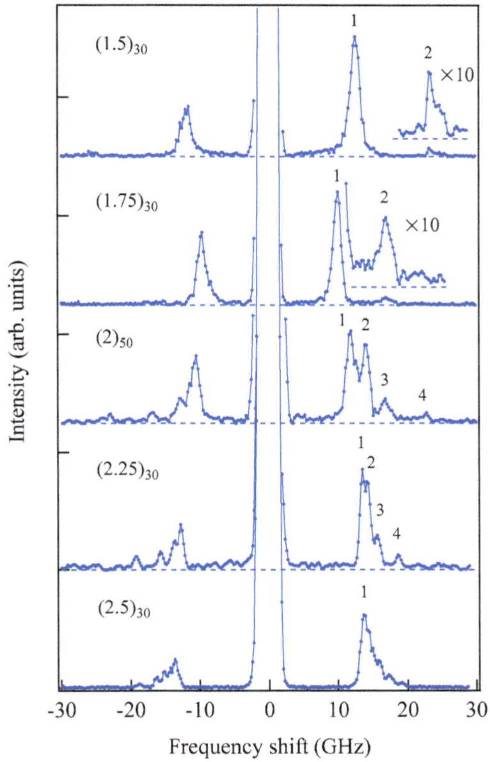

Figure 26. BLS spectra observed from $2 \pm \delta$ ML FLSLs (δ = 0.25 and 0.5) at H = 3.0 kOe. The labels 1 to 4 stand for the corresponding SSW number [95].

The even integer-type FLSLs exhibited clear SSW structures, but not the odd integer-type FLSLs. Furthermore, the noninteger-type spectra shown in Figure 26 can be smoothly connected between the $(2)_{50}$ and $(3)_{33}$ spectra. These observations indicate that the interactions leading to the occurrence of SSW systematically changed as a function of the ML number n. In order to analyze these results, we regarded the FLSLs as diluted magnetic Fe/Au alloy films with uniaxial MAEs perpendicular to the film plane and anisotropic exchange couplings. The magnetization of the alloy was assumed to be given by $M = M_{Fe}d_{Fe}/(d_{Fe} + d_{Au}) = M_{Fe}d_{Fe}/\Lambda_{SL}$. Here, d is the thickness of each layer (d_{Fe} = 3.00 Å and d_{Au} = 4.27 Å) and $M_{Fe} = 2.65 \pm 0.2\ \mu_B$ per Fe atom. The nth SSW frequency is given as

$$\Delta\nu(n) = \frac{\gamma}{2\pi}(H + H_{ex}(n))^{1/2}\left(H + H_{ex}(n) + 4\pi M_{eff}\right)^{1/2} \tag{166}$$

and

$$H_{ex}(n) = D_{//}Q_{//}^2 + D_{\perp}\left(\frac{n\pi}{L_{SL}}\right)^2 \cong D_{\perp}\left(\frac{n\pi}{L_{SL}}\right)^2 \tag{167}$$

Here, $4\pi M_{eff}$ has been already given by Equation (93), and L_{SL} is the total thickness of the FLSL, which was found to be 364 Å with XRD measurement. Figure 28 shows the SSW frequencies as a function of the external magnetic field for the $(2)_{50}$ FLSL.

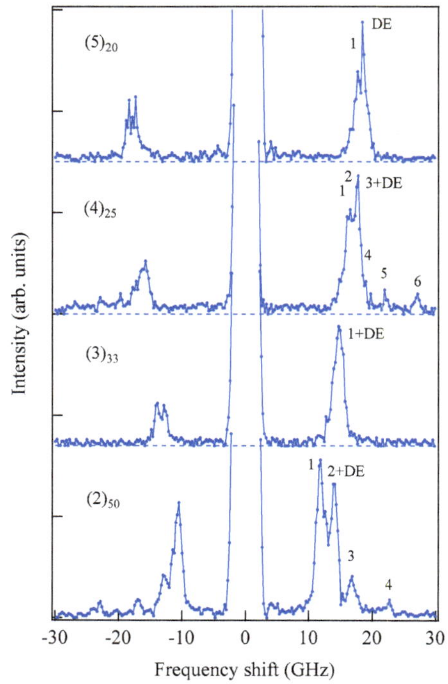

Figure 27. A comparison of BLS spectra observed from the integer-type FLSLs. The experimental conditions were the same as the ones given in Figure 25. The labels 1 to 5 and DE stand for the SSW modes and the DE mode, respectively [95].

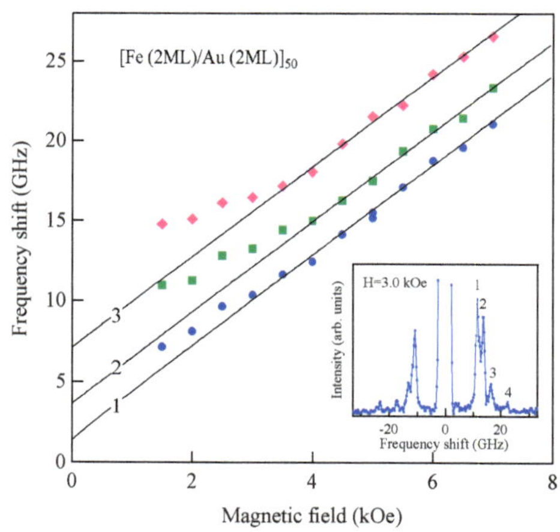

Figure 28. Magnetic field dependence of the SW frequencies for the $(2)_{50}$ FLSL. The inset shows the BLS spectrum from the $(2)_{50}$ FLSL at H = 3.0 kOe. The solid lines were calculated for the SSWs by using Equations (166) and (167) with the magnetic constants $\gamma/2\pi$ = 2.8 GHz/kOe (g = 2.0), $4\pi M_{\text{eff}}$ = 0.75 kG, and $D_{\perp}$ = 2.4 × 10^{-10} Oe·cm^2 [94].

It can be readily seen that the magnetic moment was not fully saturated below H < 4 kOe. Because we confirmed that the DE mode was superimposed on the n = 2 SSW peak through the surface dispersion examination by changing the scattering angle ϑ in Equation (1) from 15° ($Q_{//}$ = 0.61 × 10^5 cm^{-1}) to 45° ($Q_{//}$ = 1.67 × 10^5 cm^{-1}) and that the

DE mode does not contribute to the information on the exchange, the DE mode was therefore not included in the present analysis. We obtained a set of the magnetic constants given by $\gamma/2\pi = 2.8$ GHz/kOe ($g = 2.0$), $4\pi M_{\text{eff}} = 0.75$ kG, and $D_{\perp} = 2.4 \times 10^{-10}$ Oe·cm^2. Figure 29 shows the IEC constant *Ji* as a function of the atomic plane *n*.

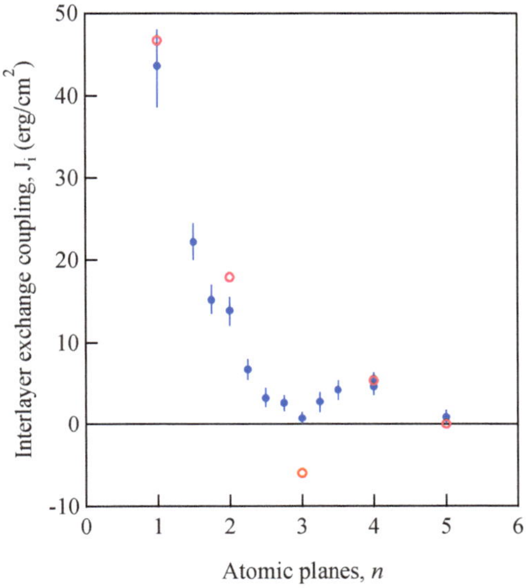

Figure 29. The IEC constant *Ji* as a function of *n*. The solid circles (•) are the BLS results, and the open circles (○) are the ab initio results [95].

We observed a pair of DE peaks for a $(1)_{30}$ FLSL but observed the SSW structure up to $n = 3$ SSW for a $(1)_{100}$ FLSL. The J_i constant was evaluated by using $Ji = 2MD_{\perp}/a_{\text{Fe}}$, in which $a_{\text{Fe}} = 2.87$ Å is the lattice constant of Fe. The open symbols stand for the ab initio results. Overall agreement between the BLS and ab initio results was fairly good, except for $n = 3$. The ab initio calculation predicted an antiferromagnetic ground state. We consider the discrepancy between the BLS and ab initio results for $n = 3$ stemmed from interface roughness. Because the IECs for $n = 2$ and 4 were ferromagnetic, the antiferromagnetic IEC for the ideal $n = 3$ FLSL may have been smeared for long-wavelength SWs detected with the BLS technique. For pure Fe films in full contact, the J_i value was expected to be 140 erg/cm^2. Hence, the present value of J_i~44 erg/cm^2 for the $(1)_{100}$ FLSL seems to be reasonable. According to the ab initio calculations, *d* electrons from the Fe atom were almost isolated even by 1 ML of the Au layer. The IEC was transmitted by itinerant *sp* electrons via second-order processes.

Figure 30 shows the perpendicular anisotropy field and the *g*-factor as a function of the atomic layer.

The solid line represents the $1/n$ dependence given by $H_A(n) = 22/n - 1.8$ (kOe). The $1/n$ dependence was expected from the interface out-of-plane anisotropy, as we have already discussed for the Fe wedge sandwiched by the Au layers. The anisotropy field $H_A(n)$ is defined as

$$H_A(n) = \frac{8\pi}{4\pi M_{\text{SL}}}\left(K_{\perp} + \frac{2k_{\perp}^{(s)}}{\Lambda_{\text{SL}} n} \times 10^8\right) \tag{168}$$

When we ignore the bulk term in Equation (168) and use $4\pi M_{\text{SL}} = 10.7$ kG, we obtain a value of $k_{\perp}^{(s)} \approx 0.34$ erg/cm^2. This value is in reasonable agreement with the values determined from the Fe wedge (see Figure 20) [79]. It is also clear that $4\pi M_{\text{eff}} = 4\pi M_{\text{SL}} - H_A(n)$ changes its sign from positive to negative around n~2. As we have already discussed for the Fe wedge, it means the in-plane magnetization state changes into the perpendicular magnetization state under zero magnetic field. The present Fe/Au FLSL and the Fe wedge sandwiched by the Au layers gave consistent results on the in-plane and out-of-plane transitions of the zero field magnetization state. Another interesting observation is the rather rapid change of the g-factor from the bulk value of 2.09 above $n \geq 4$ to the free electron value of 2.00 below $n \leq 2$. We also found that the g-factor of the Fe wedge at 2.6 Å was very close to 2.00.

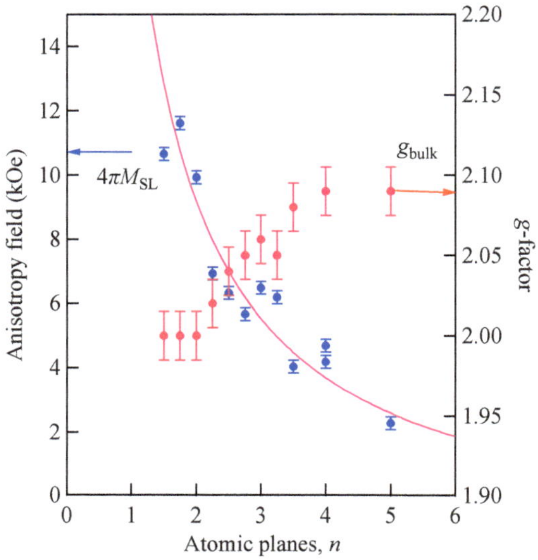

Figure 30. g-factor (•) and anisotropy field (•) as a function of n. The solid line shows the perpendicular anisotropy field $H_A(n) = 22/n - 1.8$ (kOe).

5.5. Nanogranular Films

Transition metal (TM = Fe, Co)-based granular films are higher-potential materials for various magnetic applications, for example, high-density longitudinal magnetic recording media (CoPt-SiO_2), high-frequency micromagnetic cores (CoFeB)-(SiO_2), GMR sensors (Co-Al-O), and so on [55]. Among these granular materials, TM-Al-O granular films are interesting materials for both basic magnetic research and for technological applications. Over the last twenty years, we have performed systematic studies on the magnetic and transport properties of TM-Al-O nanogranular films with the research groups of IMRAM and IMR, Tohoku University, and RIEMM, Sendai [54,55,96–101]. Readers who are interested in our results on transport and magnetization properties can refer to our references; here, I concentrate on our BLS results. It is well-known that the magnetic properties of TM-Al-O (TM = Fe, Co) granular films strongly depend on the TM composition. For example, for the Co composition x(Co) above ~70 at. %, Co-Al-O films are in a ferromagnetic (FM) ground state with a lower coercive field of $H_C > 10$ Oe. On the other hand, for x(Co) = 60~70 at %, TM-Al-O films exhibit reasonable soft magnetic properties with $H_C < 10$ Oe, whereas Co-Al-O films with x(Co) = 40~60 at % are in a superparamagnetic (SPM) state. I have already shown a BLS spectrum obtained from FM Co-Al-O films prepared at RIEMM in Figure 3 and the SW frequencies as a function of the magnetic field for a FM Fe_{64}-Al_{19}-O_{17} film prepared at RIEMM in Figure 5 [54]. The TM-Al-O nanogranular films consist of TM crystalline particles of up to several nm in diameter. The TM particles are surrounded by a nonmagnetic Al-O grain boundary of ~1 nm in thickness. In spite of the granular structure, we could observe well-defined SW spectra as shown in Figure 3, and we found that the SW frequencies, as a function of the magnetic field, are fully described by using Equations (49) and (50) as developed for uniform FM films (see Figure 5).

As already shown in Figure 5, the small exchange field term $H_{ex} = DQ^2 = 0.32$ kOe is important to reproduce the observed bulk SW frequency. As I have already mentioned, we cannot determine the SW stiffness constant D from the exchange field term. We found the resistivity ρ of the FM TM-Al-O granular films obeys the T^2 law in a wider temperature range below 200 K. Although the T^2 law can be expected from magnon scattering of conduction electrons at low temperatures, it has not been fully confirmed yet for FM metals, probably due to the much larger T^5 term by phonon scattering. Because magnon resistivity also depends on the exchange stiffness constant D, and the magnon T^2 term can be replaced by $(T/D)^2$, we can therefore expect the inverse-square law $\rho \propto (H_{ex})^{-2}$. Figure 31 shows a log ρ vs log H_{ex} plot. We obtained $\rho_{Fe} = 30.3(H_{ex})^{-2}$ μΩ·cm and $\rho_{Co} = 22.1(H_{ex})^{-2}$ μΩ·cm, respectively. Here, H_{ex} is in the kOe unit [54].

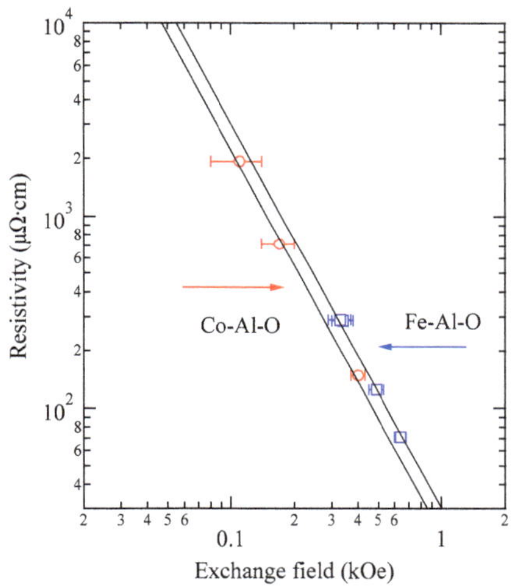

Figure 31. A log–log plot between ρ and H_E for Co-Al-O (○) and Fe-Al-O (□) granular films. The solid lines give the power law given by $\rho = a(H_{ex})^{-2}$. We obtained $a_{Fe} = 30.3$ μΩ·cm·(kOe)^2 for the Fe-Al-O films and $a_{Co} = 22.1$ μΩ·cm·(kOe)^2 for the Co-Al-O films [54].

Hereafter, I would like to concentrate on the Co-Al-O granular films prepared at RIEMM. Figure 32 shows a series of cross-polarized BLS spectra observed at room temperature from the FM and SPM Co-Al-O granular films in a magnetic field of $H = 2.0$ kOe [101].

These spectra were excited by the p-polarized 4730 Å line from a DPSS laser with an output power of 30 mW and accumulated over 5 h to improve the S/N ratio. The peak intensity of each spectrum was normalized by the total spectrum accumulation times. By virtue of the $p \to s$ polarization selection, the SAW scattering was completely suppressed. The solid squares in Figure 33 show the integrated intensity of the Stokes peak. The integrated intensity of the bulk SW in the FM state suddenly jumped in the FIM state. In order to determine the FM–SPM boundary, the exchange field H_{ex} proved a good guide. The inverse-square relation between the exchange field H_{ex} (kOe) and the resistivity ρ (μΩ·cm) gives us $H_{ex} = (22.1/\rho)^{1/2}$ kOe for FM Co-Al-O films [54]. We calculated the exchange field H_{ex} using our resistivity data ρ, and we show the calculated exchange fields by the open circles in Figure 33.

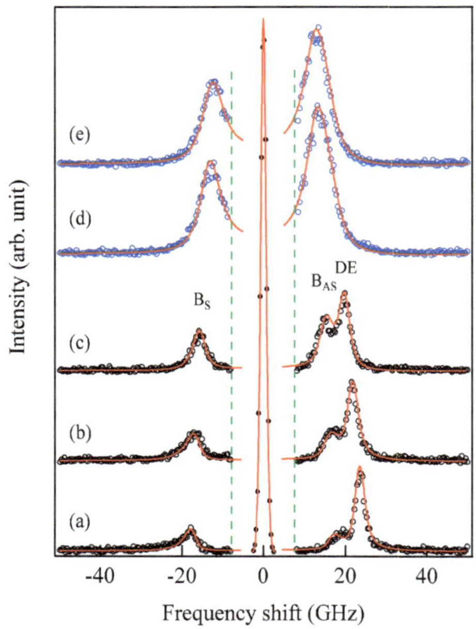

Figure 32. BLS spectrum from the SPM and FM Co-Al-O films at room temperature under an external magnetic field of 2.0 kOe. Here, (**a**) x(Co) = 72.4, (**b**) x(Co) = 67.5, (**c**) x(Co) = 63.1, (**d**) x(Co) = 50.3, and (**e**) x(Co) = 49.9, respectively. The baseline for each spectrum was properly shifted for clarity's sake. The solid lines give the numerical fits by the CM theory. The vertical broken lines indicate the frequency area in which the AOMs were activated to protect the PMT. An intensity-attenuated elastic Rayleigh peak around zero frequency is also displayed. This peak gives the instrumental function [101].

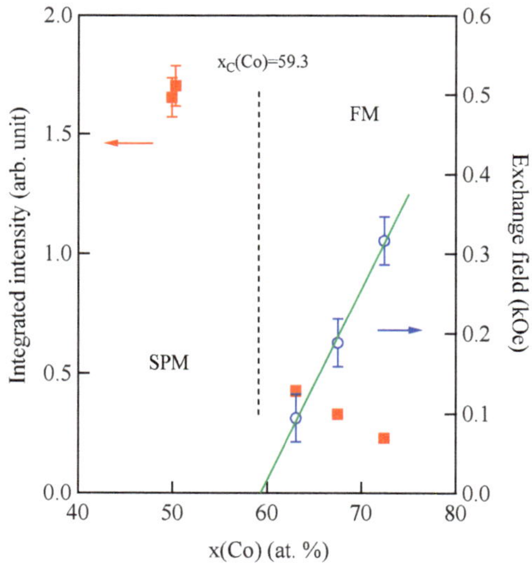

Figure 33. Integrated intensity (solid square) of the Stokes peak of BLS spectra shown in Figure 32 and the exchange field H_{ex} (○) calculated from the inverse-square law for the FM films as a function of x(Co). The green solid line gives $H_{ex} = 23.9 \times 10^{-3}$ (x(Co) − 59.3) kOe obtained from a least-squares fitting [101]. The broken line indicates the FM/SPM boundary given by H_{ex} = 0.

We also expected a linear relation $H_{ex} \propto x - x_C$, in which x_C is the FM–SPM boundary concentration. The solid line in Figure 33 displays H_{ex} = 23.9 × 10^{-3}(x(Co) − 59.3) (kOe) determined using the least-squares method. With this result, we determined the SPM-FM boundary in the Co-Al-O nanogranular system was located at x_C(Co) = 59.3 ± 1.3. The SPM–FM boundary can be characterized as a Co atomic concentration at which the exchange stiffness constant D vanishes. Therefore, we should take into account both the exchange and dipole coupling for the FM films. On the other hand, we can ignore the exchange coupling in the BLS spectrum calculation for the FIM state of the SPM films.

There is a distinct difference between the bottom three FM spectra, (a) through (c), and the top two SPM spectra, (d) and (e). The FM spectrum exhibited a characteristic dual peak structure on the positive frequency anti-Stokes (SW annihilation process) side and a single peak on the negative frequency Stokes (SW creation process) side under the present experimental conditions. These spectral features are typical for a SW spectrum from a thick FM film. An opaque magnetic film with a thickness of ~λ/2 can be treated as a semi-infinite magnet in the BLS experiment, as I have already mentioned. Here, the labels DE and B refer to the DE and bulk SW peaks, and the subscript S and AS refer to the Stokes and anti-Stokes processes. Note that the DE peak height is higher than that of the FM bulk peaks in (a) through (c). On the other hand, it is quite interesting to note that only a broad but intense peak appears on both frequency sides in the SPM spectra. It seems to be a general feature of BLS spectra from the SPM state. In fact, broad BLS peaks have been also observed in CoPt-SiO_2 granular films [102] and $(SiO_2)_{100-x}Co_x$ granular films [103]. Hereafter, for convenience's sake, let us define the magnetization-induced state under an external magnetic field in an SPM film as the field-induced magnetization (FIM) state. These FIM peak frequencies were quite sensitive to the external magnetic field H. The peak frequencies increased with the increasing magnetic field. It is also an interesting observation that the peak frequency in the anti-Stokes side was about 1 GHz higher at most than the peak frequency in the Stokes side. Damon and Eshbach discussed a dipole-coupled ferromagnetic slab and obtained a nonreciprocal DE mode in addition to the bulk SW band [24]. This means that we can expect a singlet-doublet SW structure for a BLS spectrum from an FIM slab. The peak frequency difference of ~1 GHz between the Stokes and anti-Stokes sides is probably due to the DE mode, which only appears in the anti-Stokes side in our scattering geometry (see Figure 32a–c). In order to separate the bulk and DE peaks and determine the peak frequency, peak width, and intensity from the observed broad peak, numerical spectrum analysis is strongly required. For quantitative analyses of BLS spectra beyond the peak frequency discussions we have performed so far, we employed the CM theory, which has been developed for semi-infinite magnets, by taking into account both the dipole and exchange couplings [29]. The CM theory can fully reproduce the singlet-doublet SW structure for a BLS spectrum from a ferromagnetic slab, as shown in Figure 32a–c. According to the CM theory, the SW response function $S(Q_{//}, \omega)$ is given by rather complicated formulae:

$$\begin{aligned} S(Q_{//},\omega) \propto\; & n(\omega,\ T)\cos\vartheta\,|\gamma_s(\vartheta)|^2\left(\tfrac{\omega_0}{c}\right)^4 \mathrm{Im}\int_0^\infty dz\int_0^\infty dz'\exp\left[i\Delta k_\perp z - i\Delta k_\perp^* z'\right] \\ & \times\left\{R_{zz}\chi_{zz}(z,z'|\omega) + R_{xx}\chi_{xx}(z,z'|\omega) + R_{xz}\chi_{xz}(z,z'|\omega) + R_{zx}\chi_{zx}(z,z'|\omega)\right\}. \end{aligned} \tag{169}$$

Here, $n(\omega, T)$ is the Bose–Einstein (BE) factor, R_{zz}, R_{xx}, R_{zx}, and R_{xz} are the SW-photon coupling constants, and χ_{zz}, χ_{xx}, χ_{zx}, and χ_{xz} are the dynamical susceptibilities defined in Equations (79) and (80). The observed BLS spectrum should be compared with a convoluted spectrum between the SW response function $S(Q_{//},\omega)$ and an instrumental function. We employed the intensity-attenuated Rayleigh peak as the instrumental function (see Figure 32). The solid lines in Figures 32 and 34 display a comparison between the observed spectra and the calculated spectra in the FM state and in the FIM state.

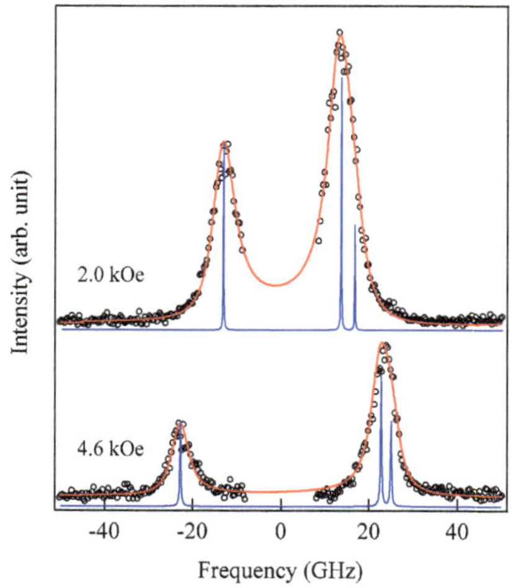

Figure 34. Comparison of BLS spectra observed from the x(Co) = 50.3 SPM film at H = 2.0 and 4.6 kOe. The solid lines on the observed spectra (○) give the numerical fits obtained via the CM theory. To show the singlet-doublet structures, the calculated response function $S(Q_{//},\omega)$ with small damping of $0.1 \leq \Gamma/2\pi \leq 0.15$ GHz is indicated by the solid lines [101].

As shown in Figures 32 and 34, we could fully reproduce the observed BLS spectrum by taking into account only the dipole coupling for SPM spectra with the damping constant of $\Gamma/2\pi$ = 2.66 GHz for the H = 2.0 kOe spectrum and 2.26 GHz for the 4.6 kOe spectrum. In order to clarify the singlet-doublet structure, we recalculated the response function $S(Q_{//},\Omega)$ with a small damping constant of $0.07 \leq \Gamma/2\pi \leq 0.15$ GHz. We adjusted the peak height of the Stokes peak of each spectrum in Figure 34. We found the doublet peaks located at 16.5 GHz and 13.4 GHz at H = 2.0 kOe, and at 25.0 GHz and 22.8 GHz at H = 4.6 kOe. Because the frequency splitting of 2~3 GHz between the doublet peaks and the peak width were comparable, the FIM doublet actually appears as a single peak in a BLS spectrum. The present numerical analysis reasonably explains why the anti-Stokes peak frequency is higher than the Stokes peak frequency.

Figure 35a is the magnetic field dependence of the bulk-type and DE-type peak frequencies in the FIM state determined by the numerical spectrum fitting with small damping constants.

Because we have confirmed no remanent magnetization at zero magnetic field, the FIM modes should be forbidden at $H = 0$. This means that the FIM mode frequencies should approach zero as the magnetic field approaches zero. On the other hand, the peak frequencies displayed in Figure 35a seem to stay finite even at zero magnetic field. As an attempt to solve this difficulty, we replaced the gyromagnetic ratio γ with a field-dependent gyromagnetic ratio $\gamma(H)$, and the magnetization M with a field-induced magnetization $M_{//}(H)$, which is given by a sum of the Langevin functions. We thus rewrote Equations (51) and (52) as follows:

$$\frac{2\pi\Delta\nu_B(H)}{\gamma(H)} = H^{1/2}[H + 4\pi M_{//}(H)]^{1/2} \tag{170}$$

and

$$\frac{2\pi\Delta\nu_{DE}(H)}{\gamma(H)} = H + 2\pi M_{//}(H) \tag{171}$$

The corrected results are shown as a function of the magnetic field in Figure 35b by filled circles for the bulk-type mode and squares for the DE-type mode. The solid lines show the right-hand side of Equations (170) and (171) with the calculated $4\pi M_{//}(H)$ values. Although agreement looks excellent as shown in Figure 35b, it should be recognized that Equations (170) and (171) are for the small-amplitude precession motion around field-induced magnetization $M_{//}(H)$.

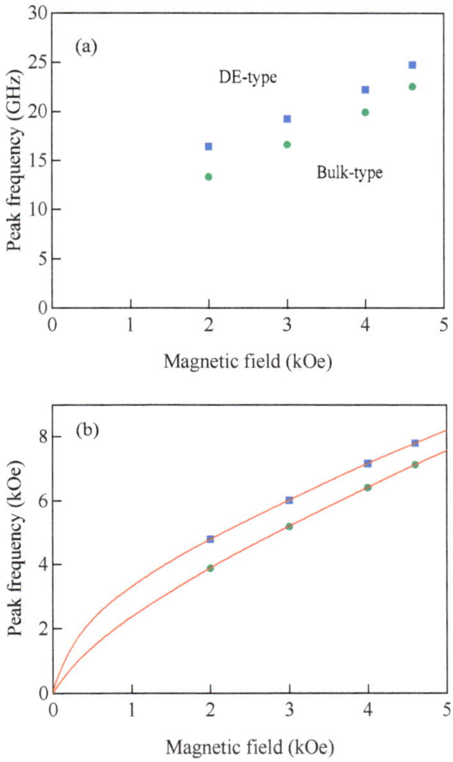

Figure 35. (**a**) FIM peak frequencies obtained from the response function $S(Q_{//},\omega)$ with a small damping constant of $0.1 \leq \Gamma/2\pi \leq 0.15$ GHz as a function of the magnetic field. (**b**) Field dependence of the FIM excitation frequencies divided by $\gamma(H)/2\pi$ as a function of the magnetic field: (•) bulk-type mode and (■) DE-type mode. The solid lines show the frequencies calculated by using the right side of Equations (170) and (171) incorporating the $4\pi M_{//}(H)$ result [101].

So far, I have shown that BLS is a unique technique for investigation of the magnetization dynamics of various opaque magnetic structures in the GHz frequency range. However, most of the BLS studies of opaque magnetic structures have been performed at room temperature, and few BLS studies have been performed at low temperature [12,104–107]. BLS studies of magnetization dynamics as a function of temperature is quite an interesting subject. As I have already pointed out, BLS intensity from opaque surfaces is much weaker compared to the phonon scattering in transparent materials, even at room temperature. Another inevitable difficulty for scattering experiments stems from the BE factor, which appears in the response function given by Equation (169). For conventional BLS studies performed above 15 K and with a narrow frequency range below 30 GHz, we can reasonably approximate the BE factor as follows:

$$n(\omega,T) = \frac{1}{\exp(\omega/k_B T) - 1} \approx \frac{k_B T}{\omega} \tag{172}$$

This approximation is known as the high-temperature approximation (HTA). Now, BLS intensity is directly proportional to the absolute temperature T. It is obvious that weak BLS intensity from opaque surfaces, even at room temperature, gets weaker at lower temperatures. Of course, we can apply an intense laser beam to increase BLS signals. However, the intense laser beam results in a local-heating effect at the beam spot. The local heating effect will be critical for phase transition studies. We must overcome these difficulties to step forward into new frontiers of BLS studies on spin dynamics or magnetization dynamics at low temperatures.

Figure 36 shows a comparison of the spectra observed under field-cooling (FC) (+) and zero-field-cooling (ZFC) (○) conditions with an external magnetic field of 4.0 kOe at 20 K [108].

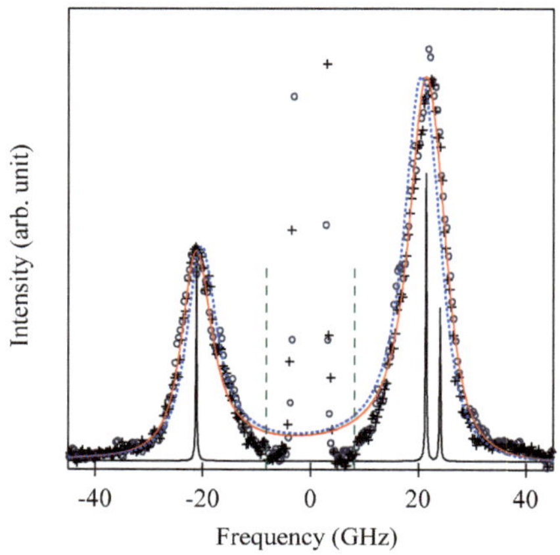

Figure 36. A comparison of the ZFC and FC BLS spectra observed at 20 K and H = 4.0 kOe. The solid line shows numerical fits by the CM theory with the ZFC fitting parameters given in the text. The broken line shows the numerical fit with the same ZFC parameters, except $H_K = 0$. To show the peak positions, the calculated response function $S(Q_{//}, \omega)$ with a small damping of $\Gamma/2\pi = 0.1$ GHz is included with a solid line [108].

After spectrum accumulation times over 6 h, we observed FC and ZFC spectra with reasonable signal to noise ratios at 20 K. We found no substantial difference between these spectra, as shown in Figure 36. For more detailed comparisons of these spectra, we properly adjusted the peak heights of the singlet Stokes peak. The vertical broken lines indicate a frequency range in which the AOMs were activated to protect the PMT from optical damage by an intense Rayleigh peak. From our ZFC and FC magnetization measurements at IMR, we estimated the blocking temperature T_B of our sample to be ~110 K. The FIM changed the temperature dependence from the SPM Langevin type above T_B to the FM power-law type below T_B. An effective magnetic anisotropy with an easy axis along the applied magnetic field appeared in SPM granular systems below T_B. For convenience's sake, we assumed a uniaxial-type anisotropy field H_K (=0.27 kOe). The solid lines in Figure 36 show the calculated ZFC spectrum with the anisotropy field, and the broken lines are the calculated ZFC spectrum without the anisotropy field term. It is obvious that the anisotropy field term improves agreement between the observed and calculated spectra. Although the H_K term was much smaller than the other fitting parameters, it plays an essential role, as we will discuss later. The inclusion of the H_K term is equivalent to replacing the external magnetic field H with the effective field $H + H_K$ in Equations (170) and (171). In order to display the SPM peak frequencies, we included a calculated spectrum for a small peak-width of $\Gamma/2\pi = 0.1$ GHz in Figure 36.

Figure 37 shows the temperature development of the BLS spectrum at 300, 100, and 15 K at $H = 4.5$ kOe.

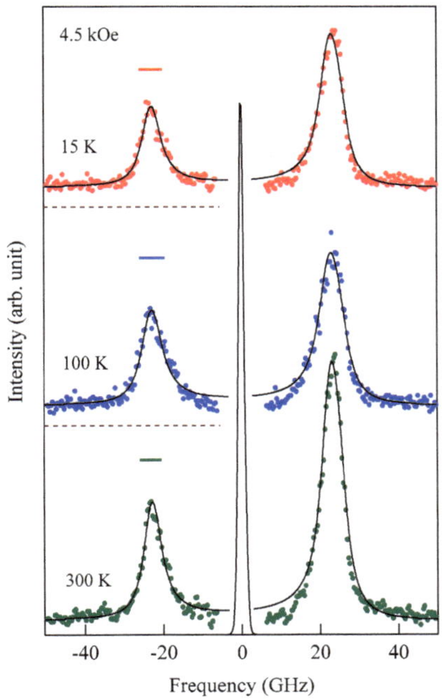

Figure 37. Temperature development of the BLS spectrum in a temperature range between 300 and 15 K under a magnetic field of 4.5 kOe. The baseline of each spectrum, indicated by the broken line, was properly shifted for clarity. The solid lines give the numerical fits obtained by the CM theory, and the horizontal bars show the damping constant of the calculated spectra [108].

I included in Figure 32 an intensity-attenuated (×1/5000) Rayleigh peak for the 300 K spectrum. The solid lines on each spectrum give the calculated BLS spectra. In the calculations, we used a common peak width of $\Gamma/2\pi$ for both the bulk and DE-type modes. In spite of this simplification, agreements between the observed and calculated spectra were reasonable. The horizontal bar on each spectrum shows the peak width for each calculated spectrum. We found that the peak width at 100 K was wider than the peak widths at 300 and 15 K. Figure 38 shows the SPM excitation frequencies and the peak width for a magnetic field H = 4.5 kOe as a function of temperature.

The labels B and DE refer to the bulk and DE-type peak frequencies, respectively. These peak frequencies were nearly insensitive to temperature but slightly increased below 50 K. The H_K term in Equations (170) and (171) gives the increasing frequencies at lower temperatures. To the contrary, the peak width $\Gamma/2\pi$ exhibited a broad peak centered at ~200 K, as shown in Figure 39. Figure 39 displays the peak width for the external fields of H = 3.0 kOe (△), 4.0 kOe (□), and 4.5 kOe (○) as a function of temperature.

The peak width clearly depends on both temperature T and the magnetic field H. We observed a narrower width for a higher magnetic field. We performed an additional BLS measurement under H = 2.0 kOe at 15 K. The inset shows a summary of the magnetic field development of the peak width at 15 K. From the results shown in the inset, we can estimate a limiting value of $\Gamma(0, 15\text{K})/2\pi$~4 GHz for the peak width at 15 K and with zero magnetic field. With these observations, it seems to be reasonable to decompose the peak width into the following terms:

$$\Gamma(H,\ T)/2\pi = \varphi/2\pi + \varsigma(H)/2\pi + \xi(T)/2\pi \tag{173}$$

Here, the $\phi/2\pi$ term describes the peak width due to the scattering of the FIM excitations by the nonuniformity of granule sizes (or granule moments) within a film, and it is expected to be temperature- and magnetic field-independent. The $\zeta(H)/2\pi$ term describes the suppression of the incoherent motion of granule moments by the external magnetic field. This term is responsible for our observation of the narrower widths for higher magnetic fields. Finally, the $\xi(T)/2\pi$ term describes the damping due to couplings between the FIM excitations and another freedom of thermally-excited magnetization dynamics. We will concentrate on this term in our following discussions.

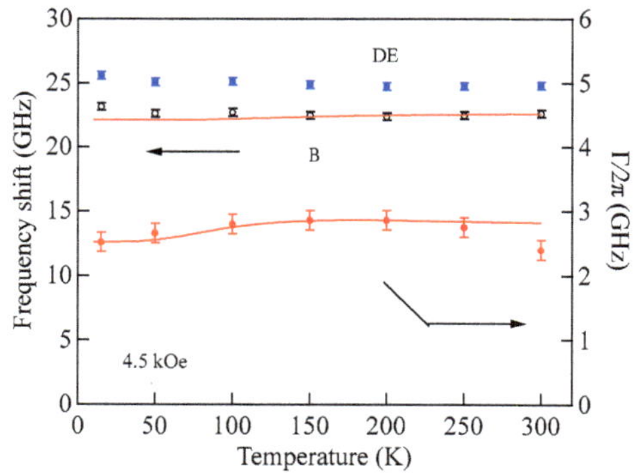

Figure 38. Temperature development of the SPM peak frequencies (Bulk-type (○) and DE-type (■) modes) and the damping constant (•) at 4.5 kOe. The SPM peak frequencies are determined from the calculated response function $S(Q_{//},\omega)$ with $\Gamma/2\pi$ = 0.1 GHz as shown in Figure 36. The solid lines show the calculated SPM peak frequency and peak width when taking into account the magnetization relaxation dynamics [108].

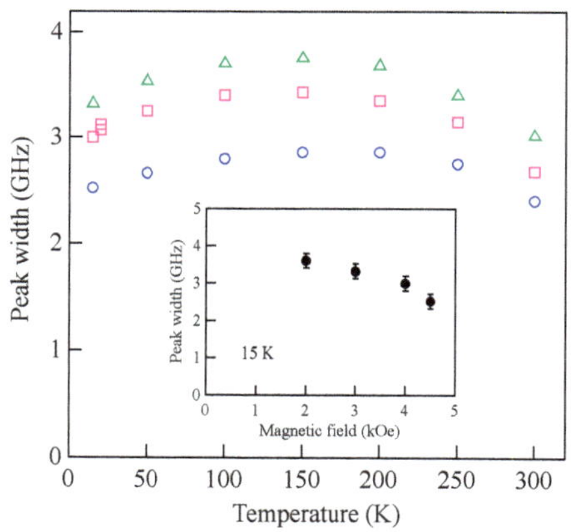

Figure 39. The peak width $\Gamma/2\pi$ as a function of temperature at H = 3.0 kOe (△), 4.0 kOe (□), and 4.5 kOe (○). The inset shows a magnetic field dependence of the peak widths at 15 K [108].

Since the pioneer work by Néel [109], the magnetization reversal dynamics in SPM nanoparticles have been intensively investigated in a wide time scale between 10^2 s and 10^{-13} s by employing various experimental techniques: magnetization and dynamical susceptibility measurements (with a time scale of 10^2 to 10^{-4} s), Mössbauer spectroscopy (with a time scale of 10^{-8}–10^{-9} s), neutron spectroscopy (with a time scale of 10^{-10}–10^{-13} s), and so on [110]. According to the Néel–Brown model, the thermally activated relaxation time is given by [111]:

$$\tau(T) = \tau_0 \exp(\Delta E/k_B T) \tag{174}$$

Here, τ_0 is the attempt relaxation time, $\Delta E = KV$ is the activation energy (or anisotropy barrier), K is the magnetic anisotropy constant per particle volume, and V is the volume of the individual particles. For convenience sake, let us introduce the Debye relaxation model [112] and replace the static FIM $4\pi M$ in Equation (170) with the dynamical FIM $4\pi M(\omega)$ as follows:

$$4\pi M(\omega) = 4\pi M(\infty) + \frac{4\pi M(0) - 4\pi M(\infty)}{1 + i\omega\tau} = 4\pi M(\infty) + \frac{4\pi \Delta M}{1 + i\omega\tau} \quad (175)$$

Here, $4\pi M(\infty)$ and $4\pi M(0)$ are the limiting high-frequency and low-frequency FIMs, and τ is the relaxation time already defined by Equation (174). Then, the bulk-type SPM excitation frequency $\Delta\nu_B$ given by Equation (170) becomes a complex frequency. For the weak relaxation case ($H + H_K + 4\pi M(\infty) >> 2\pi\Delta M$), the real and imaginary parts of the SPM excitation frequency can be readily obtained as follows:

$$\begin{aligned} \mathrm{Re}(\Delta\nu_B) = f_B = \tfrac{\gamma}{2\pi}(H + H_K)^{1/2}(H + H_K + 4\pi M(\infty))^{1/2} \\ + \tfrac{\gamma}{2\pi}(2\pi\Delta M)\left(\tfrac{H+H_K}{H+H_K+4\pi M(\infty)}\right)^{1/2}\tfrac{1}{1+\omega^2\tau^2} \end{aligned} \quad (176)$$

and

$$\mathrm{Im}(\Delta\nu_B) = \frac{\xi}{2\pi} = \frac{\gamma}{2\pi}(2\pi\Delta M)\left(\frac{H + H_K}{H + H_K + 4\pi M(\infty)}\right)^{1/2}\frac{\omega\tau}{1 + \omega^2\tau^2} \quad (177)$$

The peak width $\xi/2\pi$ exhibits a maximum at a temperature for which the condition $\omega_B\cdot\tau(T) = 2\pi f_B\cdot\tau(T) = 1$ is satisfied. In order to determine the parameters τ_0 and $\Delta E/k_B$ in Equation (174), we should perform at least two BLS measurements with different frequencies, f_1 and f_2, by simply changing the magnetic field. In the following discussions, we will use the GHz unit for these frequencies. When we observe the peak-maximum at temperatures T_1 and T_2, then we can readily calculate the relaxation parameters τ_0 and KV/k_B as follows:

$$\frac{KV}{k_B T_1} = \frac{T_2}{T_2 - T_1}\ln\left(\frac{f_2}{f_1}\right) = \frac{1}{1-\beta}\ln\left(\frac{f_2}{f_1}\right) = \ln\left(\frac{f_2}{f_1}\right)^{\frac{1}{1-\beta}} \quad (178)$$

and

$$\tau_0 = \frac{1}{2\pi f_1}\exp\left(-\frac{KV}{k_B T_1}\right) = \frac{1}{2\pi f_1}\left(\frac{f_1}{f_2}\right)^{\frac{1}{1-\beta}} = \frac{1}{2\pi}\left[\frac{f_1{}^{\beta}}{f_2}\right]^{\frac{1}{1-\beta}} \times 10^{-9}\ \mathrm{s} \quad (179)$$

Here, we have defined $\beta = T_1/T_2$. Because our BLS measurements have been performed at limited temperatures, we adapted the least-squares fourth-order polynomial fitting to determine the peak-maximum temperature. Therefore, our following discussions are limited to qualitative ones. We obtained two sets of the parameters: (f_2 = 22.1 GHz, T_2 = 178 K) from the 4.5 kOe results and (f_1 = 16.5 GHz, T_1 = 143 K) from the 3.0 kOe results. In fact, we also have the excitation frequencies: f_2 = 22.4 GHz and f_1 = 16.9 GHz. Because these frequencies involve the relaxation contribution given by Equation (176), we corrected the contributions. Utilizing Equations (178) and (179), we obtained a set of relaxation parameters: $\tau_0 = 2.2 \times 10^{-12}$ s and KV/k_B = 213 K. It is interesting to compare the present relaxation parameters with the ones for other nanogranular systems and also physically different structural relaxation systems. Neutron scattering from ~15 nm hematite particles detected a quasielastic peak due to SPM relaxation and propagative precession peaks [113]. From temperature development of the quasielastic peak, the attempt relaxation time of $\tau_0 = 7 \times 10^{-12}$ s and the anisotropy barrier $KV/k_B = 500 \pm 200$ K were determined. Linderoth et al. obtained $\tau_0 = 2 \times 10^{-12}$ s and $KV/k_B = 428 \pm 29$ K for Fe-C particles by combining the magnetization with the Mössbauer techniques [114]. These values are in the same order with the present values of $\tau_0 = 2.2 \times 10^{-12}$ s and KV/k_B = 213 K. Much longer attempt relaxation times of ~10^{-10} s have been reported on α-Fe, α-Fe_2O_3, and γ-Fe_2O_3 nanoparticle systems [110]. For comparison, we obtained τ_0~4.0×10^{-18} s and ΔE = 0.55 eV (=6380 K in the temperature scale) for P[VDF-TrFE] copolymer films [115], and τ_0~1.3×10^{-15} s and ΔE = 0.28 eV (=3250 K) for glass-former propylene glycol [116].

It is very interesting to note that the attempt time of τ_0~10^{-12} s for magnetic relaxation is three orders or more long, and that the activation energy (height of the potential barrier to jump) is one order or more lower than the barrier height for the structural relaxation. The characteristic features of SPM magnetization relaxation can be summarized as slow motion within a shallow potential minimum.

Here, we return to Equation (173). At 15 K, the magnetization relaxation time τ in Equation (174) is calculated as τ = 3.23 × 10^{-6} s. At the BLS frequency of f_B~20 GHz, the condition $2\pi f_B\tau$ >> 1 is fully satisfied. This condition means that the SPM excitation frequency is too fast to couple with the magnetization relaxation process. Therefore, we can set ξ(15 K)/2π = 0 in Equation (173). Now, assuming that the $\varphi/2\pi + \varsigma(H)/2\pi$ term in Equation (173) is independent of temperature at H = 4.5 kOe, the relaxation amplitude is given by

$$\frac{\gamma}{2\pi}(2\pi\Delta M)\left(\frac{H+H_K}{H+H_K+4\pi M(\infty)}\right)^{1/2} = 2\frac{\Delta\Gamma_R}{2\pi} = 2\left(\frac{\Gamma(178\,\mathrm{K})}{2\pi} - \frac{\Gamma(15\,\mathrm{K})}{2\pi}\right) = 2\frac{\xi(178\,\mathrm{K})}{2\pi} \tag{180}$$

When we put the available constants $\Delta\Gamma_R/2\pi$ = 0.35 GHz, $\gamma/2\pi$ = 3.11 GHz/kOe, H_K = 0 kOe, H = 4.5 kOe, and $4\pi M(\infty)$ = 6.76 kG into Equation (180), we obtain $2\pi\Delta M$ = 0.36 kG and the relaxation amplitude of 0.70 GHz. Finally, we obtain the SPM frequency and the peak width $\xi(T)/2\pi$ for the bulk-type mode as follows:

$$f_B = 22.1 + 0.70\frac{1}{1+\omega^2\tau^2} \tag{181}$$

and

$$\frac{\xi}{2\pi} = 2.52 + 0.70\frac{\omega\tau}{1+\omega^2\tau^2} \tag{182}$$

The solid lines in Figure 38 are the calculated SPM excitation frequency f_B and the peak width $\xi/2\pi$ at H = 4.5 kOe as a function of temperature by using Equations (181) and (182). The magnetic relaxation model qualitatively reproduces the observed temperature development of the peak width, except for the higher temperatures above 250 K. The calculated frequency below 100 K was lower than the observed frequencies. Because we have not included the temperature dependence of the $\gamma/2\pi$ constant and the H_K term in Equation (181), it is possible to improve agreement between the observed and calculated frequencies. We can also qualitatively reproduce the results for H = 3.0 and 4.0 kOe with the above relaxation parameters by changing the relaxation amplitude. As we have already discussed, we can ignore the $\xi/2\pi$ term at 15 K in Equation (173) because of the condition $\omega_B\tau$ >> 1. Then, we can rewrite Equation (173) as follows:

$$\Gamma(H,\ 15\,\mathrm{K})/2\pi \cong \phi/2\pi + \varsigma(H,\ 15\,\mathrm{K})/2\pi \tag{183}$$

The inset in Figure 39 shows the magnetic field dependence of the peak width Γ(H, 15 K)/2π. As the magnetic field increased, the peak width decreased. This behavior can be attributed to the magnetic field dependence of the ζ(H, 15 K)/2π term. Unfortunately, the highest magnetic field available in our BLS system with the closed-cycle refrigerator is not enough to fully separate these $\phi/2\pi$ and $\zeta/2\pi$ terms in Equation (183). However, as a rough estimation, we obtained $\phi/2\pi \approx 2$ GHz and ζ(0 kOe, 15 K)/$2\pi \approx 2$ GHz, respectively.

So far, I have demonstrated in this granular section that the BLS technique involves higher potential for investigation of fast magnetization dynamics. In the Co-Al-O system, we investigated the dynamics in the frequency range of around 20 GHz. However, it is easy to extend the frequency range below several GHz and over a few hundred GHz for the BLS technique by utilizing a tandem FPI. Furthermore, we can adjust the SPM excitation frequencies by applying an appropriate external magnetic field according to Equations (170) and (171). I would like to emphasize these advantages of the BLS technique in the magnetization dynamics study of SPM materials.

Figure 40 shows the peak intensities at H = 4.5 kOe obtained from the fittings and normalized by the total spectrum accumulation time as a function of temperature.

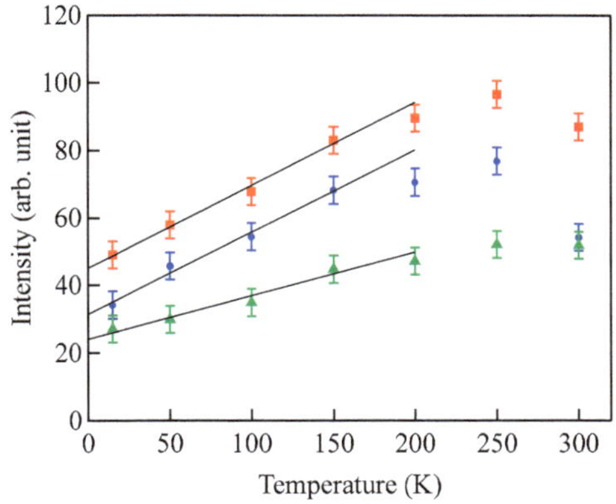

Figure 40. The peak intensities of the SPM excitations as a function of temperature: (•) bulk-type anti-Stokes peak, (■) bulk-type Stokes peak, and (▲) DE-type peak. The solid lines are the least-squares fits to the linear function $I(T) = A + B{\cdot}T$ below 150 K [108].

Although all peak intensities monotonously decreased as temperature decreased, the peak intensities at 15 K kept about 50% of the highest intensities observed at 250 K. It is important to note that the BE factor in Equation (169) is an inevitable sequence of the quantum-mechanical fluctuation-dissipation theorem [117] and independent of the details of the real physical systems. The solid lines in Figure 40 show the least-squares fits below 150 K to a linear function of temperature given by $I(T) = A + BT$. The linear function well describes the observed temperature dependence of the peak intensities. Now, we rewrite the linear function in the following form.

$$I(T)/T = A/T + B \tag{184}$$

Here, $I(T)/T$ is proportional to Equation (169) divided by the BE factor. Because the optical properties and the dynamical susceptibilities appearing in Equation (169) are essentially independent of temperature, we can explain the B term. However, we cannot explain the existence of the A/T term within the framework of the single-site magneto-optic coupling mechanism already discussed. Nevertheless, the A/T term in Equation (184) is the most crucial in our present results. When we use the least-squares parameters A = 45.0 and B = 0.246 for the Stokes peak (■), it is obvious that the A/T term dominates the BLS scattering intensity below 100 K. At 15 K, the A/T term is more than ten times larger than the B term. This can be the reason why we could observe relatively intense scattering even at 15 K.

Non-T-proportional behaviors of SW BLS intensity have already been reported in several conductive FM materials, for example, semiconductor EuS and EuO single crystals [104], a FM/AFM Co (2.8 nm)/CoO (0.7 nm) bilayer thin film [106], and so on. These materials are magnetically quite different from the present SPM Co-Al-O granular system. Because we cannot explain the A/T term in Equation (184) within the phenomenological description of the conventional single-site magneto-optical coupling theory given by Equations (14)–(18), some collective motion of electrons might contribute to the SW light scattering in conductive FM materials. In order to elucidate the light scattering mechanism, more extensive BLS studies for various conductive magnetic materials are strongly recommended.

5.6. SWs in Confined Structures and Devices

Unfortunately, I had no chance to perform BLS measurement in this research area. This area is closely related to the rapidly growing magnonics and spintronics fields, and is a promising area for the next generation of SW BLS study. For readers interested in this field, I give a few of recent references [118–120].

6. Summary

Since my SW BLS reports on CoZr and FeSi films published in 1989, I spent more than 30 years in the field of SW BLS. BLS as an optical technique to determine a set of the basic magnetic constants of a magnetic thin film was fully established during the period, as I have presented in this review. Now, the BLS technique is recognized as one of the best techniques not only to investigate the SWs and the magnetization dynamics in the GHz frequency range, but also to characterize magnetic interactions induced in various artificial magnetic structures. There is no doubt that developments both in the sample preparation techniques and the BLS technique will go hand in hand to open new frontiers of functional devices as well as basic material sciences. When I started my SW BLS research in the middle of the 1980s, quasi-two-dimensional magnetism was just a textbook subject, and I never imagined that I could measure SWs in such a thin film of a few monolayers in thickness. However, the MBE technique made it possible.

Because of the sensitivity and flexibility of the light scattering technique, the importance and usefulness of the BLS technique in materials science and engineering is still growing and growing. The downsizing of magnetic devices will result in magnetic instability due to direct or indirect interactions between magnetic elements. The microfocused BLS technique and micromagnetic simulation can be applied to characterize interactions between the magnetic elements. BLS detection of the spin current and the spin accumulation in spintronics devices will be another interesting challenge.

Through my research career in the BLS field spans over 45 years, including structural and ferroelectric phase transitions, low temperature liquids, glass formers, and SAWs and SWs from opaque surfaces, I learned many new physics, ideas, and the importance of the challenges when going into a new research field. The BLS study from opaque surface was exactly a challenge for me. When I was asked to examine BLS from SPM granular samples, to speak frankly, I had no assurance that I could observe BLS signals because of my preconception that short range interactions such as the exchange coupling were necessary to generate magnetic excitations. Fortunately, my idea was completely wrong as I have presented in this review, and BLS can be a new tool to investigate the magnetization dynamics of SPM granular materials.

Funding: This review received no external funding.

Institutional Review Board Statement: Not applicable.

Informed Consent Statement: Not applicable.

Data Availability Statement: Data are contained within the articles.

Acknowledgments: First of all, I would like to thank all of my co-authors and graduate students at Tohoku University and Ishinomaki Senshu University. I thank the technical staff of the machine shop of RISM, Tohoku University for their collaboration in assembling the multipass and tandem FPIs. Without these FPIs, I would have had no chance to write this review. I sincerely thank the late Y. Shimada (RISM and IMRAM) and H. Fujimori (IMR) of Tohoku University for their guidance to film magnetism and for their continuous interest and encouragement throughout my BLS studies. I also thank S. Nakamura (IMR) and S. Ohnuma (RIEMM) for the long period of collaborations over fifteen years on granular materials studies, and O. Kitakami (IMRAM) and K. Takanashi (IMR) of Tohoku University for their excellent sample preparations and fruitful discussions. Finally, I thank the late K-I. Kamiyoshi and T. Fujimura of RISM, Tohoku University for assigning me BLS for my Ph.D. thesis, and T. Goto of Niigata University for stimulating discussions and friendship over fifty years.

Conflicts of Interest: The author declare no conflict of interest.

References

1. Cummins, H.Z. Brillouin scattering studies of phase transitions in crystals. In *Light Scattering Near Phase Transitions*; Cummins, H.Z., Levanyuk, A.P., Eds.; North-Holland: Amsterdam, Holland, 1983; Volume 5, pp. 359–447.
2. Cooper, J.; Greig, J.R. Rapid scanning of spectral line profiles using an oscillating Fabry-Perot interferometer. *J. Sci. Instr.* **1963**, *40*, 433–437. [CrossRef]
3. Asenbaum, A. Computer-controlled Fabry-Perot interferometer for Brillouin spectroscopy. *Appl. Opt.* **1979**, *4*, 540–544. [CrossRef] [PubMed]
4. Sandercock, J.R. Brillouin scattering study of SbSi using a double-passed stabilized scanning interferometer. *Opt. Commun.* **1970**, *2*, 73–76. [CrossRef]
5. Lindsay, S.M.; Anderson, M.W.; Sandercock, J.R. Construction and alignment of a high performance multipass vernia tandem Fabry-Perot. *Rev. Sci. Instrum.* **1981**, *52*, 1478–1486. [CrossRef]
6. Dil, J.G. Brillouin scattering in condensed matter. *Rep. Prog. Phys.* **1982**, *45*, 285–334. [CrossRef]
7. Borovik-Romanov, A.S.; Kreines, M.H. Brillouin-Mandelstam scattering from thermal and excited magnons. *Phys. Rep.* **1982**, *81*, 351–408. [CrossRef]
8. Patton, C.E. Magnetic excitations in solids. *Phys. Rep.* **1984**, *103*, 251–315. [CrossRef]
9. Sandercock, J.R. Trends in brillouin scattering: Studies of opaque materials, supported films, and central modes. In *Light Scattering in Solids III*; Cardona, M., Güntherodt, G., Eds.; Springer: Berlin, Germany, 1982; pp. 121–171.
10. Grünberg, P. Light sscattering from surface magnetic excitations. *Prog. Surf. Sci.* **1985**, *18*, 1–58. [CrossRef]
11. Hillebrands, B. Brillouin Light Scattering from Layered Magnetic Structures. In *Light Scattering in Solids VII*; Cardona, M., Güntherodt, G., Eds.; Springer: Berlin, Germany, 1999; pp. 174–289.
12. Grünberg, P.; Mitawe, F. Light scattering from bulk and surface spin waves in EuO. *Phys. Rev. Lett.* **1977**, *39*, 1561–1564. [CrossRef]
13. Sandercock, J.R.; Wettling, W. Light scattering from surface and bulk thermal magnons in iron and nickel. *J. Appl. Phys.* **1979**, *50*, 7784–7789. [CrossRef]
14. Grimthditch, M.H. Brillouin scattering from metallic superlattices. In *Light Scattering in Solids V*; Cardona, M., Güntherodt, G., Eds.; Springer: Berlin, Germany, 1989; pp. 285–302.
15. Grünberg, P. Light scattering from spin waves in thin films and layered magnetic structures. In *Light Scattering in Solids V*; Cardona, M., Güntherodt, G., Eds.; Springer: Berlin, Germany, 1989; pp. 303–335.
16. Cochran, J.F. Light Scattering from Ultrathin Magnetic Layers and Bilayers. In *Ultrathin Magnetic Structures II*; Heinrich, B., Brand, J.A.C., Eds.; Springer: Berlin, Germany, 1994; pp. 222–257.
17. Hillebrands, B.; Güntherodt, G. Brillouin Light Scattering in Magnetic Superlattices. In *Ultrathin Magnetic Structures II*; Heinrich, B., Brand, J.A.C., Eds.; Springer: Berlin, Germany, 1994; pp. 258–279.
18. Slavin, A.N.; Demokritov, S.O.; Hillebrands, B. Nonlinear Spinwaves in One- and Two-Dimensional Magnetic Waveguides. In *Spin Dynamics in Confined Magnetic Structures I*; Hillebrands, B., Onudajela, K., Eds.; Springer: Berlin, Germany, 2002; pp. 35–64.
19. Demokritov, S.O.; Hillebrands, B. Spinwaves in Laterally Confined Magneti Structures. In *Dynamics in Confined Magnetic Structures I*; Hillebrands, B., Onudajela, K., Eds.; Springer: Berlin, Germany, 2002; pp. 65–92.
20. Bayer, C.; Demokritov, S.O.; Slavin, A.N.; Guslienko, K.Y.; Berkov, D.V.; Gorn, N.L.; Kostylev, M.P.; Hillebrands, B. Spin-Wave Excitations in Finite Rectangular Elements. In *Spin Dynamics in Confined Magnetic Structures III*; Hillebrands, B., Thiaville, A., Eds.; Springer: Berlin, Germany, 2005; pp. 57–104.
21. Hillebrands, B.; Hamrle, J. Investigation of Spin waves and Spin Dynamics by Optical Techniques. In *Handbook of Magnetism and Advanced Magnetic Materials, 3rd eds.*; Kronmüller, H.H., Parkin, S., Eds.; Wiley: New York, NY, USA, 2007; pp. 1566–1588.
22. Cottam, M.G.; Lockwood, D.J. *Light Scattering in Magnetic Solids*; A Wiley-Interscience Publication: New York, NY, USA, 1986; Chapter 8.
23. Farnell, G.W.; Adler, E.L. Elastic Wave Propagation in Thin Films. In *Physical Acoustics*; Mason, W.P., Thurston, R.N., Eds.; Academic Press: New York, NY, USA, 1972; Volume 9, pp. 35–127.
24. Damon, R.W.; Eshbach, J.R. Magnetostatic modes of a ferromagnetic slab. *J. Phys. Chem. Solids* **1961**, *19*, 308–320. [CrossRef]
25. Wolfram, T.; Dewames, R.E. Surface dynamics of magnetic materials. *Prog. Surf. Sci.* **1972**, *2*, 233–330. [CrossRef]
26. Rado, G.T.; Weertman, J.R. Spin-wave resonance in a ferromagnetic metal. *J. Phys. Chem. Solids* **1959**, *11*, 315–333. [CrossRef]
27. Hoffmann, F. Dynamic Pinning Induced by Nickel Layers on Permalloy Films. *Phys. Stat. Solidi b* **1970**, *41*, 807–813. [CrossRef]
28. Cottam, M.G. A magnetostatic theory of the response functions for a finite-thickness ferromagnetic slab and the application to Brillouin scattering. *J. Phys. C Solid St. Phys.* **1979**, *12*, 1709–1723. [CrossRef]
29. Camley, R.E.; Mills, D.L. Surface response of exchange- and dipole-coupled ferromagnets: Application to light scattering from magnetic surfaces. *Phys. Rev. B* **1978**, *18*, 4821–4841. [CrossRef]
30. Cottam, M.G.; Maradudin, A.A. Surface linear response functions. In *Surface Excitations*; Agranovich, V.M., Loudon, R., Eds.; North-Holland: Amsterdam, The Netherlands, 1984; pp. 1–194.
31. Mills, D.L. Surface Spin Waves on Magnetic Crystals. In *Surface Excitations*; Agranovich, V.M., Loudon, R., Eds.; North-Holland: Amsterdam, The Netherlands, 1984; pp. 379–439.
32. Camley, R.E.; Rahman, T.S.; Mills, D.L. Theory of light scattering by the spin-wave excitations of thin ferromagnetic films. *Phys. Rev. B* **1981**, *2*, 1226–1244. [CrossRef]

33. Cochran, J.F.; Dutcher, J.R. Calculation of the intensity of light scattering from magnons in thin films. *J. Mag. Mag. Mater.* **1998**, *73*, 299–310. [CrossRef]
34. Rado, G.T.; Hicken, R.J. Theory of magnetic surface anisotropy and exchange effects in the Brillouin scattering of light by magnetostatic spin waves. *J. Appl. Phys.* **1988**, *63*, 3885–3889. [CrossRef]
35. Grünberg, P. Magnetostatic spinwave modes of a ferromagnetic double layer. *J. Appl. Phys.* **1980**, *51*, 4338–4341. [CrossRef]
36. Grünberg, P. Magnetostatic spin-wave modes of a heterogeneous ferromagnetic double layer. *J. Appl. Phys.* **1981**, *52*, 6824–6829. [CrossRef]
37. Grünberg, P.; Mika, K. Magnetostatic spin-wave modes of a ferromagnetic multilayer. *Phys. Rev. B* **1983**, *27*, 2955–2963. [CrossRef]
38. Barnas, J. Spin waves in superlattices. I. General dispersion equations for exchange, magnetostatic and retarded modes. *J. Phys. C Solid State Phys.* **1988**, *21*, 1021–1036. [CrossRef]
39. Camley, R.E.; Rahman, T.S.; Mills, D.L. Magnetic excitations in layered media: Spin waves and the light-scattering spectrum. *Phys. Rev. B* **1983**, *27*, 261–277. [CrossRef]
40. Vohl, M.; Barnaś, J.; Grünberg, P. Effect of interlayer exchange coupling on spin-wave spectra in magnetic double layers: Theory and experiment. *Phys. Rev. B* **1989**, *39*, 12003–12012. [CrossRef] [PubMed]
41. Zivien, R.; Tacchi, S.; Montoncello, F.; Giovannini, L.; Nizzoli, F.; Madami, M.; Gubbiotti, G.; Carlotti, G.; Neusser, S.; Duerr, G.; et al. Bragg diffraction of spin waves from a two-dimensional antidote lattice. *Phys. Rev. B* **2012**, *85*, 012403.
42. Hayes, W.; Loudon, R. *Scattering of Light by Crystals*; A Wiley-Interscience Publication: New York, NY, USA, 1978; pp. 239–285.
43. Wettling, W.; Cottam, M.G.; Sandercock, J.R. The relation between one-magnon light scattering and the complex magneto-optic effects in YIG. *J. Phys. C* **1975**, *8*, 211–228. [CrossRef]
44. Landau, L.D.; Lifshitz, E.M. *Statistical Physics*, 2nd ed.; Pergamon Press: Oxford, UK, 1969; Chapter 12.
45. Zare, R.N. *Angular Momentum Understanding Spatial Aspects in Chemistry and Physics*; A Wiley-Interscience Publication: New York, NY, USA, 1987; Chapter 1.
46. Kittel, C. *Quantum Theory of Solids*; Wiley: New York, NY, USA, 1963; Chapter 4.
47. Fleury, P.R.; Loudon, R. Scattering of Light by One and Two-Magnon Excitations. *Phys. Rev.* **1968**, *166*, 514–529. [CrossRef]
48. Yoshihara, A.; Chida, M.; Yamato, Y.; Nakamura, F.; Iwabuchi, K.; Fujimura, T. Construction of a Piezoelectrically Scanned Brillouin Spectrometer. *Bull. Res. Inst. Sci. Meas. Tohoku Univ.* **1986**, *35*, 31–50. (In Japanese)
49. Haneda, Y.; Yoshihara, A.; Chida, M.; Yamato, Y.; Nakamura, F.; Iwabuchi, K.; Shoji, T.; Shizukuishi, T.; Fujimura, T. Construction of a Piezoelectrically Scanned Brillouin Spectrometer Part II Multipass. *Bull. Res. Inst. Sci. Meas. Tohoku Univ.* **1986**, *36*, 101–116. (In Japanese)
50. Yoshihara, A.; Haneda, Y.; Shimada, Y.; Fujimura, T. Magnon Brillouin scattering in FeSi sputtered films. *J. Appl. Phys.* **1989**, *65*, 4968–4970. [CrossRef]
51. Yoshihara, A.; Haneda, Y.; Shimada, Y.; Fujimura, T. Brillouin scattering from spin waves in sputtered CoZr films. *J. Appl. Phys.* **1989**, *66*, 328–335. [CrossRef]
52. Yoshihara, A.; Shimada, Y.; Maro, T.; Kitakami, O.; Mizushhima, K. Brillouin scattering in Fe-polyethylene co-evaporated films. *J. Magn. Magn. Mater.* **1992**, *104–107*, 1835–1836. [CrossRef]
53. Yoshihara, A. Construction of a Vernier Tandem (3+3) Pass Brillouin Spectrometer and Its Application to Surface Acoustic Wave Study in Metallic Multilayer Films. *Jpn. J. Appl. Phys.* **1994**, *33*, 3100–3109. [CrossRef]
54. Yoshihara, A.; Ohnuma, S.; Fujimori, H.; Nakamura, S.; Nojima, T. Spin Waves and Transport Properties in Ferromagnetic Co-Al-O and Fe-Al-O Granular Films: A Brillouin Scattering Study. *J. Phys. Soc. Jpn.* **2008**, *77*, 094704. [CrossRef]
55. Yoshihara, A.; Li, W.D.; Kitakami, O.; Shimada, Y. Brillouin light scattering from collective spin waves in Fe-Al-O granular films. *J. Magn. Magn. Mater.* **2004**, *268*, 257–263. [CrossRef]
56. Cottam, M.G.; Tilley, D.R. *Introduction to Surface and Superlattice Excitations*; Cambrige University Press: Cambridge, UK, 1989; Chapter 4.
57. Grimsditch, M.; Malozemoff, A.P.; Brunsch, A. Standing Spin Waves Observed by Brillouin Scattering in Amorphous Metallic $Fe_{80}B_{20}$ Films. *Phys. Rev. Lett.* **1979**, *43*, 711–714. [CrossRef]
58. Shirane, G.; Minkiewick, V.J.; Nathans, R. Spin Waves in 3d Metals. *J. Appl. Phys.* **1968**, *39*, 383–389. [CrossRef]
59. Grünberg, P.; Mayr, C.M.; Vach, W.; Grimsditch, M. Determination of magnetic parameters by means of Brillouin scattering. Examples: Fe, Ni, $Ni_{80}Fe_{20}$. *J. Magn. Magn. Mater.* **1982**, *28*, 319–325. [CrossRef]
60. Yoshihara, A. Brillouin Light Scattering Determination of Spin Wave Stiffness Constant of hcp Co Thin Film and an Attempt toward Solving the *D*-Value Problem. *J. Magn. Soc. Jpn.* **2004**, *28*, 1023–1028. (In Japanese)
61. Yoshihara, A. Brillouin Scattering from Collective Spin Waves –application to Ultrathin and Multilayer Films. *J. Magn. Soc. Jpn.* **1995**, *19*, 923–930. (In Japanese)
62. Yoshihara, A.; Takanashi, K.; Shimada, Y.; Kitakami, O. Magnon Brillouin Scattering from an Amorphous CoNbZr Thin Film. *Jpn. J. Appl. Phys.* **1994**, *33*, 3927–3928. [CrossRef]
63. Yoshihara, A.; Sato, H.; Mawatari, J.; Yoshida, K.; Kitakami, O.; Shimada, Y. Brillouin light scattering from spin waves in $Co_{100-x}Cr_x$ alloy films. *J. Magn. Magn. Mater.* **2000**, *221*, 261–267. [CrossRef]
64. Mawatari, J. Magnetic Constants of CoCr Alloy and Co Films Determined by Brillouin Light Scattering. Master's Thesis, Ishinomaki Senshu University, Ishinomaki, Japan, 2002. (In Japanese)

65. Yoshihara, A.; Haneda, Y.; Shimada, Y.; Ouchi, K. Magnetostatic Spin Waves in Sputtered Ferromagnetic Films with Perpendicular Anisotropy. *Jpn. J. Appl. Phys.* **1991**, *30*, 2010–2019. [CrossRef]
66. Srinivasan, G.; Patton, C.E. Brillouin light scattering on cobalt-chromium films. *IEEE Trans. Magn.* **1986**, *22*, 996–998. [CrossRef]
67. Heinrich, B. Ferromagnetic Resonance in Ultrathin Structures. In *Ultrathin magnetic Structures II*; Heinrich, B., Brand, J.A.C., Eds.; Springer: Berlin, Germany, 1994; Chapter 3.
68. Vernon, S.P.; Lindsay, S.M.; Stearns, M.B. Brillouin scattering from thermal magnons in a thin Co film. *Phys. Rev. B* **1984**, *29*, 4439. [CrossRef]
69. Karanikas, J.M.; Sooryakumar, R.; Printz, G.A.; Jonker, B.T. Thermal magnons in bcc cobalt-itinerancy and exchange stiffness. *J. Appl. Phys.* **1991**, *69*, 6120–6125. [CrossRef]
70. Moghadam, A.; Booth, J.G.; Lord, D.G.; Boyle, J.; Boardman, A.D. Brillouin light scattering studies of Co/GaAs [001] films. *IEEE Trans. Magn.* **1994**, *30*, 769–771. [CrossRef]
71. Liu, X.; Stamps, R L.; Sooryakumar, R.; Printz, G.A. Magnetic anisotropies in thick body centered cubic Co. *J. Appl. Phys.* **1996**, *79*, 5387–5389. [CrossRef]
72. Liu, X.; Steiner, M.M.; Sooryakumar, R.; Printz, G.A.; Farrow, R.F.C.; Harp, G. Exchange stiffness, magnetization, and spin waves in cubic and hexagonal phases of cobalt. *Phys. Rev. B* **1996**, *53*, 12166–12172. [CrossRef]
73. Grimsditch, M.; Fullerton, E.E.; Stamps, R.L. Exchange and anisotropy effects on spin waves in epitaxial Co films. *Phys. Rev. B* **1997**, *56*, 2617–2622. [CrossRef]
74. Yoshihara, A.; Mawatari, J.; Sato, H.; Okamoto, S.; Kitakami, O.; Shimada, Y. Brillouin light scattering from spin waves in epitaxial hcp Co films. *Phys. Rev. B* **2003**, *67*, 134435. [CrossRef]
75. Dimitrijević, Ž.; Kraśnicki, S.; Todorovović, J.; Wanic, A. Investigation of magnons in hexagonal cobalt. *Physica* **1967**, *37*, 501–503.
76. Kubota, T.; Hamrle, J.; Sakuraba, Y.; Gaier, O.; Oogane, M.; Sakuma, A.; Hillebrands, B.; Takanashi, K.; Ando, Y. Structure, exchange stiffness, magnetic anisotropy of $Co_2MnAl_xSi_{1-x}$ Heusler compounds. *J. Appl. Phys.* **2009**, *106*, 113907. [CrossRef]
77. Gaier, O.; Hamrle, J.; Trudel, S.; Parra, A.C.; Hillebrands, B.; Arbelo, E.; Herbort, C.; Jourdan, M. Brillouin light scattering study of $Co_2Cr_{0.6}Fe_{0.4}Al$ and Co_2FeAl Heuser compounds. *J. Phys. D Appl. Phys.* **2009**, *42*, 084004. [CrossRef]
78. Hamrle, J.; Gaier, O.; Min, S.G.; Hillebrands, B.; Sakuraba, Y.; Ando, Y. Determination of exchange constants of Heusler compounds by Brillouin light scattering spectroscopy: Application to Co_2MnSi. *J. Phys. D Appl. Phys.* **2009**, *42*, 084005. [CrossRef]
79. Yoshihara, A.; Suzuki, Y.; Katayama, T. Collective Spin Waves in an Ultrathin Fe Wedge: A Brillouin Scattering Study. *J. Phys. Soc. Jpn.* **1996**, *65*, 1469–1474. [CrossRef]
80. Geerts, W.; Suzuki, Y.; Katayama, T.; Tanaka, K.; Ando, K.; Yoshida, S. Thickness dependent oscillation of the magneto-optical properties of Au-sandwiched (001) Fe films. *Phys. Rev. B* **1994**, *50*, 12581–12586. [CrossRef] [PubMed]
81. Hillebrands, B.; Baumgart, P.; Güntherodf, G. *In situ* Brillouin scattering from surface-anisotropy-dominated Damon-Eshbach modes in ultrathin epitaxial Fe(110) layers. *Phys. Rev. B* **1987**, *36*, 2450–2453. [CrossRef]
82. Dutcher, J.R.; Heinrich, B.; Cochran, J.F.; Steigewald, D.A.; Eagelhoff, W.F., Jr. Magnetic properties of sandwiches and superlattices of fcc Fe(001) grown on Cu(001) substrates. *J. Appl. Phys.* **1988**, *63*, 3464–3466. [CrossRef]
83. Dutcher, J.R.; Cochran, J.F.; Jacob, I.; Eagelhoff, W.F., Jr. Brillouin light-scattering intensities for thin magnetic films with large perpendicular anisotropies. *Phys. Rev. B* **1989**, *39*, 10430–10432. [CrossRef]
84. Maruyama, A.; Hyomi, K.; Eickmann, J.; Falco, C.M. Underlayer-induced perpendicular magnetic anisotropy in ultrathin Co/Au/Cu) films A spin-wave Brillouin-scattering study. *Phys. Rev. B* **1998**, *58*, 8596–8604. [CrossRef]
85. Rahman, T.S.; Mills, D.L. Dipolar surface spin waves in ferromagnetic films. *J. Appl. Phys.* **1982**, *53*, 2084–2086. [CrossRef]
86. Dieny, B.; Speriosu, V.S.; Parkin, S.S.; Gurney, B.A.; Wilhoit, D.R.; Mauri, D. Giant magnetoresistive in soft ferromagnetic multilayers. *Phys. Rev B* **1991**, *43*, 1297–1300. [CrossRef] [PubMed]
87. White, R.M. *QuantumTheory of Magnetism, Second Corrected and Updated Ed.*; Springer: Berlin, Germany, 1983; Chapter 7.
88. Binasch, G.; Grunberg, P.; Saurenbach, F.; Zinn, W. Enhanced magnetoresistance in layered magnetic structures with antiferromagnetic interlayer exchange. *Phys. Rev. B* **1989**, *39*, 4828–4830. [CrossRef]
89. Grünberg, P.; Schreiber, R.; Pang, Y.; Brodsky, M.B.; Sowers, H. Layered Magntic Structures: Evidence for antiferromagnetic coupling of Fe-layers across Cr-interlayer. *Phys. Rev. Lett.* **1986**, *57*, 2442–2445. [CrossRef]
90. Nörtemann, F.C ; Stamps, R.L.; Camley, R.E.; Hillebrands, B.; Güntherodt, G. Effective-mudium theory for finite magnetic multilayer: Effect of anisotropy on dipolar modes. *Phys. Rev. B* **1993**, *47*, 3225–3230. [CrossRef] [PubMed]
91. Grimsditch, M.; Kahn, M.R.; Kueny, A.; Schuller, I.K. Collective Behavior of Magnons in Superlattices. *Phys. Rev. Lett.* **1983**, *51*, 498–500. [CrossRef]
92. Yoshihara, A.; Takanasi, K.; Obi, Y.; Fujimori, H. Brillouin scattering from spin waves in Fe/Cr multilayer films. *J. Magn. Magn. Mater.* **1993**, *126*, 333–336. [CrossRef]
93. Ashcroft, N.W.; Mermin, N.D. *Solid State Physics*; Saunders College Publishing: Fort Worth, TX, USA, 1976; Chapter 2.
94. Yoshihara, A.; Wang, Z.J.; Takanashi, K.; Grünberg, P.; Motokawa, M.; Fujimori, H. Spin wave modes in fine layered structures of Fe and Au. *J. Magn. Soc. Jpn.* **1999**, *23*, 161–163. [CrossRef]
95. Yoshihara, A.; Wang, J.T.; Takanashi, K.; Himi, K.; Kawazoe, Y.; Fujimori, H.; Grünberg, P. Interlayer exchange coupling in file-layered Fe/Au superlattices. *Phys. Rev. B* **2001**, *63*, 100405. [CrossRef]
96. Yoshihara, A.; Ohnuma, S.; Fujimori, H. Brillouin light scattering from TM-Al-O granular films (TM=Co, Fe). *J. Magn. Magn. Meter.* **2007**, *310*, 2513–2515. [CrossRef]

97. Nakamura, S.; Iwashita, T.; Nojima, T.; Yoshihara, A.; Ohnuma, S.; Fujimori, H. Electrical Resistivity in Ferromagnetic TM-Al-O (TM=Fe,Co) Granular Films: Scattering from Spin Waves and Kondo Like Behavior. *J. Phys. Soc. Jpn.* **2009**, *78*, 074708. [CrossRef]
98. Nakamura, S.; Iwashita, T.; Nojima, T.; Yoshihara, A.; Ohnuma, S.; Fujimori, H. Transport Properties of Ferromagnetic Granular Co-Al-O Films. *J. Phys. Conf. Ser.* **2010**, *200*, 012141. [CrossRef]
99. Nakamura, S.; Iwashita, T.; Nojima, T.; Yoshihara, A.; Ohnuma, S.; Fujimori, H. Transport and magnetic properties of ferromagnetic Co-Al-O and Fe-Al-O granular films. *J. Phys. Condens. Matter Conf. Ser.* **2011**, *266*, 012019. [CrossRef]
100. Nakamura, S.; Nojima, T.; Yoshihara, A.; Ohnuma, S. Temperature Dependence of the Coercive Force of Ferromagnetic TM-Al-O (TM=Fe,Co) Granular Films. *J. Korean Phys. Soc.* **2013**, *63*, 773–777. [CrossRef]
101. Yoshihara, A.; Meada, T.; Kawamura, S.; Nakamura, S.; Nojima, T.; Takeda, Y.; Ohnuma, S. Brillouin Light Scattering from Magnetic Excitations in Superparamagnetic and Ferromagnetic Co-Al-O Granular Films. *J. Phys. Soc. Jpn.* **2018**, *87*, 044707. [CrossRef]
102. Kikitsu, A.; Murayama, A.; Hyomi, K.; Falco, C. Micromagnetic study of CoPt-SiO_2 granular films by spin-wave Brillouin scattering. *J. Appl. Phys.* **2000**, *87*, 6947–6949. [CrossRef]
103. Stashkevich, A.A.; Roussigné, Y.; Djemia, P.; Billet, D.; Stognij, A.I.; Novitskii, N.N.; Wurtz, G.A.; Zayats, A.V.; Viau, G.; Chaboussant, G.; et al. Brillouin light scattering observation of the transition from the superparamagnetic to the superferromagnetic state in nanogranular (SiO_2)Co films. *J. Appl. Phys.* **2008**, *104*, 093912. [CrossRef]
104. Mirabal, M.; Grünberg, P.; Zinn, W. Observation of paramagnetic lightscattering from EuO, EuS, $Eu_{0.005}Sr_{0.995}S$, and Gd. *Solid State Commun.* **1984**, *52*, 929–931. [CrossRef]
105. Kong, S.H.; Klein, M.V.; Tsui, F.; Flynn, C.P. Brillouin-light-scattering study of long-wavelength spin waves in a single-crystal 300-Å Gadolinium film. *Phys. Rev. B* **1992**, *45*, 12297–12305. [CrossRef]
106. Ercole, A.; Lew, W.S.; Lauhoff, G.; Kernohan, E.T.M.; Lee, J.; Bland, J.A.C. Temperature-dependent spin-wave behavior in Co/CoO bilayers studied by Brillouin light scattering. *Phys. Rev. B* **2000**, *62*, 6429. [CrossRef]
107. Blachowicz, T.; Tillmanns, A.; Fraune, M.; Ghadimi, R.; Beschoten, B.; Güntherodt, G. Exchange bias in epitaxial CoO/Co bilayers with different crystallographic symmetries. *Phys. Rev. B* **2007**, *75*, 054425. [CrossRef]
108. Yoshihara, A.; Kawamura, S.; Nakamura, S.; Nojima, T.; Ohnuma, S. Brillouin Light Scattering Study on Magnetization Dynamics in Superparamagnetic Co-Al-O Granular Films Down to 15 K. *J. Phys. Soc. Jpn.* **2020**, *89*, 064701. [CrossRef]
109. Néel, L. Théorie du traînage magnétique des ferromagnétiques en grains fins avec application aux terres cuites. *Ann. Géophys.* **1949**, *5*, 99–136.
110. Mørup, S.; Hansen, M.F. Superparamagnetic Particles. In *Handbook of Magnetism and Advanced Magnetic Materials*; Kromüller, H., Parkin, S.S., Eds.; A Wiley-Interscience Publication: New York, NY, USA, 2007; Volume 6.
111. Batlle, X.; Labarta, A. Finite-size effect in fine particles: Magnetic and transport properties. *J. Phys. D Appl. Phys.* **2002**, *35*, R15–R42.
112. Yoshihara, A.; Kurosawa, M.; Morioka, Y. Brillouin Scattering from the Soft Optical Phonon in the Orthorhombic Form of a $K_3Co_{1-x}Fe_x(CN)_6$ ($x \leq 0.03$) Crystal. *Jpn. J. Appl. Phys.* **1997**, *36*, 3318–3323. [CrossRef]
113. Hansen, M.F.; Bødker, F.; Mørup, S.; Lefmann, K.; Clausen, K.N.; Lindgård, P.A. Dynamics of Magnetic Nanoparticles Studied by Neutron Scattering. *Phys. Rev. Lett.* **1997**, *79*, 4910–4913. [CrossRef]
114. Linderoth, S.; Balcells, L.; Labarta, A.; Tejada, J.; Hendriksen, P.V.; Sethi, S.A. Magnetization and Mössbauer studies of ultrafine Fe-C particles. *J. Magn. Magn. Mater.* **1993**, *124*, 269–276. [CrossRef]
115. Otani, L.; Yoshihara, A.; Ohigashi, H. Brillouin scattering study of elastic properties in ferroelectric copolymer single-crystalline films. *Phys. Rev. E* **2001**, *64*, 051804. [CrossRef]
116. Yoshihara, A.; Sato, H.; Kojima, S. Brillouin Scattering Study of Glass-Forming Propylene Glycol. *Jpn. J. Appl. Phys.* **1996**, *35*, 2925–2929. [CrossRef]
117. Moriya, T. *Spin Fluctuations in Itinerant Electron Magnetism*; Springer: Berlin, Germany, 1985; Chapter 9.
118. *Spin Wave Confinement: Propagating Waves*, 2nd ed.; Dremokritov, S.O. (Ed.) Jenny Stanford Publishing: New York, NY, USA, 2017.
119. Hirohata, A.; Yamada, K.; Nakatani, Y.; Prejbeanu, I.-L.; Dieny, B.; Pirro, P.; Hillebrands, B. Review on spintronics: Principles and device applications. *J. Magn. Magn. Mater.* **2020**, *509*, 166711. [CrossRef]
120. Karenowska, A.D.; Chumak, A.V.; Serga, A.A.; Hillebrtands, B. Magnon Spintronics. In *Handbook of Spintronicds;* Xu, Y., Awschalom, D.D., Nitta, J., Eds.; Springer: Dordrecht, The Netherlands, 2016; pp. 1505–1549.

materials

MDPI

Article

Brillouin Light Scattering Characterisation of Gray Tone 3D Printed Isotropic Materials

Fehima Ugarak, Gwenn Ulliac *, Julio Andrés Iglesias Martínez, Johnny Moughames, Vincent Laude, Muamer Kadic * and Alexis Mosset *

Institute FEMTO-ST, CNRS, University Bourgogne Franche-Comté, 25000 Besançon, France; ugarakfehima@outlook.com (F.U.); julio.iglesias@femto-st.fr (J.A.I.M.); johnny.moughames@femto-st.fr (J.M.); vincent.laude@femto-st.fr (V.L.)

* Correspondence: gwenn.ulliac@femto-st.fr (G.U.); muamer.kadic@femto-st.fr (M.K.); alexis.mosset@femto-st.fr (A.M.)

Abstract: Three-dimensional direct laser writing technology enables one to print polymer microstructures whose size varies from a few hundred nanometers to a few millimeters. It has been shown that, by tuning the laser power during writing, one can adjust continuously the optical and elastic properties with the same base material. This process is referred to as gray-tone lithography. In this paper, we characterize by Brillouin light scattering the complex elastic constant C_{11} of different reticulated isotropic polymers, at longitudinal phonon frequencies of the order of 16 GHz. We estimate the real part of the C_{11} constant to vary from 7 to 11 GPa as a function of laser power, whereas its imaginary part varies between 0.25 and 0.6 GPa. The linear elastic properties are further measured at a fixed laser power as a function of temperature, from 20 °C to 80 °C. Overall, we show that our 3D printed samples have a good elastic quality with high Q factors only ten times smaller than fused silica at hypersonic frequencies.

Keywords: 3D printed materials; Brillouin light scattering; elastic properties of polymers

Citation: Ugarak, F.; Ulliac, G.; Iglesias Martínez, J.A.; Moughames, J.; Laude, V.; Kadic, M.; Mosset, A. Brillouin Light Scattering Characterisation of Gray Tone 3D Printed Isotropic Materials. *Materials* **2022**, *15*, 4070. https://doi.org/10.3390/ma15124070

Academic Editor: Antonio Sartagata

Received: 5 May 2022
Accepted: 4 June 2022
Published: 8 June 2022

1. Introduction

Three-dimensional direct laser writing has met an exponential rise over the last decade [1,2]. In the early age, its use was mainly directed toward electromagnetic applications and it thus was the object of a quest for the highest resolution [1] and the lowest optical losses [3]. This new technology opens the way to new optical metamaterials as simple as 3D dielectrics [4–7] or to metals by using the printed sample as a mold [8]. Additional efforts have been made to convert dielectrics to conducting parts via atomic layer deposition [9] through simple thermal evaporation. More recently, 3D direct laser scanning was used for gray tone lithography [10,11]. Indeed, when changing the laser power (LP), the scanning velocity or the hatching/slicing distances, one can obviously impart different degrees of polymerisation and cross-linking during the printing process [11]. All parameters are important and the resulting properties can be more or less sensitive to them. It should be noted that the variations of electromagnetic properties such as the refractive index are rather small [12]. However, even such a small but tunable change is very useful and advantageous for manufacturing optical waveguides [6]. Micro-robotics, acoustics and elasticity have also taken advantage of 3D printing technologies for the manufacture of metamaterials via gray tone lithography [10]. One important example is the implementation of the so-called bi-material beams where one takes advantage of the very sensitive differential deformation between two rather similar materials [10,13].

A few extensive studies have been performed on different polymers to obtain the printable properties as a function of printing parameters, via Raman scattering or other characterisation techniques [14,15]. However, from one setup to another and for different reasons, characterisation must be repeated and adapted each time. In this paper, we aim at characterising via Brillouin light scattering two transparent photoresins called IP-Dip2 and IP-S that are commercially available from the Nanoscribe company [16]. Due to the isotropic elastic behavior of those photoresins and the fact that only longitudinal acoustic phonons can be probed, we limit our analysis to the dependence of the C_{11} elastic constant as a function of writing laser power, writing speed and temperature. We show how the elastic constant can be tuned and to some degree in which range of parameters one should play to obtain the largest changes. All characterisations have been done using the Brillouin Light Scattering (BLS) technique using a multipass Fabry–Pérot interferometer. Ref. [17] Hence, the information we provide regarding both the real part (determining the longitudinal phonon velocity) and the imaginary part (determining the phonon loss) of C_{11} at GHz frequencies is novel and has implications for ultrasonic MEMS applications.

2. Fabrication

In this study, we use a commercial 3D micro-printer (Photonic Professional GT+, Nanoscribe GmbH) based on two-photon absorption lithography. The photoresins investigated in this study are the commercial negative tone IP-Dip2 and IP-S (Nanoscribe GmbH) that were customized by Nanoscribe for high resolution and mesoscale applications.

We fixed the slicing and hatching distances for each resin. For the IP-Dip2 sample, they were set to $l_x = 0.3$ µm and $l_z = 0.2$ µm. For IP-S, we consistently use $l_x = 1$ µm and $l_z = 0.5$ µm. As illustrated in Figure 1, it is expected that, for the same hatching and slicing distances, a higher laser power results in a larger overlap between voxels and consequently in an enhancement of the polymerisation of photoresins. To prepare samples, a drop of IP-Dip2 (IP-S) photoresin was deposited on a fused silica (ITO-coated soda lime glass) substrate and photopolymerized using a 63X or 25X microscope objective and a femtosecond laser operating at $\lambda = 780$ nm. The reference printing parameters are a laser power of 60% and a galvanometric scanning speed of 10 mm/s for IP-Dip2 and 100% and 100 mm/s for IP-S. After printing, the samples were developed for 20 min in a propylene glycol methyl ether acetate (PGMEA) solution to remove the unexposed photoresin and rinsed for 3 min in isopropyl alcohol (IPA) to clear the developer. The printed samples are cylinders with dimensions of $[200 \times 200 \times 50]$ µm^3 and $[400 \times 400 \times 50]$ µm^3 for IP-Dip2 and IP-S, respectively, and the distance between each cylinder was set to 500 µm. This inter-cylinder distance, for both IP-S and IP-Dip2, ensures that the modification of the phonon spectrum is achieved in each cylinder. As the core study, we consider the influence of selected fabrication parameters on the elastic properties of the printed materials. Figures 2 and 3 show SEM images of samples that are used for measurements. Overall, they are homogeneous in both shape and aspect. The minor blemishes and their domed tops do not affect measurements of bulk longitudinal phonons by BLS and thus the determination of the elastic constant C_{11}.

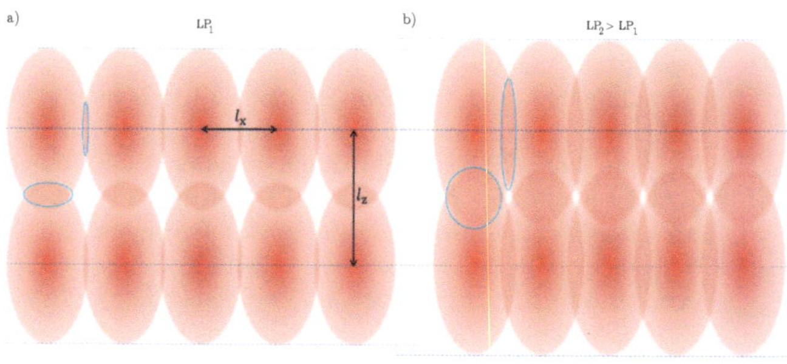

Figure 1. Principle of Direct Laser Writing voxels for different laser power. (**a**) definition of a voxel that is used in our printing and of the hatching distance l_x and slicing distance l_z; (**b**) schematic representation of the voxels for a laser power larger than in (**a**). A larger voxel is obtained, which, for the same slicing, results in a larger overlap of adjacent voxels (overlaps are depicted with blue ellipses).

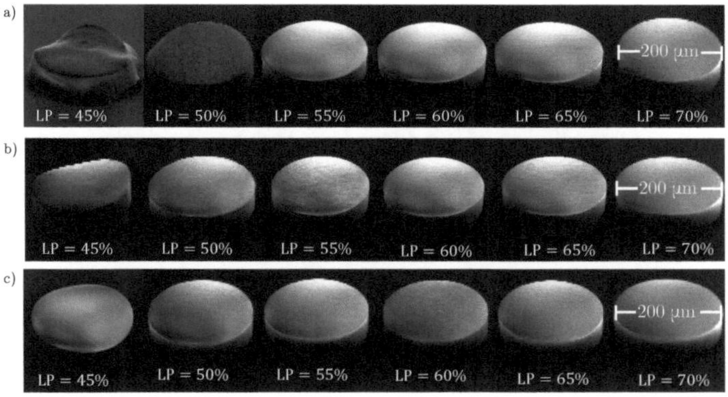

Figure 2. SEM images of IP-Dip2 samples for different laser powers and at different scanning velocities. (**a**) $v_s = 15$ mm/s; (**b**) $v_s = 10$ mm/s; (**c**) $v_s = 7$ mm/s.

Figure 3. SEM images of IP-S samples for different laser powers and at different scanning velocities. (**a**) $v_s = 200$ mm/s; (**b**) $v_s = 100$ mm/s.

3. Fundamentals of BLS for Isotropic Media

As stated in the Introduction, the BLS contact-less characterisation technique is used in order to estimate the effective elastic constant C_{11} of the samples. We remind the reader that, in our study, we only consider linear elasticity. Thus, the stress tensor σ can be related to the strain tensor ϵ by the 4th rank elasticity tensor C as:

$$\sigma_{ij} = C_{ijkl}\epsilon_{kl}. \tag{1}$$

Taking into account major and minor symmetry of the tensor, the elasticity tensor can be simplified drastically to have only 21 independent elasticity terms. However, if, in addition, we consider the material as isotropic, the simplification is even more drastic and only two independent parameters are left. It is common to express the new elasticity tensor in Voigt notation:

$$C_{\alpha\beta} = \begin{bmatrix} C_{11} & C_{12} & C_{12} & 0 & 0 & 0 \\ C_{12} & C_{11} & C_{12} & 0 & 0 & 0 \\ C_{12} & C_{12} & C_{11} & 0 & 0 & 0 \\ 0 & 0 & 0 & C_{44} & 0 & 0 \\ 0 & 0 & 0 & 0 & C_{44} & 0 \\ 0 & 0 & 0 & 0 & 0 & C_{44} \end{bmatrix}. \tag{2}$$

with C_{44} being the shear modulus and the relation $C_{12} = C_{11} - 2C_{44}$. In Voigt notation, the indices $\alpha = (ij)$ and $\beta = (kl)$ replace pairs of symmetric indices. As our material is isotropic, both elastic coefficients can be expressed by engineering moduli as follows: $C_{11} = B + 4G/3$ and $C_{44} = G$, where B and G are the bulk and shear moduli, respectively. Another common parametrization for isotropic materials uses Young's modulus E and Poisson's ratio η, with relations $C_{11} = E(1-\eta)/[(1+\eta)(1-2\eta)]$ and $C_{44} = E/[2(1+\eta)]$.

BLS relies on the inelastic scattering of light by spontaneous thermal fluctuations in a material. Those fluctuations correspond to acoustic phonons. Scattering of light in directions other than the incident one is caused by fluctuations of the material's dielectric tensor and depends both on the optical properties of the material and on the light penetration into the sample [18]. In the scattered light, there are in principle contributions from bulk phonons, through a volumetric opto-elastic mechanism, and from surface phonons, through a surface-ripple mechanism. Which of those two mechanisms is dominant in the BLS spectrum depends on the sample's thickness and on the optical properties of the material, such as its transparency at the given excitation wavelength. With our samples, bulk phonons are dominant and will only be considered in the following.

Figure 4 illustrates the general light-scattering processes involving bulk phonons in a Stokes and anti-Stokes processes. In those processes, the law of conservation of momentum is given by

$$\hbar\vec{k}_s - \hbar\vec{k}_i = \pm\hbar\vec{q} \tag{3}$$

and the conservation of energy is given by

$$\hbar\omega_s - \hbar\omega_i = \pm\hbar\omega_q. \tag{4}$$

Here, $\vec{k}_i$ and $\vec{k}_s$ are the incident and scattered wavevectors, and ω_i and ω_s are the angular frequencies of inci dent and scattered light, respectively. The positive and negative signs on the right-hand sides correspond to 'anti-Stokes' and 'Stokes' processes, respectively. The anti-Stokes process is a process of annihilation (or absorption) of a phonon, whereas the Stokes process represents a creation (or emission) of a phonon during scattering. Angle θ between $\vec{k}_i$ and $\vec{k}_s$ is the scattering angle that depends on the scattering geometry [19]. Our measurements were performed in the back-scattering geometry where the angle $\theta = \pi$. This backscattering arrangement provides a comparably larger interaction volume, allowing for a higher temporal resolution and Brillouin spectra that can be recorded faster, compared to other common scattering geometries [20]. The connection between the frequency shift f of Brillouin peaks and the phase velocity v_L of the longitudinal bulk acoustic phonon is

$$v_L = \frac{2\pi f}{q_L} = \frac{\lambda f}{2n}\sin\left(\frac{\theta}{2}\right) \tag{5}$$

where $\vec{q}_L = 2n\vec{k}_i$ is the phonon wavevector with $||\vec{k}_i|| = 2\pi/\lambda$, where λ is the wavelength of the incident light. From the measurement of the frequency of the longitudinal bulk acoustic phonons, the linewidth Γ_b (FWHM) of the Brillouin peak, and the mass density of the material, one can estimate the real and imaginary parts of the longitudinal modulus C_{11} [19]. The real part of the longitudinal modulus is obtained through the relation

$$\Re(C_{11}) = \rho v_L^2 = \frac{\rho\lambda^2}{4n^2}f^2. \tag{6}$$

Imaginary part of the longitudinal modulus $\Im c_{11}$ depends on the linewidth Γ_b (FWHM) of the Brillouin peak through

$$\Im(C_{11}) = \frac{\rho\lambda^2}{4n^2}f\frac{\Gamma_b}{2\pi}. \tag{7}$$

As an illustration, Figure 5 shows a typical BLS recorded spectrum. The central peak is the Rayleigh elastic scattering of the probe laser. The Stokes and anti-Stokes frequency shifts are visible respectively to the left and to the right side of the central peak. The peaks around $\pm$33 GHz are induced by the longitudinal bulk phonon of the substrate. Peaks for photoresins appear around $\pm$16 GHz.

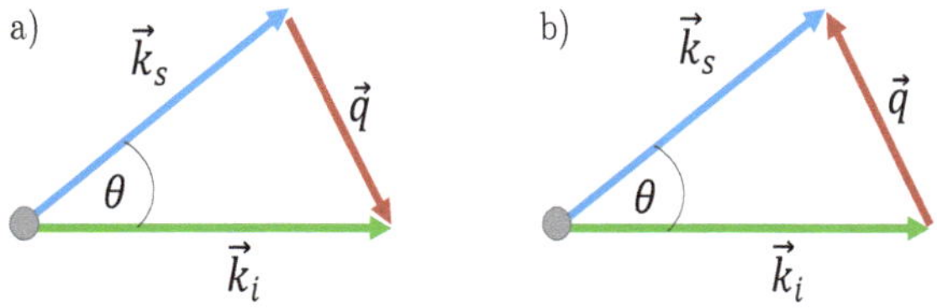

Figure 4. Vector representation of the light-scattering processes via bulk phonons (**a**) Stokes process; (**b**) Anti-Stokes process.

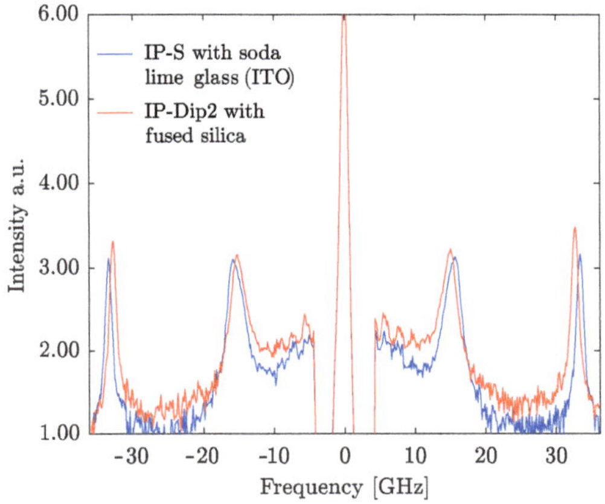

Figure 5. Example of BLS spectra measured directly from the setup for two different photoresins on two different substrates. Clear anti-Stokes and Stokes peaks are observed in both cases.

4. Experimental Setup

Figure 6 presents the main components of our experimental setup. We use a continuous single longitudinal mode Nd-YAG laser, doubled at 532 nm, vertically polarized, to illuminate the sample with an incident power of 20 mW. A small amount of the laser beam ($\sim$1%) is diverted by a beam splitter (BS) in order to obtain a reference beam (green dashed line), giving the Rayleigh elastic peak. The incident laser beam is controlled by mirrors M1 and M2. Lens L1 ($f = 100$ mm, N.A. $= 0.1$) focuses the incident laser beam, giving a spot diameter smaller than 50 µm, and collects backscattered light. Mirror M2 is very small compared to the area of the pupil of lens L1, in order to reduce the losses of collected scattering light in comparison with a standard 50/50 beam splitter. We illuminate the sample under quasi-normal incidence ($\sim 2^\circ$ with respect to the sample surface) to avoid dazzling the detector with reflected light. We assume that this small angular detuning does not affect the Brillouin frequency shift because the wavenumber of bulk acoustic phonons in isotropic media is independent of the incidence angle with respect to the normal to the surface. Backscattered light is focused by achromatic lens L2 ($f = 400$ mm) in a Sandercock-type 3+3-pass tandem Fabry–Pérot interferometer (TFP-2 HC) with respect to its numerical aperture. The two steering mirrors M3 and M4 can be adjusted so the backscattered light beam enters the entrance pinhole of the interferometer. The very weak signal produced by spontaneous inelastic scattering of light by bulk acoustic phonons is filtered with a high contrast ratio and detected by a synchronized photomultiplier. In order to observe and position the laser spot onto the sample, it was inevitable to install a removable confocal-like setup with a beam splitter, a white light source and a CMOS camera, which are not represented here. Furthermore, to perform measurements as a function of the temperature of the sample, the sample was inserted inside a Linkam head (not represented) with a thin window transparent to visible light. The temperature change was performed from 30 °C to 80 °C in steps of 10 °C and back.

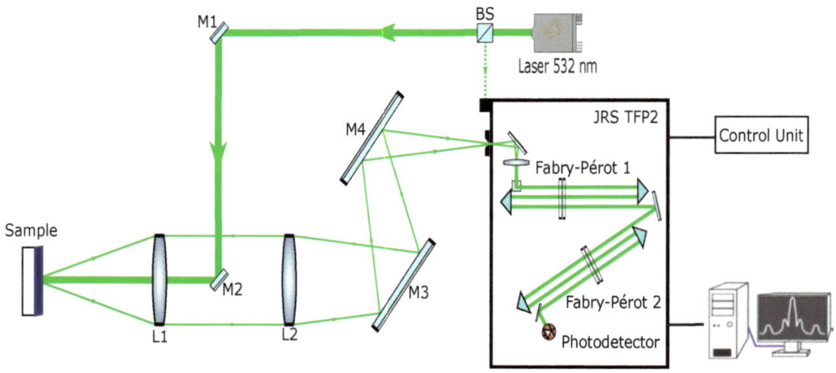

Figure 6. Brillouin light back-scattering setup.

5. Results for Isotropic Materials IP-Dip2 and IP-S

In this section, we analyze the dependence of the complex-valued elastic constant C_{11} for different writing laser powers and scanning velocities v_s. Both IP-S and IP-Dip2 samples are considered. In order to retrieve the elastic properties, we assume that IP-S is a typical amorphous polymer with mass density $\rho = 1280\ \mathrm{kg/m^3}$ that, at room temperature ($T = 293$ K), has a refractive index $n = 1.52$ [21]. IP-Dip2 is a photoresin with a refractive index $n = 1.55$ [21] at room temperature and mass density $\rho = 1170\ \mathrm{kg/m^3}$.

As explained before, one can in principle change the effective properties of the printed material via at least three different controls: the laser power, the hatching/slicing distance and the scanning velocity. In this study, we keep the hatching/slicing distances constant since these parameters are crucial for maintaining low surface roughness as well as high writing resolution.

In Figure 7, we show the measured elastic constant C_{11} for IP-Dip2 (panel (a)) and for IP-S (panel (b)). The study is done for both photoresins as a function of the relative laser power. In both cases, we show results for different scanning velocities v_s. For IP-Dip2, we limit the laser power to LP = 70% in order to avoid burning the photoresin. For both photoresins, we observe a stiffening process as a function of the laser power. As the laser power increases, the value of the real part of the longitudinal modulus C_{11} increases before reaching a maximum. Simultaneously, the imaginary part of C_{11} decreases to reach a minimum. Hence, viscoelastic damping reduces with the laser power.

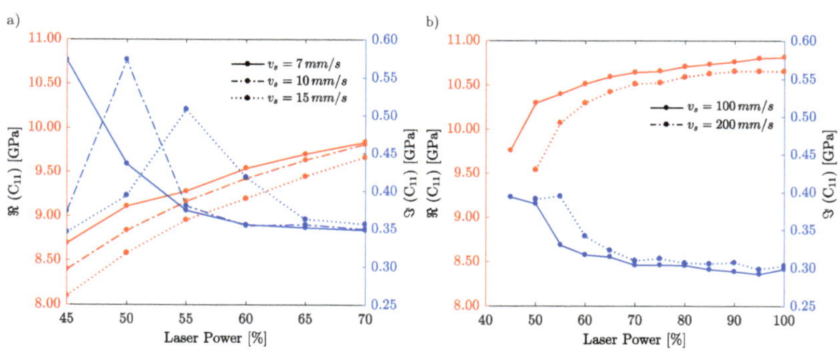

Figure 7. Measured elastic constant c_{11} for different writing laser powers, and for (**a**) IP-Dip2 and (**b**) IP-S. The red color represents the real part of the longitudinal modulus and blue represents its imaginary part. Each type of line corresponds to a different scanning velocity.

When the writing velocity increases, the effective dose received by each voxel becomes smaller and polymerisation via diffusion decreases, resulting in a softer material (lower elastic constant). We attribute this effect to the enhancement of the cross-linking and to the fact that two-photon absorption induces a saturation of the polymerisation process [11]. For IP-S, this aspect is more pronounced and we reach a plateau very fast. It is often observed in the literature that faster polymerisation combined with better diffusion results in more homogeneous sample properties and thus in a better surface and volume quality [14]. Here, we furthermore emphasize the importance of damping that increases for both photoresins when the laser power decreases. As observed in previous studies, for low laser power, polymerisation can in principle be rather inhomogeneous and some parts can contain monomers (not polymerised). However, we also note a decrease in damping in the case of IP-Dip2 for the lowest laser power. This surprising phenomenon was checked with additional samples and repeating measurements; results have always been consistent. We attribute this effect to a process of grain-like formation in the samples when the quality deteriorates [22]. This deterioration of quality is often attributed to domain formation.

Interestingly, we observed that in general for both photoresins and for all printing conditions the imaginary part is rather low and thus bulk longitudinal phonons can propagate in these polymers in the GHz regime with reasonable loss. The quality factor times frequency product, Qf, which is expected to be a constant in the ultrasonic range for most viscoelastic materials, is at best of the order of 50 10^{10} Hz, only smaller by a factor 10 compared to Qf measured for fused silica optical fibers [23].

The datasheet for the IP-S resin indicates that $E = 5$ GPa from static measurements (at very low mechanical frequencies) [16]. In the MHz range, Young's modulus can be estimated from the analysis of the resonances of beams. The reported value for Young's modulus is $E = 10$ GPa for the IP-S photoresin for frequencies between 1 and 3 MHz [7]. For example, converting to elastic constants, this yields to $Re(c_{11}) = E(1-\nu)/[(1+\nu)(1-2\nu)] \approx 15$ GPa assuming a Poisson's ratio $\nu = 0.3$. Hence, we observe similar values in the GHz regime.

6. Temperature Dependence of Elastic Constants

Lastly, we examine the dependence of elastic properties with temperature. For that purpose, we insert the samples into a Linkam chamber installed on the sample holder to control its temperature. Figure 8 shows the temperature dependence of the elastic constant C_{11} for IP-Dip2 and for a single laser power of LP = 60%, but for three different scanning velocities, $v_s = 7$ mm/s in panel (a), $v_s = 10$ mm/s in panel (b), and $v_s = 15$ mm/s in panel (c). We first note that the temperature dependence is rather linear and that the initial properties are recovered for all samples after a full cycle. For the lowest scanning velocity, a small difference appears between the heating cycle and the cooling cycle. As expected for many polymers, as temperature increases, the material becomes softer: the real part of the longitudinal modulus decreases whereas the imaginary part increases, as often for viscous media. However, we note that the imaginary part is mostly two orders of magnitude smaller than the real part of C_{11} and that the polymer is indeed in a very stiff and limited lossy state, making it suitable for acoustic experiments without detrimental absorption over the full range of temperature study.

Figure 9 proposes a similar study but for IP-S. The laser power is set to LP=100%. Two different scanning velocities are considered, $v_s = 100$ mm/s in panel (a) and $v_s = 200$ mm/s in panel (b). The hysteresis deviation between increasing and decreasing temperature cycles is now quite small. IP-S is thus more stable and less sensitive to ambient temperature variations as compared to IP-Dip2.

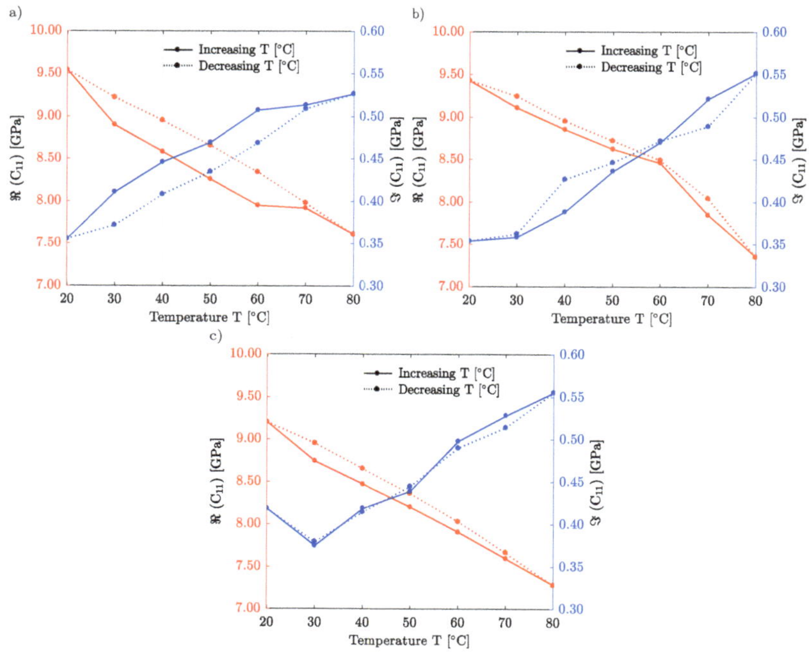

Figure 8. Temperature dependence of the elastic constant C_{11} for IP-Dip2. The scanning speed is (**a**) $v_s = 7$ mm/s; (**b**) $v_s = 10$ mm/s; and (**c**) $v_s = 15$ mm/s.

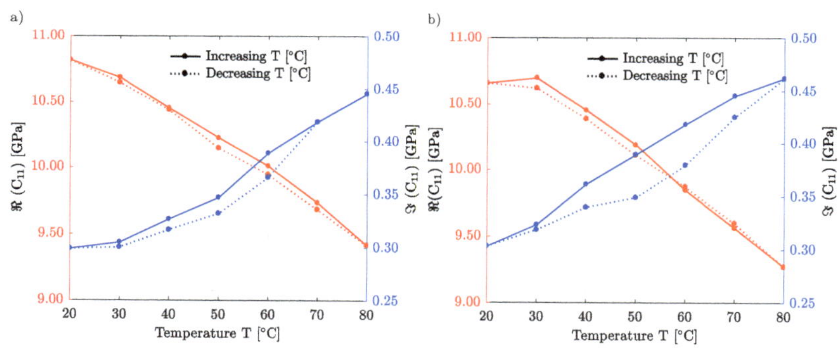

Figure 9. Temperature dependence of the elastic constant C_{11} for IP-S. The scanning velocity is (**a**) $v_s = 100$ mm/s, (**b**) $v_s = 200$ mm/s.

7. Conclusions

In this paper, we have characterized using Brillouin light scattering the elastic properties of two commercially available photoresins used for two-photon lithography. Specifically, the Brillouin shift of longitudinal phonons was recorded as a function of writing laser power, scanning velocity, and sample temperature. Measurements show that the real part of the elastic constant C_{11} changes relatively by about 10 to 15%, whereas the imaginary part changes by about the same amount but in the opposite direction. The elastic constant can thus be tuned continuously between 7 and 11 GPa at phonon frequencies of the order of 16 GHz. Concurrently, viscoelastic loss can be tuned between 0.25 and 0.6 GPa. We hope that our findings can be useful for the design of smart structures by gray-tone lithography and can motivate the use of 3D printing at the microscale for acoustics in the GHz regime.

Author Contributions: F.U. performed experimental characterisations. G.U. performed fabrications. J.A.I.M. helped process the data and set-up the instrumentation. All authors contributed to conceptualization and to methodology; software J.A.I.M. and V.L.; validation, M.K., V.L. and A.M.; formal analysis, all authors; investigation, all authors; resources, all authors; data curation, F.U.; writing—original draft preparation, A.M., M.K. and V.L.; writing—review and editing; supervision, M.K., V.L. and A.M.; project administration, M.K.; funding acquisition, M.K. and V.L. All authors have read and agreed to the published version of the manuscript.

Funding: This work was supported by the EIPHI Graduate School [Grant No. ANR-17-EURE-0002]; the French Investissements d'Avenir program, project ISITEBFC [Grant No. ANR-15-IDEX-03].

Data Availability Statement: All data are available upon reasonable request from the corresponding authors.

Acknowledgments: This work was partly supported by the french RENATECH network and its FEMTO-ST technological facility.

Conflicts of Interest: The authors declare no conflict of interest.

Sample Availability: Samples are available from the authors.

References

1. Fischer, J.; Wegener, M. Three-dimensional optical laser lithography beyond the diffraction limit. *Laser Photon. Rev.* **2013**, *7*, 22–44. [CrossRef]
2. Mueller, J.B.; Fischer, J.; Mayer, F.; Kadic, M.; Wegener, M. Polymerization Kinetics in Three-Dimensional Direct Laser Writing. *Adv. Mater.* **2014**, *26*, 6566–6571. [CrossRef] [PubMed]
3. Moughames, J.; Porte, X.; Thiel, M.; Ulliac, G.; Larger, L.; Jacquot, M.; Kadic, M.; Brunner, D. Three-dimensional waveguide interconnects for scalable integration of photonic neural networks. *Optica* **2020**, *7*, 640. doi: 10.1364/OPTICA.388205. [CrossRef]
4. Moughames, J.; Jradi, S.; Chan, T.M.; Akil, S.; Battie, Y.; Naciri, A.E.; Herro, Z.; Guenneau, S.; Enoch, S.; Joly, L.; et al. Wavelength-scale light concentrator made by direct 3D laser writing of polymer metamaterials. *Sci. Rep.* **2016**, *6*, 33627. doi: 10.1038/srep33627. [CrossRef] [PubMed]
5. Chen, X.; Moughames, J.; Ji, Q.; Martínez, J.A.I.; Tan, H.; Adrar, S.; Laforge, N.; Cote, J.M.; Euphrasie, S.; Ulliac, G.; et al. Optimal isotropic, reusable truss lattice material with near-zero Poisson's ratio. *Extrem. Mech. Lett.* **2020**, *41*, 101048. [CrossRef]
6. Porte, X.; Dinc, N.U.; Moughames, J.; Panusa, G.; Juliano, C.; Kadic, M.; Moser, C.; Brunner, D.; Psaltis, D. Direct (3+1)D laser writing of graded-index optical elements. *Optica* **2021**, *8*, 1281. doi: 10.1364/OPTICA.433475. [CrossRef]
7. Iglesias Martínez, J.A.; Moughames, J.; Ulliac, G.; Kadic, M.; Laude, V. Three-dimensional phononic crystal with ultra-wide bandgap at megahertz frequencies. *Appl. Phys. Lett.* **2021**, *118*, 063507. [CrossRef]
8. Kaschke, J.; Gansel, J.K.; Wegener, M. On metamaterial circular polarizers based on metal N-helices. *Opt. Express* **2012**, *20*, 26012–26020. [CrossRef] [PubMed]
9. Kern, C.; Kadic, M.; Wegener, M. Experimental evidence for sign reversal of the hall coefficient in three-dimensional metamaterials. *Phys. Rev. Lett.* **2017**, *118*, 016601. [CrossRef] [PubMed]
10. Ji, Q.; Moughames, J.; Chen, X.; Fang, G.; Huaroto, J.J.; Laude, V.; Martínez, J.A.I.; Ulliac, G.; Clévy, C.; Lutz, P.; et al. 4D Thermomechanical metamaterials for soft microrobotics. *Commun. Mater.* **2021**, *2*, 93. [CrossRef]
11. Yang, L.; Münchinger, A.; Kadic, M.; Hahn, V.; Mayer, F.; Blasco, E.; Barner-Kowollik, C.; Wegener, M. On the schwarzschild effect in 3D two-photon laser lithography. *Adv. Opt. Mater.* **2019**, *7*, 1901040. [CrossRef]
12. Gissibl, T.; Wagner, S.; Sykora, J.; Schmid, M.; Giessen, H. Refractive index measurements of photo-resists for three-dimensional direct laser writing. *Opt. Mater. Express* **2017**, *7*, 2293–2298. [CrossRef]
13. Qu, J.; Kadic, M.; Naber, A.; Wegener, M. Micro-Structured Two-Component 3D Metamaterials with Negative Thermal-Expansion Coefficient from Positive Constituents. Available online: http://www.nature.com/articles/srep40643 (accessed on 3 June 2022).
14. Iglesias Martínez, J.A.; Groß, M.F.; Chen, Y.; Frenzel, T.; Laude, V.; Kadic, M.; Wegener, M. Experimental observation of roton-like dispersion relations in metamaterials. *Sci. Adv.* **2021**, *7*, eabm2189. [CrossRef] [PubMed]
15. Dudek, K.; Martínez, J.I.; Ulliac, G.; Kadic, M. Micro-Scale Auxetic Hierarchical Mechanical Metamaterials for Shape Morphing. *Adv. Mater.* **2022**, *34*, 2110115. [CrossRef] [PubMed]
16. Nanoscribe. Available online: http://www.nanoscribe.com (accessed on 20 May 2022).
17. Sandercock, J. Brillouin scattering study of SbSI using a doubled-passed stabilised scanning interferometer. *Opt. Commun.* **1970**, *2*, 73–76. [CrossRef]
18. Speziale, S.; Marquardt, H.; Duffy, T.S. Brillouin scattering and its application in geosciences. *Rev. Mineral. Geochem.* **2014**, *78*, 543–603. [CrossRef]
19. Fioretto, D.; Scarponi, F. Dynamics of a glassy polymer studied by Brillouin light scattering. *Mater. Sci. Eng. A* **2009**, *521*, 243–246. [CrossRef]

20. Philipp, M.; Müller, U.; Sanctuary, R.; Seck, P.; Krüger, J.K. Scanning Brillouin Microscopy: Acoustic microscopy at gigahertz frequencies. *Arch. Des Sci. Nat. Phys. Math.* **2012**, *NS46*. Available online: http://hdl.handle.net/10993/13482 (accessed on 20 May 2022).
21. Schmid, M.; Ludescher, D.; Giessen, H. Optical properties of photoresists for femtosecond 3D printing: Refractive index, extinction, luminescence-dose dependence, aging, heat treatment and comparison between 1-photon and 2-photon exposure. *Opt. Mater. Express* **2019**, *9*, 4564–4577. [CrossRef]
22. Menczel, J.D.; Prime, R.B. *Thermal Analysis of Polymers: Fundamentals and Applications*; John Wiley & Sons: Hoboken, NJ, USA, 2009.
23. Beugnot, J.C.; Laude, V. Electrostriction and guidance of acoustic phonons in optical fibers. *Phys. Rev. B* **2012**, *86*, 224304. [CrossRef]

materials

Review

Brillouin Scattering and First-Principles Studies of $BaMO_3$ (M = Ti, Zr, and Cu) Perovskites

Md Al Helal [1,2,*] and Seiji Kojima [2]

1 Department of Physics, Begum Rokeya University, Rangpur 5400, Bangladesh
2 Graduate School of Pure and Applied Sciences, University of Tsukuba, Ibaraki 305-8573, Japan
* Correspondence: helalphy82@gmail.com; Tel.: +880-1-727231188

Abstract: Perovskite oxides with the general formula ABO_3 comprise a large number of families among the structures of oxide-based materials, and currently, several perovskite structures have been identified. From a variety of compositions and structures, various functions are observed in perovskite compounds, and therefore, they became very useful for various applications in the electronic and medical industries. One of the most puzzling issues for perovskite compounds is the understanding of the vibration and relaxation dynamics in the gigahertz range. In that sense, the micro-Brillouin scattering system is a very effective tool to probe the gigahertz dynamics, and also, first-principles calculations can be used to describe the phonon structure with different atomic contributions. The micro-Brillouin scattering system and first-principles calculations provide the fundamental information on a variety of vibration and relaxation processes related to structural phase transitions under different external conditions such as temperature, electric field, and pressure. This review article summarizes the Brillouin scattering and first-principles studies on $BaMO_3$ (M = Ti, Zr, and Cu). Through a detailed analysis of the existing results, we summarize the existing limitations and future perspectives in these research areas, which may propel the development of different perovskite ferroelectrics and extend their practical application areas.

Keywords: Brillouin scattering; first-principles calculation; perovskite; barium oxide; ferroelectric

Citation: Helal, M.A.; Kojima, S. Brillouin Scattering and First-Principles Studies of $BaMO_3$ (M = Ti, Zr, and Cu) Perovskites. *Materials* **2022**, *15*, 6747. https://doi.org/10.3390/ma15196747

Academic Editor: Pavel Lukáč

Received: 12 August 2022
Accepted: 22 September 2022
Published: 28 September 2022

1. Introduction

In solid-state physics, crystals with the ABO_3-type (A: Large cation with different valence state and B: Transition metal) perovskite family structure are being widely studied because they put up most of the metal ions in the periodic table owing to its substantial number of different ions [1]. These types of oxides are significant in the field of various technological applications in modern electronics, non-linear optics, catalysis, refractories, geophysics, particle accelerators, etc. [2–6]. One of the major benefits of perovskite-structured oxides is that they can accept considerable substitutions in one or both cationic sites (A and B sites) while retaining their original crystal structures [7]. Currently, these types of compounds have several outstanding applications in the electronic industries such as piezoelectric devices, transducers, random access memories, tunable microwave device displays, sensors, capacitors, and wireless communications [8–12].

The ideal structure of perovskite is cubic, which is shown in Figure 1. Although few compounds have this pure cubic phase, many perovskites have a slightly distorted phase with lower symmetry (e.g., tetragonal, orthorhombic, or rhombohedral). The lattice distortion in a crystal results from the off-center and/or displacement of the A or B site cations, which cause a ferroelectric phase transition owing to the polar phonon mode softening at a Γ point of the Brillouin zone. Meanwhile, the softening of the phonon mode can also appear due to the rotation of the oxygen octahedron at the *R* or *M* point of the Brillouin zone boundary [13]. Furthermore, many oxygen- or cation-deficient structures have been observed. The different categories of distortion in the perovskite materials are significantly related to their physical properties, in particular their ferromagnetic or ferroelectric properties.

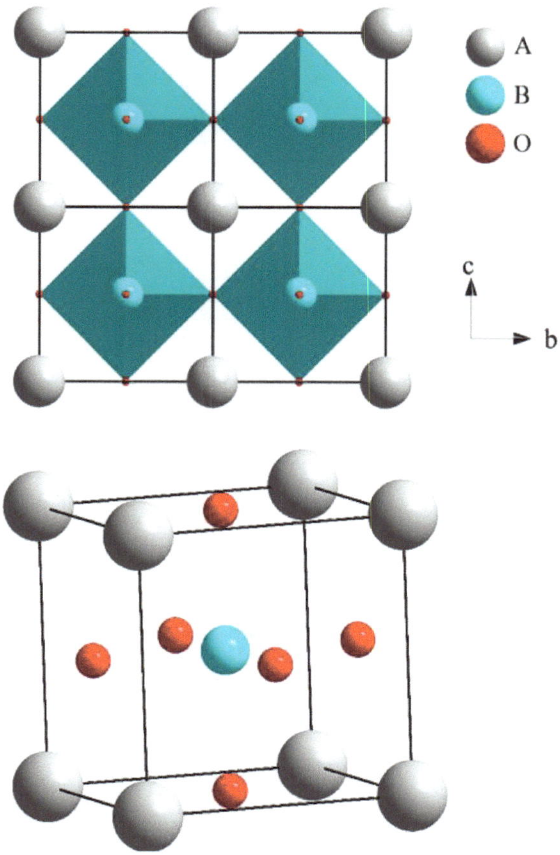

Figure 1. The 2D and 3D view of ABO_3-type cubic perovskites.

Since its discovery, barium titanate ($BaTiO_3$ (BTO)) has been studied extensively due to its remarkable ferroelectric properties, and it has been the subject of many studies for more than 50 years. Barium zirconate ($BaZrO_3$ (BZO)) has attracted much attention from the scientific community due to its possible applications in hydrogen sensors and fuel cells. Barium cuprate ($BaCuO_3$ (BCO)) is another promising perovskite material, which can be used to improve catalytic activity. The process of phase transition is different in these three crystals.

BTO is stable in a paraelectric cubic phase (space group $Pm\overline{3}m$) at high temperatures with five atoms per unit cell. Upon cooling, BTO exhibits three successive phase transitions from the paraelectric cubic phase to ferroelectric tetragonal (space group *P4mm*), orthorhombic (space group *Amm2*), and rhombohedral (*R3m*) phases at 393, 278, and 183 K, respectively. There exist numerous experimental and theoretical studies explaining the ferroelectric behavior of BTO single crystals. In spite of numerous reports, still, the true nature of the ferroelectric phase transition in BTO is controversial. The ferroelectricity and concomitant cubic to tetragonal distortion were initially discussed on the basis of displacive-type phase transition from the condensation of the soft optic mode [14,15]. However, with the lowering of temperature, the subsequent phase transitions of the orthorhombic and rhombohedral phases are inconsistent with the displacive nature [16,17].

This anomaly together with the results of low-temperature diffuse X-ray scattering [18] results in the development of a phenomenological order–disorder model [19]. In addition, the deviation of the refractive index from high-temperature linear behavior [20], first-order Raman spectra [21], the Curie-Weiss law for the permittivity [22], and significant softening of the longitudinal acoustic phonon [23,24] in the paraelectric cubic phase were observed in BTO crystals, which were ascribed to the existence of local polar clusters. Although BTO single crystals were studied more than 50 years, there is a significant lack of some decisive properties concerning the existence of the local polar cluster and the effect of external stress on this cluster, which must be attained to unlock the full potential of BTO for possible future technological applications.

The lattice dynamics and structural stability of BZO are found to be more complex, which has drawn great attention owing to its low-temperature structure and properties. As other perovskites, so far, no phase transition has been noticed in BZO at ambient pressure at any temperature. BZO has a cubic structure up to 2K according to neutron and X-ray diffraction studies [25]. However, these structural data disagree with the first-principles calculations [26,27], where the instability of the R or M phonon mode is observed due to the rotations of oxygen octahedra. In addition, a new transverse optic mode at about 2 THz was observed by the terahertz time-domain spectroscopy technique, which was forbidden in the consideration of cubic symmetry according to the infrared (IR) selection rule [28]. The origin of this peak was attributed to an IR active mode of the $P\bar{1}$ symmetry caused by local distortion. In BZO crystals, the first-order Raman active modes are observed, which is unfavorable for the cubic $Pm\bar{3}m$ symmetry, indicating the lowering of the cubic symmetry [29,30]. However, the observed Raman spectra in BZO are consistent with the local tetragonal and orthorhombic symmetry [30–32].

One of the most interesting properties of perovskite materials is superconductivity. The discovery of superconductivity in La-Ba-Cu-O perovskite oxide with the K_2NiF_4-type layered structure containing the CuO_2 plane was observed by Bednorz and Müller in 1984 [33]. After their report, significant attention has been paid to new types of high-temperature superconductors, mainly Cu-based perovskite oxides due to their excellent physical properties. It has been found that the presence of Cu on the B site cation is needed for superconductivity to occur in such perovskites. Due to the tremendous effort, a high value of transition temperatures (T_c) was found in the $YBa_2Cu_3O_7$ [34] and $Bi_2Sr_2Ca_2Cu_3O_{10}$ [35] systems in 1987 and 1988, respectively, and currently, T_c has been further increased to 130–155 K in the $HgBa_2Ca_2Cu_3O_{8+\delta}$ system [36]. So far, all the high-temperature superconductors are Cu-based oxides, and T_c is clearly associated with the Cu-O chain. In addition to superconductivity, Cu-based perovskites exhibit high electrical conductivity and catalytic activity, which are close to those of metals such as Cu. Very recently, another typical Cu-based superconducting $BaCuO_3$ (BCO) perovskite has been reported [37,38], and T_c was found to be around 10 k.

In conventional dielectric spectroscopy measurement, it is very difficult to measure the broadband spectra between 1 and 1000 GHz. However, Brillouin scattering techniques enable the acquisition of such broad spectra by a single scan. To observe the dynamics of local polarization fluctuation in perovskite-type ferroelectrics, Brillouin scattering has been used as a very significant tool due to its sensitivity to local symmetry breaking related to the heterogeneity among a few methods, which can be used to directly identify the local structure. Another advantage of Brillouin scattering is that the spatial resolution is comparable to the wavelength of an excited laser, and it is possible to measure the acoustic phonon and optic soft mode in different types of perovskites.

In several applications, perovskites have been proven to be of great interest because of their meaningful physical properties for different device applications. In the present paper, we review the recent systematic studies on $BaMO_3$ (M = Ti, Zr, and Cu) perovskites and discuss the existing limitations and future perspectives and research in these areas.

2. Methodology

2.1. Experimental Section

In solid-state physics, the determination of the velocity of an acoustic wave is given by the elastic constant and density. There are many methods to determine the elastic constants. Among them, the ultrasonic pulse-echo method and the mechanical resonance method have been well established [39,40]. However, as a non-contact and non-destructive method, Brillouin scattering spectroscopy is a powerful tool to measure the elastic constants with different external conditions, for example pressure, temperature, and electric field. Such a spectroscopic technique covers a large frequency range between 1 and 1000 GHz, and recently, it has been used to study not only the acoustic phonons, but also the optical soft mode [41], magnetic excitation [42], and localized excitation, such as fracton [43].

The frequency shift υ_B is proportional to the phase velocity V of the sound wave, which is measured by the following equation:

$$\Delta\upsilon_B = \pm\frac{2n_m V}{\lambda_0}\sin\left(\frac{\theta}{2}\right) \quad (1)$$

where λ_0, n_m, and θ are the wavelength of incident light, the refractive index of the material to be studied, and the scattering angle, respectively.

The optical arrangement of a micro-Brillouin scattering apparatus is shown in Figure 2 [44]. The tandem multi-pass Fabray-Pérot interferometer, (FP1 and FP2), was designed by Sandercock (JRS, Zwillikon, Switzerland). The laser is focused on the sample with the help of a small mirror in the optical microscope arrangement. The scattered light enters FP1 via pinhole P1 and aperture A1 and is collected by the objective lens. The laser light is then directed via mirror M3 to FP2. The light strikes the 90°-oriented prism PR1 after the transmission through FP2, then is reflected backwards and returns to FP2. It continues via aperture A2 to FP1. The light is then passed via L1 and is focused onto mirror M4 with mirror M1 underneath, after the transmission through FP1. Mirror M4 then returns the laser light back to lens L1, and the light is projected to FP1 through M2 after collimation. The light is directed to PR2 through M5 after the final pass through FP2. A bandpass filter is formed by the combination of lens L2, prism PR2, and output pinhole P2. The light is projected onto the photomultiplier (PM) from mirror M6 via aperture A3, lens L2, and output pinhole P2. Finally, a conventional photon counter records the scattered signal.

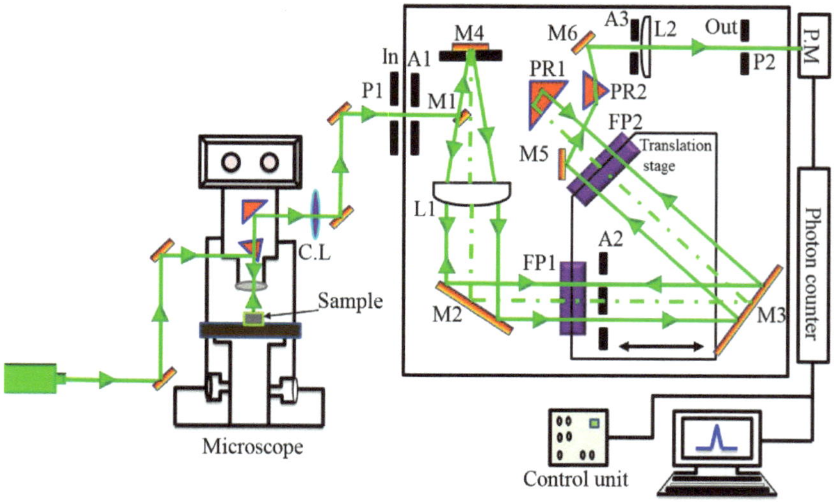

Figure 2. A schematic illustration of micro-Brillouin scattering setup.

To measure the micro-Brillouin scattering spectra, a 3 + 3 pass Sandercock-type tandem Fabray–Pérot interferometer (JRS TEP-1, JRS, Zwillikon, Switzerland), was used with a photon counting system. All the Brillouin scattering spectra were collected in a backwards scattering geometry [44,45]. The objective lens of the microscope focuses an incident laser beam with a wavelength of 532 nm and a power of about 100 mW onto the samples, and simultaneously, the scattered light is collected. The mirror space was set at 2 mm with a free spectral range of 75 GHz. The temperature of the sample was controlled by a heating and/or a cooling stage (THMS600, Linkam, Salfords, UK) with a stability of ± 0.1 K. Two surfaces of the sample were coated with silver paste, and gold contact wires were attached to the electrodes to apply the electric field.

2.2. Theoretical Section

The density functional theory (DFT) formalism is the basis of first-principles calculations upon which the different properties of a material can be derived. Using DFT, without solving the multibody Schrödinger equation, the problem can be described with an equivalent non-interacting problem. The ground state structure is obtained after solving the Kohn–Sham equations [46]. The ultrasoft pseudopotential, given by Vanderbilt, was used to optimize the geometry, because it allows for a lower cut-off energy, without sacrificing transferability too much [47]. The ultrasoft pseudopotential method performs the calculations of the material by considering the electron density in the valance as the soft part and the core region as the hard part, where the cut-off energy is dramatically reduced and the computational efficiency increased appreciably. The bulk modulus, independent elastic constants, and shear modulus were estimated using the stress–strain method contained in the Cambridge Serial Total Energy Package (CASTEP) code. On the other hand, the norm-conserving pseudopotential was used for the phonon calculations. The system geometry was optimized under the energy convergence at a total energy of 0.5×10^{-5} eV/atom, the maximum interatomic force 0.01 eV/Å, the maximum stress 0.02 GPa, and an ionic displacement of 5×10^{-4} Å. Understanding the energy-dependent optical parameters is very crucial to explore the potential application of a new compound. The optical properties are very important from the viewpoint of a possible optoelectronic device applications. The optical constants' spectra were extracted from the frequency-dependent dielectric function, $\varepsilon(\omega) = \varepsilon_1(\omega) + i\varepsilon_2(\omega)$. The real part of the dielectric constant, $\varepsilon_1(\omega)$, is calculated from the imaginary part, $\varepsilon_2(\omega)$, using Kramers-Kronig transformations. The imaginary dielectric function, $\varepsilon_2(\omega)$, can be calculated through the following equation:

$$\varepsilon_2(\omega) = \frac{2e^2\pi}{\Omega\varepsilon_0}\sum_{k,v,c} k\Psi_k^c|\hat{u}.\hat{r}|\Psi_k^v|^2\delta(E_k^c - E_k^v - E) \tag{2}$$

Here, $\hat{r}$ is the electron radius vector, $\hat{u}$ is the polarization vector of the incident electric field, e is the charge of the electron, ω is the frequency of light, and Ψ_k^c and Ψ_k^v are the conduction and valence band wave functions at k, respectively.

3. Typical Phenomena in $BaMO_3$ (M = Ti, Zr, and Cu) Perovskites

3.1. Precursor Dynamics and Domain Switching in $BaTiO_3$ Single Crystal

$BaTiO_3$ (BTO) is the first produced and most studied perovskite material with its superior dielectric, ferroelectric, and piezoelectric properties [48]. The high dielectric constants of BTO results from its crystal structure. In BTO crystals, three phonon frequencies were obtained from the FTIR measurements [49], located at around 40, 180, and 482 cm^{-1}. The lowest frequency phonon mode is called the soft mode and is overdamped in a paraelectric phase. From the measured frequency, Luspin et al. [49] estimated the dielectric constant using the Lyddane-Sachs-Teller (LST) relation. Due to the soft mode dynamics, the dielectric constant increases markedly in the vicinity of the phase transition temperature upon cooling. The obtained dielectric constant gives a diverging value by approaching the Curie temperature, while the soft mode frequency remains nearly constant. To explain this discrepancy, two possible origins can be invoked. The first one comes from the contribution of impurities, which gives rise to the rapid increase of the dielectric constant due to the coupling with the soft mode. Another possibility is the existence of an additional relaxational mechanism in the microwave region. However, they had no direct conclusion about this point. Therefore, it is necessary to observe this relaxational mechanism by considering another approach. In BTO crystals, the size of the Ba ion is too large compared with the Ti ion, which is quite small and remains stable in the octahedral position. The Ti ion having a charge of +4 has the minimum energy position outside the center, in the direction of each of the O_6 ions surrounding it. Due to the off-centering position of the Ti ions, the mechanism of the phase transition, in particular, the order-disorder- and/or displacive-type, is still under discussion. It was shown by Takahasi [50] and Comes et al. [18] that Ti ions go to the off-center position along eight equivalent [111] directions inside the cubic matrix with a non-zero local dipole moment. Therefore, this leads to a local polar nanoregion (PNR) in a cubic phase, which appears a few hundred degrees above the cubic to tetragonal phase transition temperature (T_c), called the Burns temperature (T_B). It is believed that PNRs play an important role in the appearance of the outstanding piezoelectric performance in electrical devices. In order to identify such types of nano-metric local polar regions, a low-frequency inelastic light scattering spectrum can be a very effective tool, because it can be used to directly probe the local symmetry and structure. Figure 3 shows a group of Brillouin scattering spectra at 398 and 403 K without any applied electric fields.

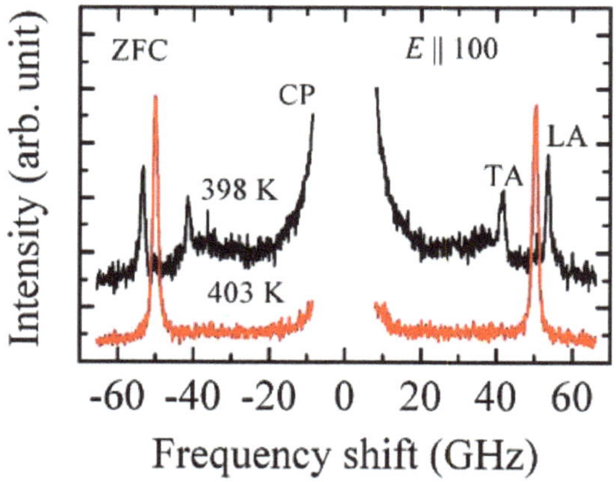

Figure 3. Brillouin scattering spectra at 398 and 403 K of (100)-oriented $BaTiO_3$ single crystals.

It is pertinent to note that a longitudinal acoustic (LA) mode and a weak central peak (CP) appear in the Brillouin scattering spectrum above T_c (402 K). It is now believed that the CP appears in low-frequency light scattering spectra owing to the appearance of random fields, which originate from the displacement of Ti ions inside the oxygen octahedra [23,51]. It was also suggested from different experiments [52,53] that the motions of off-center Ti ions are correlated and create PNRs in which all the Ti ions are oriented in the same direction and fluctuate together. Above T_c, previous theoretical reports also supported the creation of PNRs with local symmetry breaking [54,55]. Therefore, the presence of the CP in BTO crystals can be attributed to the polarization fluctuation of PNRs, which has also been observed in BTO-based ferroelectrics. However, to make further conclusions, it is very important to measure the broad CP by means of different experimental techniques. Ko et al. [56] measured the temperature dependence of the broad CP and discussed the critical slowing down behavior of the CP. They explained their result by the existence of polar clusters and the growth of these clusters with longer correlation lengths a few ten degrees above T_c. To make still further conclusions, besides the temperature dependence, the electric field dependence of the broad CP is very important, which can reveal some unanswered questions. So far, there is no available result on the electric field dependence of the broad CP. Therefore, future experiments will test the growth of PNRs by measuring the broad CP by applying an electric field in different directions.

Another approach to the effect of PNRs is to observe the behavior of LA phonons in the zero field cooling (ZFC) and field cooling (FC) process. At first, the Brillouin spectra are measured without applying the electric field, called the ZFC process. To apply the electric field, the sample is heated above T_B and short circuited to remove any kind of memory effects. Then, an electric field of 1 kV/cm at a high temperature is applied, and the Brillouin spectra upon cooling are collected, called the FC process. After measuring the Brillouin scattering spectra in ZFC and FC, they are fit by a Lorentzian function convoluted by a Gaussian instrumental function from which the LA frequency shift (ϑ_{LA}) and the FWHM (Γ_{LA}) can be obtained. Figure 4a and b shows the ϑ_{LA} and Γ_{LA} in the ZFC and FC processes, respectively. A sudden change of both ϑ_{LA} and Γ_{LA} at around 402 K reflects the transition from the cubic to tetragonal phase.

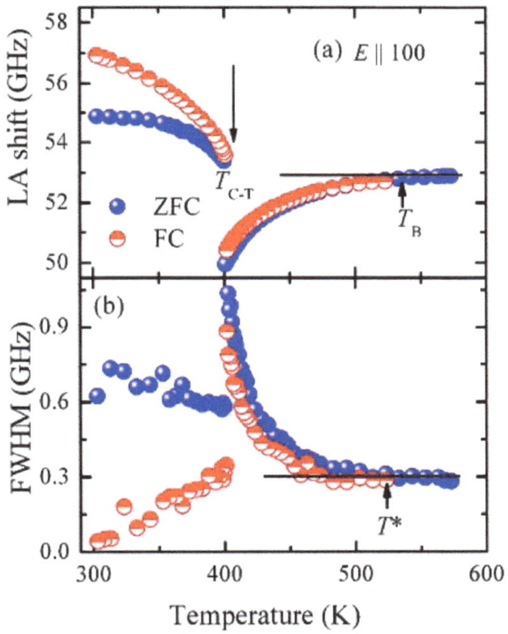

Figure 4. The temperature dependences of (**a**) frequency shift (υ_{LA}) and (**b**) FWHM (ΓLA) of LA phonon mode of 100-BT single crystals measured in ZFC (blue solid symbols) and FC (red solid half-filled symbols) processes [24].

Here, the interesting feature is that the LA frequency shift begins to deviate from its linear temperature dependence at around 533 K. It has been reported that the local strain fluctuation is created by the local polarization fluctuation inside a PNR. The macroscopic elastic anomaly and the piezoelectric coupling in a PNR occur due to the coupling between the local strain fluctuation in a PNR and acoustic phonons [57]. Therefore, we assigned this temperature as the Burns temperature, T_B, where the dynamic PNRs begin to appear. On the other hand, the sound attenuation, related to Γ_{LA}, begins to increase abruptly towards the transition temperature at around 503 K, where the permanent PNRs arise, and the temperature is known as the intermediate temperature, T^*. The observed T_B and T^* in this study were somewhat higher than previously found by the Brillouin scattering results [56] carried out on BTO single crystals. The discrepancies of these results may appear because of the quality of the sample, because the authors noticed T_c at 373 K, which is far behind the actual T_c in BTO crystals. Interestingly, these two temperatures were also measured by the acoustic emission measurement [58] carried out on a single crystal of BTO, and the reported values of T_B were approximately between 530–570 K, while T^* was 506 K, which is in good agreement with the present Brillouin scattering results. Very recently, T_B was also measured by the resonant ultrasound spectroscopy (RUS) technique [59]. The onset of the frequency vs. temperature curve in the RUS method shows the deviation from the exponential softening towards a weaker temperature dependence near 560 K, which they denoted as T_B. These two characteristic temperatures have been noticed in many relaxor ferroelectrics [18,60]. However, so far, this temperature has not been clearly identified [58].

The electric field has a serious effect on the crystal structure and polarization properties of BTO. Under the applied electric field, Ti ions shift from a random position to aligned positions, obtaining high polarization and a high dielectric constant [61]. To analyze the effect of an electric field, we applied a 1 $kVcm^{-1}$ electric field at a higher temperature (above T_B), and then, the Brillouin spectra were taken in the field cooling process. A significant difference was noticed between ϑ_{LA} and Γ_{LA} in both the cubic and tetragonal phases. It has already been discussed and reported that in BTO crystals, the Ti ions fluctuate along eight equivalent [111] directions inside a PNR [53]. Owing to the application of an electric field along the [100] direction, the density of dynamic/static PNRs can be decreased, which is expected to weaken the LA phonons' scattering by PNRs, resulting in a higher frequency shift and lowering the damping of the LA phonons. However, it is significant to note that T_B is not influenced by the applied electric field, and T^* and T_c change after applying electric field. This observation is in good agreement with acoustic emission results [58]. According to acoustic emission results, the electric field dependence is stronger in T^* than T_c due to the possible field ordering of PNR dipoles, which could favor their freezing. In the tetragonal phase, a significant change between ϑ_{LA} and Γ_{LA} has been observed. Due to the sufficiently strong electric field, the correlation among the PNRs becomes sluggish, and the alignment of nanodomains/static PNRs is facilitated and enables complete switching into the macro/single-domain state at T_c, giving rise to the formation of the ferroelectric state. In addition, the large differences in ϑ_{LA} and Γ_{LA} between ZFC and FC indicate the transformation from the ferroelectric [100] direction to the [001] direction in the tetragonal phase. Upon the subsequent heating (heating after field cooling), the crystals transform back to their initial state, which results in hysteresis below T_c.

In ferroelectrics, to understand the domain switching process, many research works have been performed and remarkable progress has been achieved [62–66]. Previously, it was reported that under an applied electric field antiparallel to the polar direction, only 180° domain switching can be possible in ferroelectrics. However, non-180° domain switching was also suggested by both the theoretical and experimental observations [67,68]. Under the application of an external electric field, now, it has been established that two types of domains can nucleate: 90° domains with polarization perpendicular to each other. which minimizes strain via the formation of twin walls, and 180° domains having anti-parallel spontaneous polarization, which diminishes the depolarization fields [69]. To observe the domain switching effect, a set of typical Brillouin scattering spectra in the tetragonal phase at 303 K is shown in Figure 5.

When there is no electric field, the transverse acoustic (TA) mode is denoted as the *a*-domain, since the sample is cut along the [100]-oriented plate. When the field is increased up to 2.94 kV/cm, another TA mode at around 30 GHz begins to appear and persists up to 3.9 kV/cm. The additional TA mode in BTO crystals means the appearance of the *c*-domain. Within 2.94 and 3.9 kV/cm, both domains co-exist. Above 3.9 kV/cm, the intensity of the additional TA mode goes to the maximum and saturates, suggesting that the 90° domain switching has completed. The 90° domain switching was also noticed by Li et al. [70] and Salje et al. [58]. Moreover, whether the enhanced polar properties are related to the engineered domain structures or whether the increased density of the domain boundaries between domains plays an important role in BTO crystals is an open question.

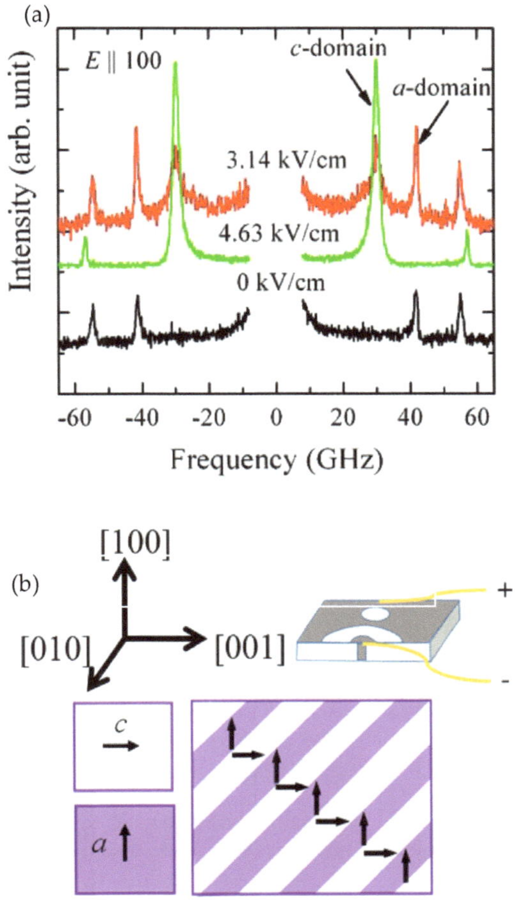

Figure 5. (**a**) A set of typical Brillouin scattering spectra BTO single crystals in the ferroelectric tetragonal phase at some selected electric fields at 303 K; (**b**) shows the orientation of a and c domains.

3.2. Local Lattice Disorder in $BaZrO_3$ Single Crystals

The perovskite structure of BZO single crystals, long known to maintain their cubic symmetry at room temperature without any structural phase transition with wide temperature variations, has recently been exhibited to have various ground state structures with low symmetries involving octahedral rotation. However, the ground state structure in BZO is questionable. Several first-principles calculations reported that the true structural state of BZO may be tetragonal *I4/mcm* because of the unstable antiphase antiferrodistortive phonons of oxygen octahedra [69,71]. These local tetragonal distortions can be averaged out and may not be possible to detect using X-ray or neutron diffraction experiments, which results in the cubic symmetry. From the quantum mechanical viewpoint, Brillouin scattering is assumed as one of the interactions of light photons with acoustic or vibrational quanta such as phonons. No coupling between the ultrasonic transducer and other devices with samples is needed to produce acoustic excitation and detection in Brillouin scattering, which leads to accurately measuring the acoustic properties. Figure 6a shows a set of Brillouin scattering spectra at some selected temperatures, and a contour color map is displayed in Figure 6b.

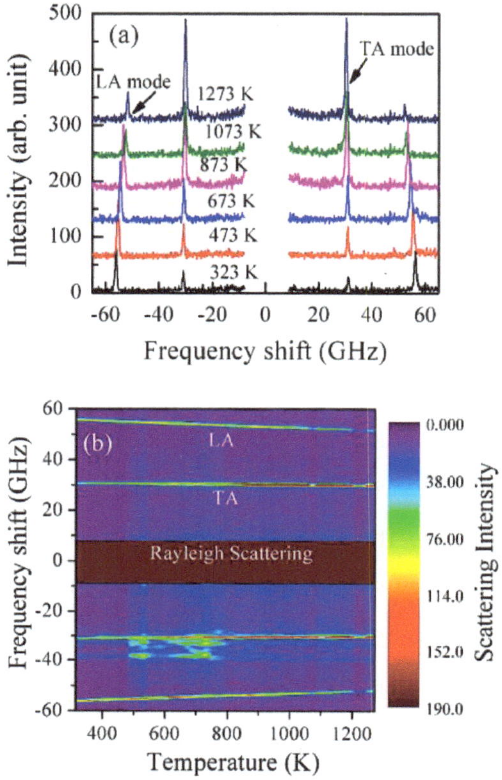

Figure 6. (**a**) A set of typical Brillouin scattering spectra at some selected temperatures and (**b**) a contour color map of intensity versus temperature and frequency of BZO single crystals [13].

We observed clearly an LA and a TA phonon mode. From the temperature dependence of these phonon modes, no sharp discontinuity related to a phase transition was observed. Therefore, macroscopically, it remains as the ground state structure at room temperature. However, the most striking feature of this Brillouin spectrum is the appearance of the TA mode, which exists up to 1273 K. The Brillouin spectra were recorded in a backward scattering geometry, and only the LA mode should propagate along the [100] direction based on the selection rules in the cubic symmetry [72]. Due to this, several previous experiments noticed only the LA phonon mode in the cubic symmetry [73]. Therefore, the strong TA mode, not allowed in the cubic symmetry, reflects the presence of local lattice distortion, which can destroy the center of the symmetry in a cubic matrix. This local lattice distortion is also evident from terahertz time-domain spectra (THz-TDS). Figure 7a,b shows the time-domain waveform and transmission power spectra converted from the time-domain waveform of a BZO single crystal at 8 K, where a clear absorption peak at about 1.94 THz is observed.

In cubic perovskite with the space group $Pm\bar{3}m$, three IR active transverse optic (TO) modes are allowed according to the selection rule. Two former measurements on the IR phonon spectra have been reported [74,75]. In addition, the IR phonon frequencies have been reported by first-principles calculations [76,77]. In comparison with those published in BZO ceramics by IR and first-principles calculations, the observed TO mode was observed in THz-TDS at a very low frequency, which reflects the additional new mode in a BZO crystal. The FTIR spectra of BZO ceramics [74] show three narrow phonon peaks, as expected for a simple cubic perovskite structure. Among three phonon modes, the

TO2 mode was the strongest one and was assigned as the Slater mode (Zr-O_6 stretching vibrations). The eigenvector of the lowest-frequency TO1 mode was close to the last mode, playing the role od the ferroelectric soft mode. The highest-frequency mode above 15 THz was due to the O_6 deformation and called the Axe mode. In addition, to fit the FTIR spectra, Nuzhnyy et al. [74] used the nine oscillator models and concluded that not all the oscillators have a clear meaning. However, the appearance of a new mode at 2 THz in a BZO crystal and an additional mode at 9 THz in BZO ceramics [74] indicated that BZO has a lower symmetry than cubic Pm$Pm\bar{3}m$ and/or these modes come from the local lattice distorted structure. Very recently, D.H. Gim et al. measured the Raman and IR spectra of both the ceramics and single crystals of BZO [75]. They also observed the additional phonon modes in both samples and reported that these additional phonon modes are due to the presence of local distortion or disorder. Therefore, the additional phonon modes in Brillouin scattering, THz-TDS, Raman spectra [75], as well as IR spectra [74] can be considered evidence of the presence of the local lattice distorted structure in BZO.

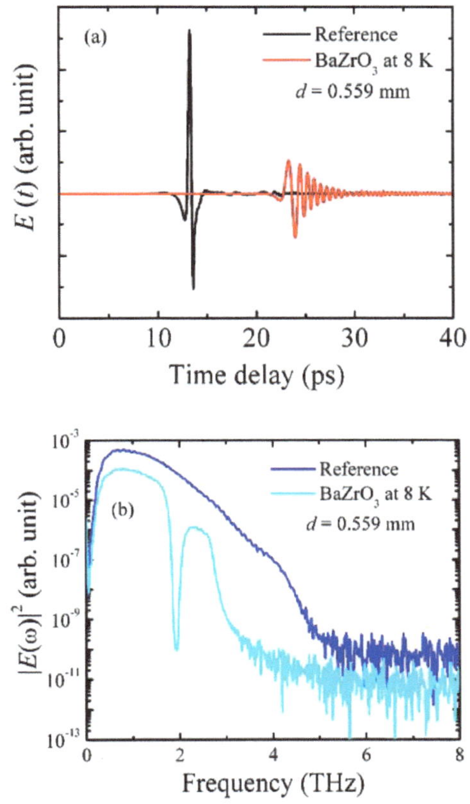

Figure 7. (**a**) Time-domain waveform and (**b**) the power spectra of a $BaZrO_3$ single crystal converted from time-domain waveforms at 8 K.

3.3. *Lattice Instability and IR Spectra of $BaZrO_3$ Crystals*

Tolerance factor t is a well-established parameter, which is used to identify the stable structure. The formula for t of ABO_3-type perovskite is given by

$$t = \frac{R_{A-O}}{R_{B-O}\sqrt{2}} \quad (3)$$

Here, R_{A-O} is the sum of A and O ionic radii and R_{B-O} is the sum of B and O ionic radii. Generally, $t < 1$ leads to the rotation and expansion of B-O_6 octahedra. A low-temperature anti-ferroelectric phase is generated due to the octahedral rotations. If $t > 1$, the B-O_6 octahedra are stretched from their preferred B-O bond lengths, resulting in the distortions of the B cation by creating room for the B cations to move to the off-center position. In BZO crystals, the values of the ionic radii of Ba and Zr atoms are 1.61 Å and 0.72 Å, respectively, which leads to $t = 1$, so the cubic structure should have no driving force to deform to a lower symmetry. However, the calculation of the phonon spectrum of BZO using DFT calculations is unfavorable to achieve the stable structure of BZO. The phonon dispersion relation for the five-atom unit cell of BZO is shown in Figure 8. The very important characteristics of the phonon spectra are related to the lowest-frequency acoustic phonon branch, which goes to negative at the R- and M-points in the Brillouin zone. However, theses unstable phonon modes were absent in. [78,79], but appeared in [69], where all the calculations were performed using the DFT method. The negative phonon frequencies correspond to the imaginary frequencies, which implies the change of the crystal symmetry. Due to the anharmonic phonon interaction, the negative phonon frequencies may also appear, which are responsible for a structural phase transition of BZO with respect to some external conditions. Therefore, BZO is dynamically unstable, which can change in phase under pressure or an electric field. This observation is consistent with the present Brillouin scattering and THz-TDS results. Indeed, under the applied pressure, very recently, the cubic to rhombohedral $R\bar{3}c$ and rhombohedral to tetragonal *I4/mcm* phase transitions at 8.4 and 19.2 GPa, respectively, were observed [75]. A few years before, Yang et al. also observed the pressure dependence of the phase transition from the cubic to tetragonal *I4/mcm* at 17.2 GPa [70]. We also observed the pressure dependence of the phase transition by first-principles calculations [13].

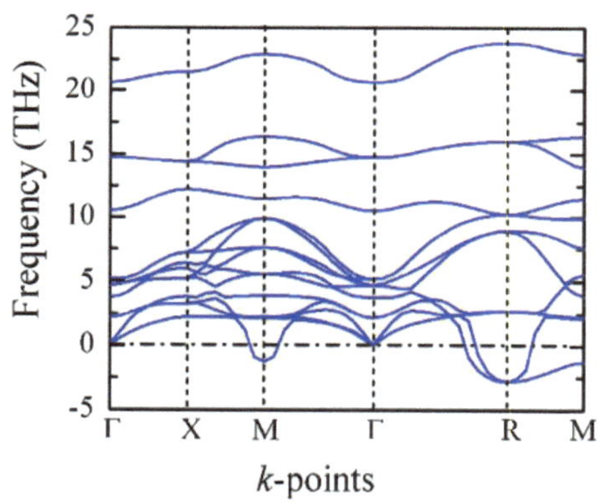

Figure 8. Phonon dispersion relation of BZO crystals [13].

The atomic contribution in the phonon spectra can be seen from the partial phonon density of states (PPDOS), which is shown in Figure 9. It can be noticed from the PPDOS that the negative phonon frequency appears due to the vibration of Zn and O atoms. Therefore, it can be concluded that within the frequency range from 0.5 to 3.0 THz, an unstable phonon mode at the *R*- and/or *M*-points of the Brillouin zone is responsible for the phase transition and the superior dielectric and piezoelectric responses of BZO, which correspond to the tilts of the ZrO_6 octahedra. Among the perovskite ferroelectric oxides such as PZT [80], this type of unstable phonon mode is observed commonly at the *R*- or *M*-points of the Brillouin zone.

Finally, we also estimated the IR spectra of BZO using first-principles calculations. According to the group theoretical analysis, due to the cubic symmetry, three phonon branches of the T_{1u} symmetry is IR active, and one branch of the T_{2u} symmetry is not IR, nor

Raman active. The estimated IR spectra of the BZO single crystal are displayed in Figure 10 where the frequencies of the three TO modes are 2.4, 5.3, and 14.8 THz, respectively.

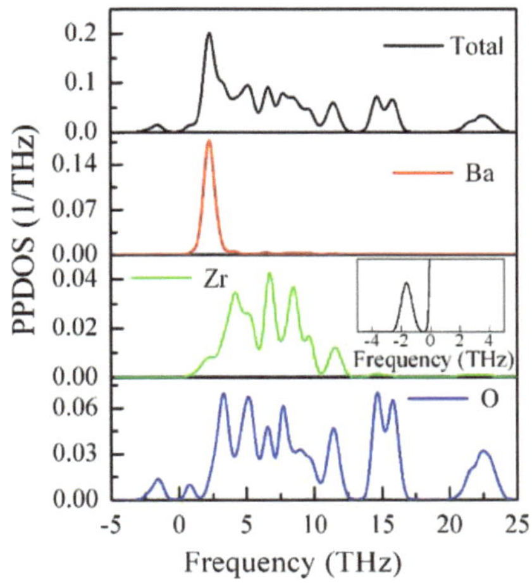

Figure 9. Partial and total phonon density of states of BZO crystals. Inset shows the contribution of Zr atom.

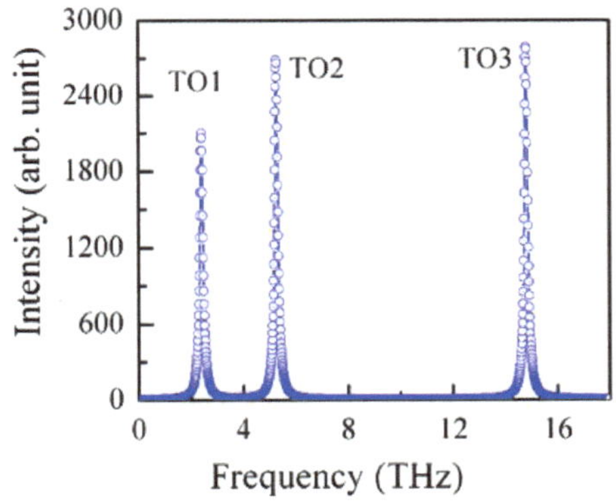

Figure 10. Calculated IR active TO mode frequencies of BZO crystals in the cubic phase.

These values of IR active modes are in good agreement with previous theoretical results [72,76]. However, Nuzhnyy et al. [74] obtained the values of these IR active modes to be 3.48, 6.4, and 15.6 THz, respectively, which are higher than the present calculated values. It is well known that the mode frequencies obtained from DFT calculations may vary due to the choice of different functionals and pseudopotentials, as shown for BZO, as well as for BTO. The origin of the additional modes observed by Nuzhnyy et al. and our THz-TDS has not been discussed anywhere using first-principles calculations. J. W. Bennett et al. estimated the values of IR modes by considering the $P\bar{1}$ symmetry [71]. However, their reported values also cannot explain the experimentally observed additional modes in BZO. Therefore, further theoretical calculations are not only necessary, but will also be interesting.

3.4. First-Principles Calculations of $BaCuO_3$ Crystals

Very recently, another important type of perovskite, BCO, was discovered [36]. BCO belongs to the family of chemically simple perovskite oxides. At a fundamental level, to know the dynamical stability of BCO over the whole Brillouin zone boundary, the phonon dispersion relation is very important. As discussed above, the observation of negative phonon frequencies corresponds to the instability of the crystal symmetry; otherwise, they are considered as dynamically stable. The calculated phonon spectra and PPDOS of BCO at absolute zero temperature along the principal symmetry direction of the Brillouin zone are shown in Figures 11 and 12, respectively. As for BZO, negative phonon frequencies are absent in BCO. On the other hand, at the *G*-point, the positive value of the low-frequency acoustic phonon (indicated by red color) clearly reflects that BCO single crystals are dynamically stable. So far, there have been no available data, neither experimental, nor theoretical, to compare the present results. From the PPDOS, it is seen that the low-frequency region (1–4 THz) is mainly due to the vibration of Ba atoms, while the intermediate-frequency region (4–8 THz) arises because of the vibration of both the Cu and O atoms. The high-frequency region (above 8 THz) is composed of the vibration of O atoms due to the lower mass.

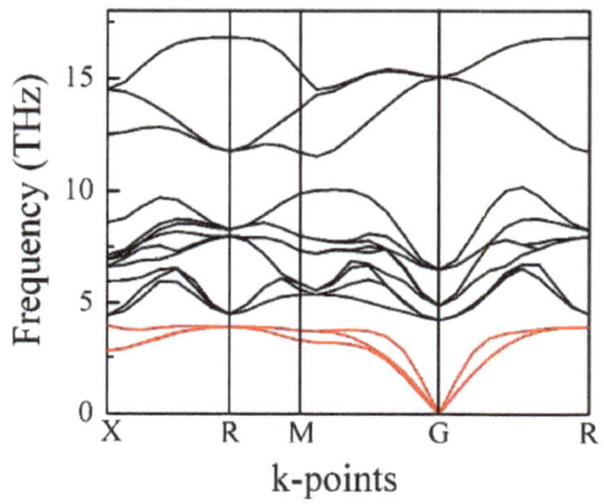

Figure 11. Phonon spectra of $BaCuO_3$ at zero pressure.

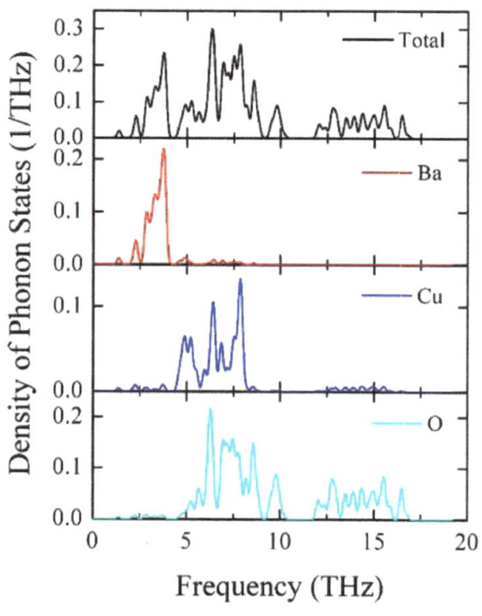

Figure 12. Density of phonon states of $BaCuO_3$ at zero pressure.

The elastic properties of a material are essential to study because of its response under different external loads. this also helps predict the numerous applications of a solid in different sectors by exploring the outstanding physical properties, such as elastic stiffness, bulk, modulus, compressibility, Young's modulus, ductility, etc. The independent elastic constants for cubic symmetry, namely C_{11}, C_{12}, and C_{44}, of BCO have been calculated using the stress–strain method [81,82]. All the estimated elastic constants are positive and satisfy the mechanical stability criteria $(C_{11} - C_{12}) > 0$, $(C_{11} + 2C_{12}) > 0$, and $C_{44} > 0$. Therefore, BCO crystals are stable from the viewpoint of both mechanics and thermodynamics.

For the material containing the cubic phase, the equations for the calculation of B and G are

$$B_V = B_R = \frac{1}{3}(C_{11} + 2C_{12}) \tag{4}$$

$$G_V = \frac{1}{5}(C_{11} - C_{12} + 3C_{44}) \tag{5}$$

$$G_R = \frac{5(C_{11} - C_{12})C_{44}}{C_{44} + 3(C_{11} - C_{12})} \tag{6}$$

$$B = \frac{B_V + B_R}{2} \tag{7}$$

$$G = \frac{G_V + G_R}{2} \tag{8}$$

where B_V and G_V represent the Voigt approximation, B_R and G_R indicate the Reuss approximation, and B and G have been calculated following the Voigt–Reuss–Hill approximation. The hardness of a material strongly depends on C_{44} and G. In order to check the hardness of the other well-known perovskites, we compared our results with BTO, BZO and $SrZrO_3$, which is shown in Figure 12. From Figure 13, it can be noticed that BCO has a smaller value of not only B and C_{44}, but also the other elastic moduli, which should result in BCO being a soft material. Therefore, in comparison with those perovskites, BCO is more suitable for use as a thermal barrier coating substance.

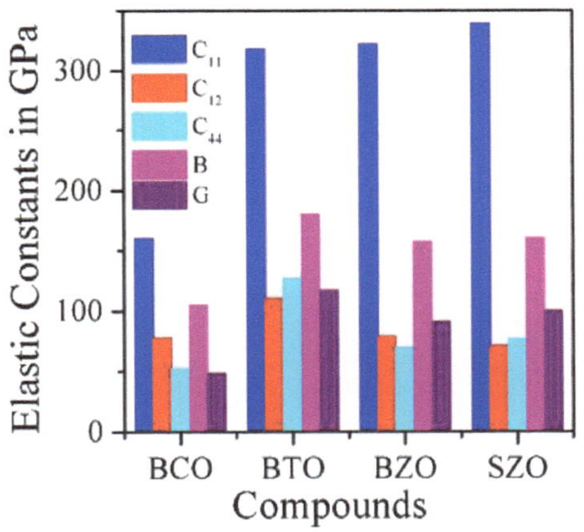

Figure 13. Comparison of different elastic moduli of $BaCuO_3$ with other perovskites.

The most interesting phenomena observed in BCO is that it exhibits superconductivity at low temperatures. Recently, for Ba_2CuO_{4-y}, T_c was found to be around 73 K [83]. We also estimated T_c by using the Allen-Dynes modified McMillan equation [84] as

$$k_B T_c = \frac{\hbar\omega_{log}}{1.2} \exp\left[-\frac{1.04(1+\lambda)}{\lambda - \mu^*(1+0.62\lambda)}\right] \tag{9}$$

where λ is the electron–phonon coupling constant, defined as $\lambda = N(E_F)V_{e-ph}$, with V_{e-ph} denoting the electron–phonon interaction energy responsible for the Fermi surface instability and Cooper pairing, and μ^* is the repulsive Coulomb pseudopotential. The values of ω_{log}, λ, and μ^* were obtained as 121, 0.84, and 0.01, respectively, using the calculation by the quantum ESPRESSO code. These values were used to calculate T_c using the above equation, and we obtained T_c = 9.96 K. The large difference arose because of the oxygen defect. It has already been mentioned that the oxygen defect can increase the T_c of a material [85,86]. However, here, we report the T_c for the pure BCO structure.

At some selected pressures, the real ($\varepsilon_1(\omega)$) and imaginary ($\varepsilon_2(\omega)$) parts of the dielectric constants are displayed in Figure 14a,b, respectively. The low-energy optical response is dominated by $\varepsilon_2(\omega)$ for metallic compounds. From Figure 14a, it is seen that $\varepsilon_1(\omega)$ crosses zero from the bottom level at 37.9 eV, and at the same energy, $\varepsilon_2(\omega)$ approaches zero, which reflects the metallic nature of BCO. It also suggests that the material becomes transparent when the energy of the incident photon exceeds 37.9 eV. Due to the intraband transition, the prominent peaks are observed in both $\varepsilon_1(\omega)$ and $\varepsilon_2(\omega)$ at around 27 and 31 eV. The peak at around 27 eV almost diminishes, when the applied pressure is 18 GPa.

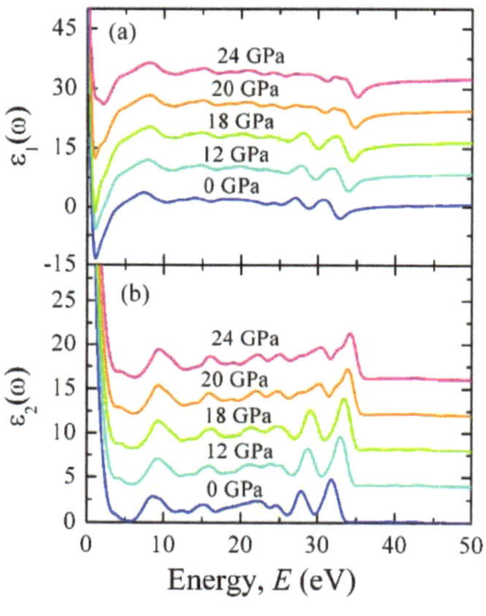

Figure 14. The energy dependence of the (**a**) real and (**b**) imaginary part of the dielectric constant of $BaCuO_3$ at some selected pressures. Different offset values are set to visualize the optical spectra clearly.

To explain these observations, we calculated the pressure dependence of the partial density of electronic states, which is shown in Figure 15. It can be noticed from the pressure dependence of the partial density of states that there is a prominent hybridization between the neighboring Cu-$d_{x^2-y^2}$ and O-2p_x orbitals at 18 GPa, which is supposed to screen the Cu core holes, resulting in the decrease of the peak intensity.

Significant absorption at zero photon energy was observed in BCO, as shown in Figure 16a, indicating the metallic nature. Many peaks were observed in the absorption spectrum. However, the intensities of these peaks in the IR and visible regions were small compared to the UV region. At around 27 and 31 eV, two intense peaks were observed. As a result, the solar spectrum in UV regions can be absorbed by BCO. The real part of the photoconductivity (σ) of the BCO crystal is displayed in Figure 16b. It is seen that σ has a finite value at zero photon energy, which reflects the metallic conductivity of BCO, totally consistent with the results of the absorption and dielectric constant. The finite value of σ can be increased owing to the absorption of the photon with sufficient energy. The maximum value of σ was obtained in the UV region. A sudden jump and decrease of intensity of a peak around 27 eV can be noticed from Figure 16a,b with increasing pressure, which might be associated with the transition from the initial state $2p^6 3d^9$ to the final state $2p^5 3d^{10}$. Overall, there are no studies on pure BCO to compare the result. Many studies are required for the superconducting BCO to understand the basic mechanism of its superconductivity.

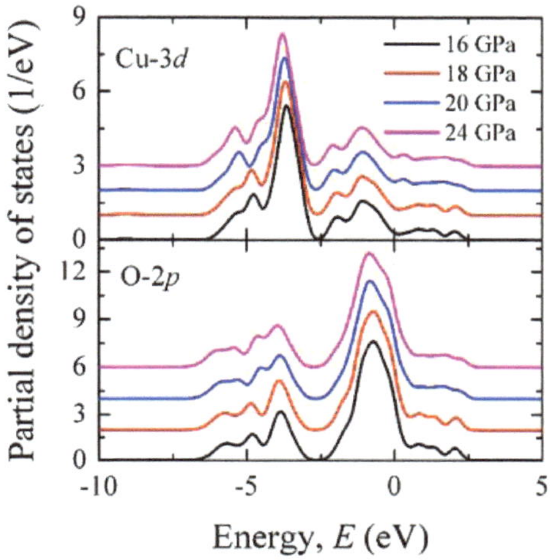

Figure 15. Partial DOS of Cu-3d and O-2p orbitals at 16, 18, 20, and 24 GPa.

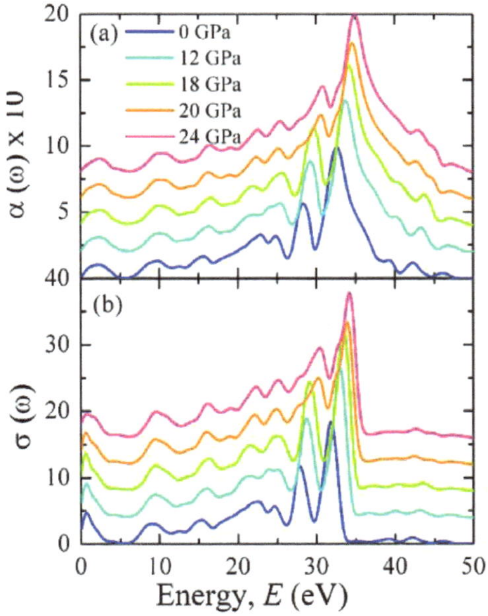

Figure 16. The energy dependence of (**a**) absorption coefficient and (**b**) conductivity of $BaCuO_3$ at some selected pressures. Different offset values are set to visualize the optical spectra.

4. Summary

The crystal structure of perovskite oxide has a variety of compositions and component elements, which give rise to a wide range of applications due to its unique properties. Here, three different types of perovskites were reviewed by means of Brillouin scattering and first-principles studies.

In BTO single crystals, a significant softening of the frequency shift and FWHM of the LA phonon is observed in the vicinity of a cubic–tetragonal phase transition temperature. The Burns temperature, T_B = 533 K, and intermediate temperature, T^* = 503 K, were identified, which show very close agreement with recent acoustic emission measurement result. When the sample was cooled down with an electric field, an increase of the frequency shift and a decrease of the FWHM in both the cubic and tetragonal phase were observed due to the reduced density of the twin domain walls and static/dynamic PNRs. A complete 90° domain switching process was noticed under the application of an electric filed in the ferroelectric tetragonal phase. In the BZO single crystal, no phase transition was observed from 8 K to 1273 K, reflecting that the BZO system is macroscopically cubic. The presence of a TA phonon in the Brillouin spectra suggests local lattice distortion, which destroys the center of symmetry in the cubic matrix. This observation is also justified with the observation of an additional absorption peak near 2 THz in the THz-TDS experiment. The imaginary part of the frequency in the phonon dispersion spectra suggests structural instability with respect to oxygen octahedral rotations. In BCO, the dynamical and mechanical stabilities were confirmed from the calculation of the phonon spectra and elastic constants using first-principles calculations. The magnitude of different elastic constants of BCO compared with other related perovskites shows that BCO is a mechanically stable soft perovskite with improved ductility. The superconducting transition temperature, T_c = 9.96 K, was obtained from phonon calculations. The change of the peak position and the reduction of the peak intensity in various optical constants with pressure were in good agreement with the observation of a pressure-dependent partial DOS. The shift of the peak position at 18 GPa in different optical parameters was attributed to the transition from the initial state $2p^63d^9$ to the final state $2p^53d^{10}$ of the Cu atom. Since there are no experimental reports on cubic BCO available, the results discussed here will be very helpful, not only for future experimental and theoretical research, but also for the assessment of the suitability of BCO in different optoelectronic devices.

Author Contributions: M.A.H. performed the experiment and theoretical calculations and wrote the manuscript. S.K. analyzed the results and reviewed the manuscript. All authors have read and agreed to the published version of the manuscript.

Funding: This research received no external funding.

Institutional Review Board Statement: Not applicable.

Informed Consent Statement: Not applicable.

Data Availability Statement: Not applicable.

Acknowledgments: The authors are thankful to Tatsuya Mori for his kind help in measuring the THz-TDS. We are also grateful to Mijanur Rahman and Aftabuzzaman for the fruitful discussions.

Conflicts of Interest: The authors declare that they have no known competing financial interests or personal relationships that could have appeared to influence the work reported in this paper.

References

1. Nagata, S.; Katsui, H.; Hoshi, K.; Tsuchiya, B.; Toh, K.; Zhao, M.; Shikama, T.; Hodgson, E. Recent research activities on functional ceramics for insulator, breeder and optical sensing systems in fusion reactors. *J. Nucl. Mater.* **2013**, *442*, S501–S507. [CrossRef]
2. Seyfi, B.; Baghalha, M.; Kazemian, H. Modified $LaCoO_3$ nano-perovskite catalysts for the environmental application of automotive CO oxidation. *Chem. Eng. J.* **2009**, *148*, 306–311. Available online: https://www.cheric.org/research/tech/periodicals/doi.php?art_seq=768853 (accessed on 15 May 2009). [CrossRef]
3. Kreisel, J.; Glazer, A.; Jones, G.; Thomas, P.; Abello, L.; Lucazeau, G. An X-ray diffraction and Raman spectroscopy investigation of A-site substituted perovskites compounds: The $(Na_{1-x}K_x)_{0.5}Bi_{0.5}TiO_3$ (0 x1) solid solution. *J. Phys. Condens. Matter* **2000**, *12*, 3267–3280. [CrossRef]
4. Fu, P.; Shan, Q.; Shang, Y.; Song, J.; Zeng, H.; Ning, Z.; Gong, J. Perovskite nanocrystals: Synthesis, properties and applications. *Sci. Bull.* **2017**, *62*, 369–380. [CrossRef]

5. Protesescu, L.; Yakunin, S.; Bodnarchuk, M.I.; Krieg, F.; Caputo, R.; Hendon, C.H.; Yang, R.X.; Walsh, A.; Kovalenko, M.V. Nanocrystals of cesium lead halide perovskites ($CsPbX_3$, X = Cl, Br, and I): Novel optoelectronic materials showing bright emission with wide color gamut. *Nano Lett.* **2015**, *15*, 3692–3696. [CrossRef] [PubMed]
6. Kawamura, Y.; Mashiyama, H.; Hasebe, K. Structural study on cubic–tetragonal transition of $CH_3NH_3PbI_3$. *J. Phys. Soc. Jpn.* **2002**, *71*, 1694–1697. Available online: https://ui.adsabs.harvard.edu/link_gateway/2002JPSJ...71.1694K/doi:10.1143/JPSJ.71.1694 (accessed on 13 December 2001). [CrossRef]
7. Xu, B.; Yin, K.; Lin, J.; Xia, Y.; Wan, X.; Yin, J.; Bai, X.; Du, J.; Liu, Z. Room temperature ferromagnetism and ferroelectricity in Fe-doped BaTiO3. *Phys. Rev. B* **2009**, *79*, 134109. [CrossRef]
8. Vanderbilt, D. First-principles based modelling of ferroelectrics. *Curr. Opin. Solid State Mater. Sci.* **1997**, *2*, 701. [CrossRef]
9. Kotecki, D.E.; Baniecki, J.D.; Shen, H.; Laibowitz, R.B.; Saenger, K.L.; Lian, J.J.; Shaw, T.M.; Athavale, S.D.; Cabral, C.; Duncombe, P.R.; et al. (Ba,Sr)TiO_3 dielectrics for future stacked-capacitor DRAM. *IBM J. Res. Dev.* **1999**, *43*, 367. [CrossRef]
10. Tinte, S.; Niguez, J.; Rabe, K.M.; Vanderbilt, D. Quantitative analysis of the first-principles effective Hamiltonian approach to ferroelectric perovskites. *Phys. Rev. B* **2003**, *67*, 064106. [CrossRef]
11. Rao, F.; Kim, M.; Freeman, A.J. Structural and electronic properties of transition-metal/$BaTiO_3$(001) interfaces. *Phys. Rev. B* **1997**, *55*, 13953. [CrossRef]
12. Mitchell, R.H. Perovskite: Modern and Ancient. *Acta Cryst.* **2002**, *58*, 1075. [CrossRef]
13. Helal, M.A.; Kojima, S. Structural instability and phase transition in $BaZrO_3$ single crystals: Brillouin scattering and DFT study. *Mater. Sci. Eng. B* **2021**, *271*, 115314. [CrossRef]
14. Cochran, W. Crystal stability and the theory of ferroelectricity. *Adv. Phys.* **1960**, *9*, 387. [CrossRef]
15. Shirane, G.; Frazer, B.C.; Minkiewicz, V.J.; Leake, J.A.; Linz, A. Soft Optic Modes in Barium Titanate. *Phys. Rev. Lett.* **1967**, *19*, 234. [CrossRef]
16. Kay, H.F.; Vousden, P. XCV. Symmetry changes in barium titanate at low temperatures and their relation to its ferroelectric properties. *Philos. Mag.* **1949**, *40*, 1019. [CrossRef]
17. Rhodes, R.G. Structure of $BaTiO_3$ at low temperatures. *Acta Crystallogr.* **1949**, *2*, 417–419. [CrossRef]
18. Comes, R.; Lambert, M.; Guinier, A. The chain structure of $BaTiO_3$ and $KNbO_3$. *Solid State Commun.* **1968**, *6*, 715. [CrossRef]
19. Comes, R.; Lambert, M.; Guinier, A. Désordre linéaire dans les cristaux (cas du silicium, du quartz, et des pérovskites ferroélectriques). *Acta Crystallogr. Sect.* **1970**, *A26*, 244. [CrossRef]
20. Burns, G.; Dacol, F.H. Glassy polarization behavior in ferroelectric compounds $Pb(Mg_{1/3}Nb_{2/3})O_3$ and $Pb(Zn_{1/3}Nb_{2/3})O_3$. *Solid State Commun.* **1983**, *48*, 853. [CrossRef]
21. Li, L.-M.; Jiang, Y.-J.; Zeng, L.-Z. Temperature Dependence of Raman Spectra in $BaTiO_3$. *J. Raman Spectrosc.* **1996**, *27*, 503. [CrossRef]
22. Rusek, K.; Kruczek, J.; Szot, K.; Rytz, D.; Gorny, M.; Roleder, K. Non-Linear Properties of $BaTiO_3$ above T_C. *Ferroelectrics* **2008**, *375*, 165. [CrossRef]
23. Ko, J.-H.; Kojima, S.; Koo, T.-Y.; Jung, J.H.; Won, C.J.; Hur, N.J. Elastic softening and central peaks in $BaTiO_3$ single crystals above the cubic-tetragonal phase-transition temperature. *Appl. Phys. Lett.* **2008**, *93*, 102905. [CrossRef]
24. Helal, M.A.; Kojima, S. Effect of electric field on elastic properties of $BaTiO_3$ single crystals: A micro-Brillouin scattering study. *Jpn. J. Appl. Phys.* **2022**, *61*, SG1016. [CrossRef]
25. Levin, I.; Amos, T.J.; Bell, S.M.; Farber, L.; Vanderah, T.A.; Roth, R.S.; Toby, B.H. Phase equilibria, crystal structures, and dielectric anomaly in the $BaZrO_3$-$CaZrO_3$ system. *J. Solid State Chem.* **2003**, *175*, 170. [CrossRef]
26. Lebedev, A.I.; Sluchinskaya, I.A. Structural instability in $BaZrO_3$ crystals: Calculations and experiment. *Phys. Solid State* **2013**, *55*, 1941. [CrossRef]
27. Bennett, J.W.; Grinberg, I.; Rappe, A.M. Effect of symmetry lowering on the dielectric response of $BaZrO_3$. *Phys. Rev. B* **2006**, *73*, 180102. [CrossRef]
28. Helal, M.A.; Mori, T.; Kojima, S. Softening of infrared-active mode of perovskite $BaZrO_3$ proved by terahertz time-domain spectroscopy. *Appl. Phys. Lett.* **2015**, *106*, 182904. [CrossRef]
29. Helal, M.A.; Mori, T.; Kojima, S. Terahertz time-domain spectroscopy and Raman scattering studies of incipient ferroelectric $BaZrO_3$. *Ferroelectrics* **2016**, *499*, 107. [CrossRef]
30. Chemarin, C.; Rosman, N.; Pagnier, T.; Lucazeau, G. A High-Pressure Raman Study of Mixed Perovskites $BaCe_xZr_{1-x}O_3$ ($0 \leq x \leq 1$). *J. Solid State Chem.* **2000**, *149*, 298–307. [CrossRef]
31. Giannici, F.; Shirpour, M.; Longo, A.; Martorana, A.; Merkle, R.; Maier, J. Long-Range and Short-Range Structure of Proton-Conducting Y:$BaZrO_3$. *Chem. Mater.* **2011**, *23*, 2994. [CrossRef]
32. Kim, D.Y.; Patrik, E.D.C.; Miyoshi, S.; Tsuchita, T.; Yamaguchi, S. Local Bond Structure of $BaZrO_3$ Doped with Various Dopant Probed by Raman Spectroscopy. *Meet. Abstr.* **2012**, *MA2012-02*, 2098. [CrossRef]
33. Bednorz, J.B.; Müller, K.A. Possible highT_c superconductivity in the Ba−La−Cu−O system. *Z. Phys. B* **1986**, *64*, 189–193. [CrossRef]
34. Hor, P.H.; Meng, R.L.; Wang, Y.Q.; Gao, L.; Huang, Z.J.; Bechtold, J.; Forster, K.; Chu, C.W. High-pressure study of the new Y-Ba-Cu-O superconducting compound system. *Phys. Rev. Lett.* **1987**, *58*, 1891. [CrossRef]
35. Maeda, H.; Tanaka, Y.; Fukutomi, M.; Asano, T. A New High-T_c Oxide Superconductor without a Rare Earth Element. *Jpn. J. Appl. Phys.* **1998**, *27*, L209. [CrossRef]

36. Gao, L.; Xue, Y.Y.; Chen, F.; Xiong, Q.; Meng, R.L.; Ramirez, D.; Chu, C.W.; Eggert, J.H.; Mao, H.K. Superconductivity up to 164 K in $HgBa_2Ca_{m-1}Cu_mO_{2m+2+\delta}$ (m = 1, 2, and 3) under quasihydrostatic pressures. *Phys. Rev. B* **1994**, *50*, 4260. [CrossRef] [PubMed]
37. Seko, A.; Ishiwata, S. Prediction of perovskite-related structures in A CuO_{3-x}(A = Ca, Sr, Ba, Sc, Y, La) using density functional theory and Bayesian optimization. *Phys. Rev. B* **2020**, *101*, 134101. [CrossRef]
38. Helal, M.A.; Farid Ul-Islam, A.K.M.; Liton, M.N.H.; Kamruzzaman, M. Hydrostatic pressure dependent structural, elastic, vibrational, electronic, and optoelectronic properties of superconducting $BaCuO_3$: A DFT insight. *J. Phys. Chem. Solids* **2022**, *161*, 110452. [CrossRef]
39. Eros, S.; Reitz, J.R. Elastic Constants by the Ultrasonic Pulse Echo Method. *J. Appl. Phys.* **1958**, *29*, 683. [CrossRef]
40. Sumino, Y.; Ohno, I.; Goto, T.; Kumazawa, M. Measurement of the elastic constant and internal frictions on single-crystal MgO by rectangular parallelepiped resonance. *J. Phys. Earth* **1976**, *24*, 263. [CrossRef]
41. Maczka, M.; Hanuza, J.; Ko, J.-H.; Kojima, S. High-resolution broadband Brillouin scattering study of antiferroelectric phase transition in $CsBi(MoO_4)_2$. *Phys. Rev. B* **2003**, *68*, 174101. [CrossRef]
42. Grunberg, P. Brillouin scattering from spin waves in thin ferromagnetic films. *J. Magn. Mater.* **1980**, *766*, 15–18. [CrossRef]
43. Courtens, E.; Pelous, J.; Phalippou, J.; Vacher, R.; Woignier, T. Brillouin-scattering measurements of phonon-fracton crossover in silica aerogels. *Phys. Rev. Lett.* **1987**, *58*, 128. [CrossRef]
44. Kojima, S. 100th Anniversary of Brillouin Scattering: Impact on Materials Science. *Materials* **2022**, *15*, 3518. [CrossRef] [PubMed]
45. Kojima, S. Gigahertz Acoustic Spectroscopy by Micro-Brillouin Scattering. *Jpn. J. App. Phys.* **2010**, *49*, 07HA01. [CrossRef]
46. Kohn, W.; Sham, L.J. Self-consistent equations including exchange and correlation effects. *Phys. Rev.* **1965**, *140*, A1133. [CrossRef]
47. Vanderbilt, D. Soft self-consistent pseudopotentials in a generalized eigenvalue formalism. *Phys. Rev. B* **1990**, *41*, 7892–7895. [CrossRef] [PubMed]
48. Nikulin, N. *Fundamentals of Electrical Materials*; Mir Publishers: Moscow, Russia, 1988.
49. Luspint, Y.; Servoin, J.L.; Gervais, F. Soft mode spectroscopy in barium titanate. *J. Phys. C Solid St. Phys.* **1980**, *13*, 3761. [CrossRef]
50. Takahasi, H. A Note on the Theory of Barium Titanate. *J. Phys. Soc. Jpn.* **1961**, *16*, 1685. [CrossRef]
51. Helal, M.A.; Aftabuzzaman, M.; Tsukada, S.; Koima, S. Role of polar nanoregions with weak random fields in Pb-based perovskite ferroelectrics. *Sci. Rep.* **2017**, *7*, 44448. [CrossRef] [PubMed]
52. Liu, Y.; Withers, R.; Nguyen, B.; Elliott, K. Structurally frustrated polar nanoregions in $BaTiO_3$-based relaxor ferroelectric systems. *Appl. Phys. Lett.* **2007**, *91*, 152907. [CrossRef]
53. Tsuda, K.; Sano, R.; Tanaka, M. Nanoscale local structures of rhombohedral symmetry in the orthorhombic and tetragonal phases of $BaTiO_3$ studied by convergent-beam electron diffraction. *Phys. Rev. B* **2012**, *86*, 214106. [CrossRef]
54. Grinberg, I.; Rappe, A.M. Local structure and macroscopic properties in PMN-PT and PZN-PT solid solutions. *Phys. Rev. B* **2004**, *70*, 220101. [CrossRef]
55. Xu, G.; Wen, J.; Stock, C.; Gehring, P.M. Phase instability induced by polar nanoregions in a relaxor ferroelectric system. *Nat. Mater.* **2008**, *7*, 562–566. [CrossRef]
56. Ko, J.H.; Kim, T.H.; Roleder, K.; Rytz, D.; Kojima, S. Precursor dynamics in the ferroelectric phase transition of barium titanate single crystals studied by Brillouin light scattering. *Phys. Rev. B* **2011**, *84*, 094123. [CrossRef]
57. Tsukada, S.; Kojima, S. Time-reversal operation of tunneling dynamics of Bose-Einstein condensates in optical lattices. *Phys. Rev. B* **2008**, *78*, 144106. [CrossRef]
58. Dul'kin, E.; Petzelt, J.; Kamba, S.; Mojaev, E.; Roth, M. Relaxor-like behavior of $BaTiO_3$ crystals from acoustic emission study. *Appl. Phys. Lett.* **2010**, *97*, 032903. [CrossRef]
59. Sale, E.K.H.; Carpenter, M.A.; Nataf, G.F.; Picht, G.; Webber, K.; Weerasinghe, J.; Lisenkov, S.; Bellaiche, L. Elastic excitations in $BaTiO_3$ single crystals and ceramics: Mobile domain boundaries and polar nanoregions observed by resonant ultrasonic spectroscopy. *Phys. Rev. B* **2013**, *87*, 014106. [CrossRef]
60. Helal, M.A.; Kojima, S. Brillouin scattering and elastic constants of relaxor ferroelectric $0.44Pb(Mg_{1/3}Nb_{2/3})O_3$-$0.56PbTiO_3$ single crystals. *Ferroelectrics* **2017**, *512*, 36. [CrossRef]
61. Richerson, D.W. *Modern Ceramic Engineering: Properties, Processing, and Use in Design*, 2nd ed.; Marcel Dekker, Inc.: New York, NY, USA, 1992.
62. Cruz, M.P.; Chu, Y.H.; Zhang, J.X.; Yang, P.L.; Zavaliche, F.; He, Q.; Shafer, P.; Chen, L.Q.; Ramesh, R. Strain Control of Domain-Wall Stability in Epitaxial $BiFeO_3$ (110) Films. *Phys. Rev. Lett.* **2007**, *99*, 217601. [CrossRef]
63. Merz, W.J. Domain Formation and Domain Wall Motions in Ferroelectric $BaTiO_3$ Single Crystals. *Phys. Rev.* **1954**, *95*, 690–698. [CrossRef]
64. Rojac, T.; Kosec, M.; Damjanovic, D. Large Electric-Field Induced Strain in $BiFeO_3$ Ceramics. *J. Am. Ceram. Soc.* **2011**, *94*, 4108–4111. [CrossRef]
65. Jiang, B.; Bai, Y.; Chu, W.Y.; Su, Y.J.; Qiao, L.J. Direct observation of two 90° steps of 180° domain switching in $BaTiO_3$ single crystal under an antiparallel electric field. *Appl. Phys. Lett.* **2008**, *93*, 152905. [CrossRef]
66. Potnis, P.; Huber, J. In-situ observation of needle domain evolution in barium titanate single crystals. *J. Eur. Ceram. Soc.* **2013**, *33*, 327–333. [CrossRef]
67. Nelson, C.T.; Gao, P.; Jokisaari, J.R.; Heikes, C.; Adamo, C.; Melville, A.; Baek, S.H.; Folkman, C.M.; Winchester, B.; Gu, Y.J.; et al. Domain Dynamics During Ferroelectric Switching. *Science* **2011**, *334*, 968–971. [CrossRef]

68. Baek, S.H.; Jang, H.W.; Folkman, C.M.; Li, Y.L.; Winchester, B.; Zhang, J.X.; He, Q.; Chu, Y.H.; Nelson, C.T.; Rzchowski, M.S.; et al. Ferroelastic switching for nanoscale non-volatile magnetoelectric devices. *Nature Mater.* **2010**, *9*, 309–314. [CrossRef]
69. Bilić, A.; Gale, J.D. Ground state structure of $BaZrO_3$: A comparative first-principles study. *Phys. Rev. B* **2009**, *79*, 174107. [CrossRef]
70. Li, Y.W.; Li, F.X. The effect of domain patterns on 180° domain switching in $BaTiO_3$ crystals during antiparallel electric field loading. *Appl. Phys. Lett.* **2014**, *104*, 042908. [CrossRef]
71. Kersch, A.; Fischer, D.J. Phase stability and dielectric constant of ABO_3 perovskites from first principles. *Appl. Phys.* **2009**, *106*, 01405. [CrossRef]
72. Vacher, R.; Boyer, L. Brillouin Scattering: A Tool for the Measurement of Elastic and Photoelastic Constants. *Phys. Rev. B* **1972**, *6*, 639. [CrossRef]
73. Ko, J.-H.; Kim, D.H.; Kojima, S. Central peaks, acoustic modes, and the dynamics of polar nanoregions in $Pb[(Zn_{1/3}Nb_{2/3})_xTi_{1-x}]O_3$ single crystals studied by Brillouin spectroscopy. *Phys. Rev. B* **2008**, *77*, 104110. [CrossRef]
74. Nuzhnyy, D.; Petzelt, J.; Savinov, M.; Ostapchuk, T.; Bovtun, V.; Kempa, M.; Hlinka, J.; Buscaglia, V.; Buscaglia, M.T.; Nanni, P. Broadband dielectric response of $Ba(Zr,Ti)O_3$ ceramics: From incipient via relaxor and diffuse up to classical ferroelectric behavior. *Phys. Rev. B* **2012**, *86*, 014106. [CrossRef]
75. Gim, D.-H.; Sur, Y.; Lee, Y.H.; Lee, J.H.; Moon, S.; Oh, Y.S.; Kim, K.H. Pressure-Dependent Structure of $BaZrO_3$ Crystals as Determined by Raman Spectroscopy. *Materials* **2022**, *15*, 4286. [CrossRef] [PubMed]
76. Evarestov, R.A. Hybrid density functional theory LCAO calculations on phonons in $Ba(Ti,Zr,Hf)_3$. *Phys. Rev. B* **2011**, *83*, 014105. [CrossRef]
77. Sundell, P.G.; Björketun, M.E.; Wahnström, G. Thermodynamics of doping and vacancy formation in $BaZrO_3$ perovskite oxide from density functional calculations. *Phys. Rev. B* **2006**, *73*, 104112. [CrossRef]
78. Akbarzadeh, A.R.; Kornev, I.; Malibert, C.; Bellaiche, L.; Kiat, J.M. Combined theoretical and experimental study of the low-temperature properties of $BaZrO_3$. *Phys. Rev. B* **2005**, *72*, 205104. [CrossRef]
79. Muller, K.A.; Burkard, H. $SrTiO_3$: An intrinsic quantum paraelectric below 4 K. *Phys. Rev. B* **1979**, *19*, 3593. [CrossRef]
80. Glazer, A.M. The classification of tilted octahedra in perovskites. *Acta Cryst. B* **1972**, *28*, 3384–3392. [CrossRef]
81. Liton, M.N.H.; Roknuzzaman, M.; Helal, M.A.; Kamruzzaman, M.; Islam, A.K.M.F.U.; Ostrikov, K.; Khan, M.K.R. Electronic, mechanical, optical and photocatalytic properties of perovskite $RbSr_2Nb_3O_{10}$ compound. *J. Alloys Compd.* **2021**, *867*, 159077. [CrossRef]
82. Helal, M.A.; Islam, A.K.M.A. Elastic, electronic, and optical properties of hypothetical $SnNNi_3$ and $CuNNi_3$ in comparison with superconducting $ZnNNi_3$. *Phys. B* **2011**, *406*, 4564–4568. [CrossRef]
83. Li, W.M.; Zhao, J.F.; Cao, L.P.; Hu, Z.; Huang, Q.Z.; Wanga, X.C.; Liu, Y.; Zhao, G.Q.; Zhang, J.; Liu, Q.Q.; et al. Superconductivity in a unique type of copper oxide. *Proc. Natl. Acad. Sci. USA* **2019**, *116*, 12156–12160. [CrossRef] [PubMed]
84. Allen, P.B.; Dynes, R.C. Transition temperature of strong-coupled superconductors reanalyzed. *Phys. Rev. B* **1975**, *12*, 905. [CrossRef]
85. Casais, M.T.; Alonso, J.A.; Rasines, I.; Hidalgo, M.A. Preparation, neutron structural study and characterization of $BaNbO_3$: A Pauli-like metallic perovskite. *Mater. Res. Bull.* **1995**, *30*, 201–208. [CrossRef]
86. Gasparov, V.A.; Ermolov, S.N.; Khassanov, S.S. Superconducting, surface and anomalous electron transport properties of $BaNbO_{3-x}$ films. *Phys. B* **2000**, *284*, 1119–1120. [CrossRef]

materials

MDPI

Article

Quasielastic Light Scattering in the Broadband Brillouin Spectra of Relaxor Ferroelectric $PbMg_{1/3}Nb_{2/3}O_3$

Nikita K. Derets [1], Alexander I. Fedoseev [1], Jae-Hyeon Ko [2], Seiji Kojima [3] and Sergey G. Lushnikov [1,*]

1 Division of Physics of Dielectrics and Semiconductors, Ioffe Institute, Politekhnicheskaya 26, 194021 St. Petersburg, Russia
2 School of Nano Convergence Technology, Nano Convergence Technology Center, Hallym University, 1 Hallymdaehakgil, Chuncheon 24252, Gangwondo, Republic of Korea
3 Division of Materials Science, University of Tsukuba, Tsukuba 305-8573, Ibaraki, Japan
* Correspondence: sergey.lushnikov@mail.ioffe.ru; Tel.: +7-(921)-3936869

Abstract: In this paper, the behavior of quasielastic light scattering (QELS) in a $PbMg_{1/3}Nb_{2/3}O_3$ (PMN) crystal under broadband Brillouin light scattering in a temperature range from 750 K to 80 K was studied. It was shown that QELS consists of two components: narrow (0.9 GHz to 11 GHz) and wide (80 GHz to 600 GHz). The dependencies of the intensity, I, of these components on the frequency, ν, are well described by the power law $I \sim e\nu^{\alpha}$, with different α, and are determined by the distribution of the relaxation times. The analysis of the Brillouin spectra showed that the behavior of the relaxation time of both the components of QELS with temperature change is well described by the Arrhenius law. Additionally, in the vicinity of the intermediate temperature $T^* \approx 380$ K, a critical relaxation time behavior for the narrow component of QELS was detected. In the vicinity of the same temperature, a maximum in the integral intensity of both the components of QELS was observed, which is adjacent to another maximum in the region of the Vogel–Fulcher temperature $T_{VF} \approx 250$ K corresponding to the transformation of the crystal to a nonergodic state.

Keywords: Brillouin scattering; relaxations; quasielastic light scattering; relaxors ferroelectrics

Citation: Derets, N.K.; Fedoseev, A.I.; Ko, J.-H.; Kojima, S.; Lushnikov, S.G. Quasielastic Light Scattering in the Broadband Brillouin Spectra of Relaxor Ferroelectric $PbMg_{1/3}Nb_{2/3}O_3$. *Materials* **2023**, *16*, 346. https://doi.org/10.3390/ma16010346

Academic Editor: Ingo Dierking

Received: 26 November 2022
Revised: 20 December 2022
Accepted: 26 December 2022
Published: 30 December 2022

1. Introduction

Quasielastic light scattering (QELS) is an intriguing subject in the physics of the vibrational spectrum of dielectric crystals. This phenomenon is observed in many compounds during phase transitions or transformations, which are studied by optical spectroscopy [1]. The nature of this contribution to elastic scattering at an unshifted frequency in the spectra of inelastic light scattering has been studied in detail. Several mechanisms are described that lead to the appearance of QELS in light-scattering spectra, including the influence of defects, entropy, fluctuation, and others, which differ in their dependence of the QELS width on the wave vector, the behavior of the intensity, the width itself, and the region of existence, with the change in the external thermodynamic parameters (temperature, pressure) [1]. The existence of QELS has also been observed in disordered structures, such as glasses, supercooled liquids, etc. [2–4]. In recent years, particular interest in this phenomenon has arisen in the study of the vibrational spectrum of partially disordered crystals and primarily in relaxor ferroelectrics with a perovskite structure. This is due to the fact that the existence of QELS in relaxor ferroelectrics was associated with polar nanoregions (PNRs), which dramatically change the dynamics of the crystal lattice of perovskite ferroelectrics. Indeed, in relaxor ferroelectrics with the general formula $AB'_{1-x}B''_{x}O_3$, a frequency-dependent anomaly of the dielectric response is observed, which is "stretched" in temperature by hundreds of degrees. This anomaly is not associated with a structural phase transition. Anomalies in the dielectric response correspond to frequency-dependent anomalies in the speed of sound, damping, and several other physical properties [5,6]. For the first time, polarized QELS in relaxors was observed in a $PbMg_{1/3}Nb_{2/3}O_3$ (PMN) crystal in the Raman scattering spectra [7] within a temperature region where an anomalous dielectric response was observed, called the diffuse phase transition region. Later, it became clear that QELS is a characteristic feature of the low-frequency vibrational spectrum of relaxors and is observed in experiments using optical spectroscopy [8,9]. Experimental studies have shown that QELS is observed in a wide frequency range (more than three orders of magnitude in frequency) and has a complex temperature dependence [10–13]. At the moment, it is assumed that its appearance is associated with the appearance of PNRs (or polar clusters) in the vicinity of the Burns temperature (T_B). A description of various ideas about PNRs can be found in a number of papers on this topic [14–16]. The behavior of QELS with temperature change is determined by the dynamics of these nanoregions [8]. In the current approach describing the dynamics of PNRs, there are two main mechanisms that contribute to the dielectric dispersion and, accordingly, to the anomaly of the dielectric response of relaxors. The first mechanism is related to the dynamical flipping of the cluster polarization. This mechanism determines mainly the high-frequency contribution to the dispersion of the dielectric response. Below the freezing temperature, T_{VF}, the clusters freeze out and the flipping of the polarization disappears. The second mechanism is determined by the vibration (breathing) of the polar cluster boundaries. Both of these mechanisms determine a broad distribution function for relaxation time, which expands with decreasing temperature. It is the vibrations of the walls of the polar nanoclusters that determine the dispersion of the dielectric response at $T < T_{VF}$ [9].

Studies into QELS in relaxors have shown that it is polarized and can be complex, i.e., it can consist of several contributions, as was observed for $Na_{1/2}Bi_{1/2}TiO_3$ (NBT) [17]. Studies of the polarized spectra of Brillouin and Raman scattering in PMN-56%PT made it possible to relate the QELS component observed in the vertical–vertical (VV) polarization configuration to the longitudinal component of polarization fluctuations and that observed in a vertical–horizontal (VH) polarization configuration to the transverse component of polarization fluctuations [12,13]. An analysis of the intensity distribution of QELS in PMN crystals in the range of three orders of magnitude of frequency allowed the authors [11] to show its selfsimilarity. The intensity distribution is described as a power-law spectrum using the expression $I \sim \nu^{\alpha}$, where I, ν, and α are the intensity, frequency, and power exponent, respectively. The power-law spectrum implies a Lorentzian superposition (Debye relaxators) with different half-widths, that is, with different relaxation times, defined as the reciprocal of the Lorentzian half-width. Thus, in the frequency range under consideration, we have the power-law distribution for relaxation time [11]. Another consequence of such a description of QELS is the conclusion of the authors from [11] on the possible use of the fractal approach to describe the dynamics of the PNRs of a PMN crystal. This conclusion is consistent with the results of the analysis of the generalized density of states function [18] and the behavior of heat capacity [19] in PMN obtained within the framework of the fractal approach. Recent studies on the spatial organization of nanoregions in relaxors [16] support the assumption that fractals exist in PMN crystals. A detailed analysis of dielectric dispersion over a wide frequency range based on infrared reflectivity (IR) and terahertz time-domain spectra combined with high -frequency dielectric data above the megahertz range did not support the picture of fractal dynamics proposed in [20]. Additional studies into QELS in PMN crystals have not been carried out ever since. This motivated us to investigate QELS using Brillouin light scattering in a PMN crystal over a wide range of frequencies and temperatures.

PMN is a model for studying the properties of relaxors. For PMN, the anomaly in the dielectric response in the vicinity of maximum temperature, $T_m \sim 270$ K, at a frequency of 10 kHz is not associated with a macroscopic change in symmetry, which remains cubic (space group $Pm\bar{3}m$) down to helium temperatures [21]. The lattice dynamics of the PMN crystal exhibit complex behavior in this case. In the vicinity of 900 K, the appearance of a polarized Raman spectrum of the first order is observed [22]. This effect is called a dynamic phase transition and is associated by the authors (in a number of works) with the nonequivalence of oxygen octahedra at B′ and B″ cations, which begin to manifest themselves at these temperatures [22]. A further decrease in temperature leads to a deviation in refractive index from its high-temperature linearity, which is due to the formation of PNRs ($T_B \approx 640$ K) [23]. In the vicinity of this temperature, quasielastic scattering also occurs in the spectra of inelastic neutron scattering [5]. Its intensity increases with decreasing temperature and reaches a maximum simultaneously with permittivity (at the GHz frequency) in the vicinity of the intermediate temperature, $T^* \sim 380$ K [24]. The anomalies in several of the physical properties of PMN are associated with this temperature [25], including a sharp increase in the intensity of the elastic component in inelastic neutron scattering [24]. The observed features in this temperature range are associated with a transition from a cubic isotropic to a cubic anisotropic state [24] or with the formation of static nanoregions [9,14,25]. Another characteristic point in the discussed pattern of phase transformations in PMN is the temperature of the ferroelectric phase transition induced by an applied electric field in the vicinity of 210 K [5]. In the absence of an electric field, a low-temperature region is isolated at $T_{VF} < 250$ K, where the dynamics of the PNRs "freeze" and a nonergodic state appears [14].

The dynamics of the PNRs in PMN in a wide frequency range have been studied in sufficient detail using dielectric, IR, and terahertz time-domain spectroscopy [5,9,20]. Broadband Brillouin scattering is a complementary method for studying the dynamics of PNRs in a frequency range from 10^9 to 10^{12} GHz, which allows one to study the relaxation mechanisms in the gigahertz frequency domain. This motivated us to investigate the temperature behavior of the low-frequency region of the vibrational spectrum of a PMN crystal using Brillouin light scattering in a frequency range from 0.6 GHz to 800 GHz within the temperature range, which includes the regions of appearance, transformation, and freezing of the PNRs.

2. Materials and Methods

In this work, the low-frequency region of the vibrational spectrum of a PMN crystal was studied in detail by means of Brillouin light scattering. The PMN crystal was grown by the Czochralski method at the SI Vavilov State Optical Institute (JSC SI Vavilov State Optical Institute, St. Petersburg, Russia) in the form of a large boule with a diameter of about 30 mm and a length of more than 60 mm under the direction of [100]. Part of the boule was used to prepare large samples (up to 8 cm^3) for experiments on inelastic neutron scattering (see references to our works in [5]). Part of this boule was used to prepare samples for experiments on inelastic light scattering (see references to our works in [5]). PMN samples from this boule were tested before experiments on neutron scattering, including at single-crystal neutron diffraction facilities in PSI, Villigen, Switzerland. It was shown that there are no deviations from the stoichiometry, composition, and structure of perovskite; the mosaic is less than 20′. From a boule, a sample in the form of a parallelepiped oriented as [001] with sizes $1 \times 3 \times 5$ mm was prepared. The sample was oriented by using single-crystal X-ray diffraction.

A high-contrast 3 + 3 pass tandem Fabry–Perot interferometer combined with an optical microscope (Olympus BH-2) was used in the micro-Brillouin experiments. Scattering was excited by a green YAG laser with a wavelength of $\lambda = 532$ nm and power of 100 mW. A backscattering geometry, with a vertically polarized incident light and either a vertically or horizontally polarized scattered light, was employed. The phonon wave vector ($\mathbf{q}_{ph}$) was oriented along the [100] axis. The sample was put in a heat/cooling cell (THMS 6000) with temperature variations from −190 to 600 °C and stability of ±0.1 °C. When scanning the low-frequency region of the vibrational spectrum, several free spectral ranges (FSRs) of 6, 12, 75, 200, and 800 GHz were used. All observed spectra were polarized. To obtain the combined spectrum of Brillouin light scattering in PMN over a wide frequency range (from 0.6 to 800 GHz), the experimental Brillouin spectra obtained at different FSRs were merged together. The normalization was carried out for the spectrum obtained at FSR = 800 GHz. The measurements were carried out when the sample was cooled from 750 to 80 K at each of the FSRs indicated above. In the used geometry of the scattering, a longitudinal acoustic phonon was observed, the velocity of which is determined by the C_{11} elastic stiffness constant. The temperature behavior of this phonon is discussed in [26].

3. Results and Discussion

3.1. Broadband Brillouin Spectra in PMN

Figure 1 shows the combined Brillouin scattering spectra, which reflect the evolution of the low-frequency region of the vibrational spectrum of the PMN crystal with temperature. The graph was plotted with the use of a double logarithmic scale for the convenience of describing QELS, the existence of which in PMN crystals is known from previous works [7,11]. Koreeda et al. [11] showed that QELS in the combined light-scattering spectra of PMN is well described by a power-law function (which is a superposition of a set of Lorentzians) centered at a zero frequency. On a double logarithmic scale, the power-law function is described as the linear dependence of intensity on frequency. The set of Lorentzians in QELS corresponds to the distribution of the relaxation times [11]. It is clearly seen from Figure 1 that the combined spectrum of inelastic light scattering consists of at least two linear sections corresponding to two QELS components: a narrow, low-frequency component in the range from 0.9 to 11 GHz and a wide, high-frequency component from 80 to 600 GHz. The high-frequency component continues into the THz range, being part of the QELS in the Raman spectra of PMN [7]. It is clearly seen that the spectra presented in Figure 1 cannot be described using the expression $I \sim e\nu^{\alpha}$, as was conducted in [11]. An analysis of the intensity distribution of the spectra showed that the narrow and wide components of QELS can be described by two linear dependencies that correspond to two power-laws with different values for the exponents α–α_N and α_B, corresponding to the narrow and wide components, respectively. This indicates the existence of two relaxation time distribution functions in the considered frequency range and is consistent with the bimodal nature of the relaxation time distribution function obtained by analyzing the dielectric dispersion in PMN [27]. At a temperature of 285 K, we obtained the values $\alpha_N \approx -0.5$ and $\alpha_B \approx -0.32$. The differences in the combined spectra obtained and shown in [11] are apparently associated with different experimental conditions. Koreeda et al. [11] performed measurements using PMN with a horizontally polarized light incident along the [110] crystallography direction of the crystal and vertically polarized scattered light, whereas, in our experiments, a vertically polarized incident and scattered light along the [100] crystallography direction were used. In this case, apparently, various mechanisms which determine the distribution function of relaxation times manifest.

The intermediate region of the combined spectrum that separates the two QELS components (between 11 and 80 GHz) is highlighted since it contains a longitudinal acoustic phonon (LA) and an additional intensity value, which, in our opinion, is not related to either the wide or narrow QELS components (Figure 1). Let us consider this region of the combined spectrum and its temperature evolution (Figure 1). Figure 1, which was obtained at a temperature of 600 K, clearly shows that the linear function describing the broad QELS component is replaced by an intensity minimum in the vicinity of 60 GHz. This minimum passes into the asymmetric contour of the LA phonon. In the low-frequency region of the considered part of the spectrum, the excess intensity turns into a linear dependence for the intensity of the narrow QELS component. It should be noted that the shape of the LA phonon line is not described by the Lorentz function or by a damped harmonic oscillator. The characteristic distortion of the LA phonon line shape and the "dip" in the intensity of the high-frequency phonon wing in the vicinity of 60 GHz corresponds to the appearance of Fano resonance [28,29]. In our case, the Fano resonance effect means the existence of an interaction between the LA phonon and some broadband excitation, the frequency of which is comparable to but is less than the frequency of the LA phonon. The excess intensity, which apparently indicates the excitation interacting with the LA phonon, is weak and extended over a wide frequency range (about 80 GHz). The nature of this additional contribution to the combined spectrum in the vicinity of 60 GHz is not clear. Perhaps this is a manifestation of a soft overdamped optical mode, the existence of which is indicated by the results of experiments using neutron, hyper-Raman, and IR spectroscopy [5,9,30]. The only thing that does not match this assumption is the frequency of this excitation (<49 GHz), which is less than the experimentally obtained values of the soft optical mode frequency. Unfortunately, it is not possible to carry out the calculations, describe the shape of the phonon line, or determine the coupling constant in the framework of the Fano model due to the weak excitation intensity, which manifests itself as an excess intensity in the considered range of the combined spectrum. If this might be a soft, overdamped optical mode, then it apparently interacts with an acoustic phonon. It should be noted that the interaction of QELS and LA phonons can also take place. Lowering the temperature below 400 K changes the picture in the frequency region considered for the combined scattering spectrum; it is no longer possible to distinguish an additional intensity in the vicinity of 60 GHz, and the asymmetry of the LA phonon line contour, including the intensity minimum near the high-frequency region of the phonon line, disappears (the line shape is well described by the Lorentzian function) (Figure 1). Additional experiments on other scattering geometries and other crystallographic directions should clarify the reason for the distortion of the LA phonon line.

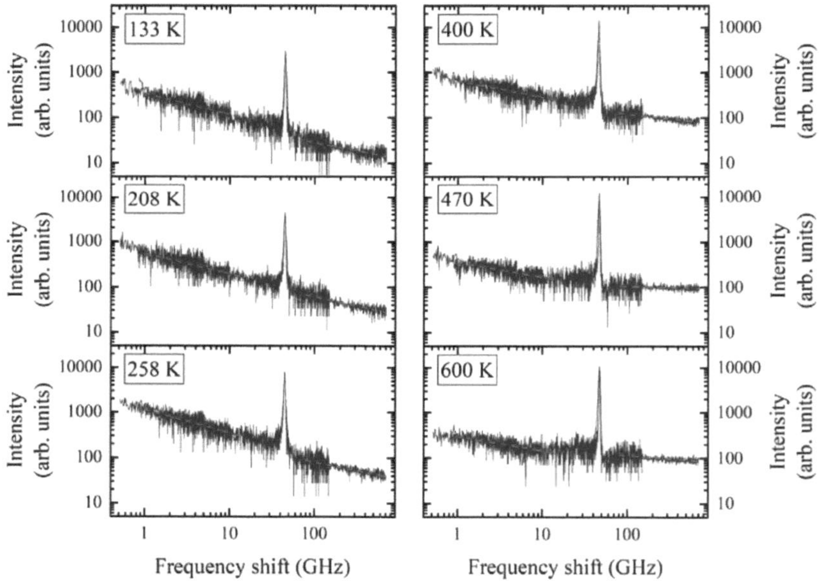

Figure 1. Combined spectra of Brillouin scattering in a PMN crystal obtained by "merging" the experimental spectra with different free spectral intervals (see text). Red dashed line is the result of a linear approximation.

The study of the possible relaxation processes in PMN that are associated with the narrow and wide QELS components will be considered in the next section when analyzing the temperature behavior of the Brillouin spectra. It should be noted, again, that the attempt to build a combined spectrum for VH polarization was unsuccessful since the intensities of these spectra with different FSRs are close to the background level. This greatly distinguishes PMN from the other examples of the studied relaxors, in which QELS was observed in the VH polarization [8,12,13].

3.2. Narrow and Broad Components of QELS in Brillouin Spectra

Let us now return to the analysis of the narrow and broad QELS components in the PMN scattering spectra. These QELS components in the combined spectra most closely correspond to the experimental spectra of Brillouin light scattering with FSR = 12 GHz and

800 GHz. An analysis of these spectra with varying temperatures makes it possible to study the temperature evolution of the narrow and broad QELS components in PMN in detail. The quasielastic component contour is well described by a Lorentzian function centered at zero frequency, and the excitation is, thus, a Debye relaxator. By measuring the half-width at the half-height $\delta\nu$ of the Quasielastic component, we can estimate the relaxation time, τ, as

$$\tau = \frac{1}{2\pi\delta\nu} \tag{1}$$

According to the fluctuation-dissipation theorem, the spectral function $S(\omega)$ is proportional to the imaginary part of the susceptibility $\chi''(\omega)$

$$S(\omega) \approx \frac{k_b T}{\hbar\omega}\chi''(\omega) \text{ for } \hbar\omega << k_B T \tag{2}$$

$$\frac{T}{I_{CP}} \propto \left(\int_0^\infty \frac{\chi''(\omega)}{\omega} d\omega\right)^{-1} \propto \chi\prime(0)^{-1} \tag{3}$$

where k_B is the Boltzmann constant, and T is the temperature in Kelvins. Thus, the integral intensity of the quasielastic scattering, I_{CP}, in the Brillouin spectra is proportional to the real part of the static susceptibility $\chi'(0)$. Thus, the determination of the experimental Brillouin spectra of the integrated intensity and half-width at half-height QELS (by fitting) makes it possible to obtain (expressions 1 and 3) the temperature dependencies of the susceptibility and the relaxation time of the narrow and wide QELS components.

Let us consider the behavior of the Brillouin scattering spectra with FSR = 12 GHz. Figure 2 shows examples of the polarized spectra at various temperatures. The LA phonon is absent in the spectra with VV polarization since its frequency $\omega \approx 49$ GHz lies outside the existing frequency window. In Figure 2, for the VV polarization, one can clearly see the quasielastic scattering component centered at zero frequency, which is practically absent in the VH polarization. This suggests that the considered QELS component is polarized. It is well demonstrated in Figure 2 that QELS changes with temperature. The fitting of the experimental spectra, the examples of which are shown in Figure 2, was carried out using the sum of the Gaussian function describing the instrumental function, the Lorentzian function describing QELS, and the background intensity represented by the constant.

$$I = A_{Gauss}\exp\left(-\frac{\omega_{Gauss}{}^2}{2\Gamma_{Gauss}{}^2}\right) + \frac{2A_{QELS}}{\pi}\frac{\Gamma_{QELS}}{4\omega^2 + \Gamma_{QELS}{}^2} + I_B \tag{4}$$

As a result of the calculations, we obtained the temperature dependencies of the integral intensity of the corresponding QELS component and the relaxation time (Figure 3). The intensity of the QELS in PMN in the Brillouin spectra with FSR = 12 GHz increases upon cooling, and a broad maximum with a complex structure is observed in a temperature range from 450 to 200 K. In this case, the intensity value increases by a factor of five in the vicinity of 250 K relative to its high-temperature values. Figure 3 shows that the maximum temperature dependence for intensity consists of two components: the first being a weak maximum in the region of 380 K, with the second being intense at 250 K, masking the first. At $T < 250$ K, the intensity of the QELS decreases. The temperature dependence of the relaxation time (τ_s) demonstrates a weak temperature dependence during cooling, against the background of which the anomaly in the vicinity of 380 K is clearly visible. In this case, the value of τ_s changes by a factor of 2.5 from 100 ps to 250 ps at $T \approx 380$ K. The maximum in the observed anomaly τ_s corresponds to a "shoulder" in the temperature dependence of the intensity and may reflect the critical dynamics of the corresponding relaxation process. The background change in relaxation time can be described by using the Arrhenius law $\tau_s = \tau_0 \exp(E_a/k_B T)$ with the activation energy E_a = 6.41 meV, and τ_0 = 85.5 ps. These values do not agree with the values for the activation energy and the pre-exponential factor known to us from the literature and determined in the framework of the existing models. Therefore, we could not choose an appropriate relaxation mechanism.

The anomalous behavior of the narrow QELS component in the vicinity of 380 K may be due to the interaction of the polarization fluctuations with a soft transverse optical mode. This is possible provided that the order parameter does not have a unique wave vector and is made up of a range of wave vectors around the center of the Brillouin zone, as it is assumed in [31]. Then, the occurrence of the interplay of the soft mode with the polarization fluctuations and the manifestation of the anomalous behavior of QELS in the vicinity of T^* looks reasonable.

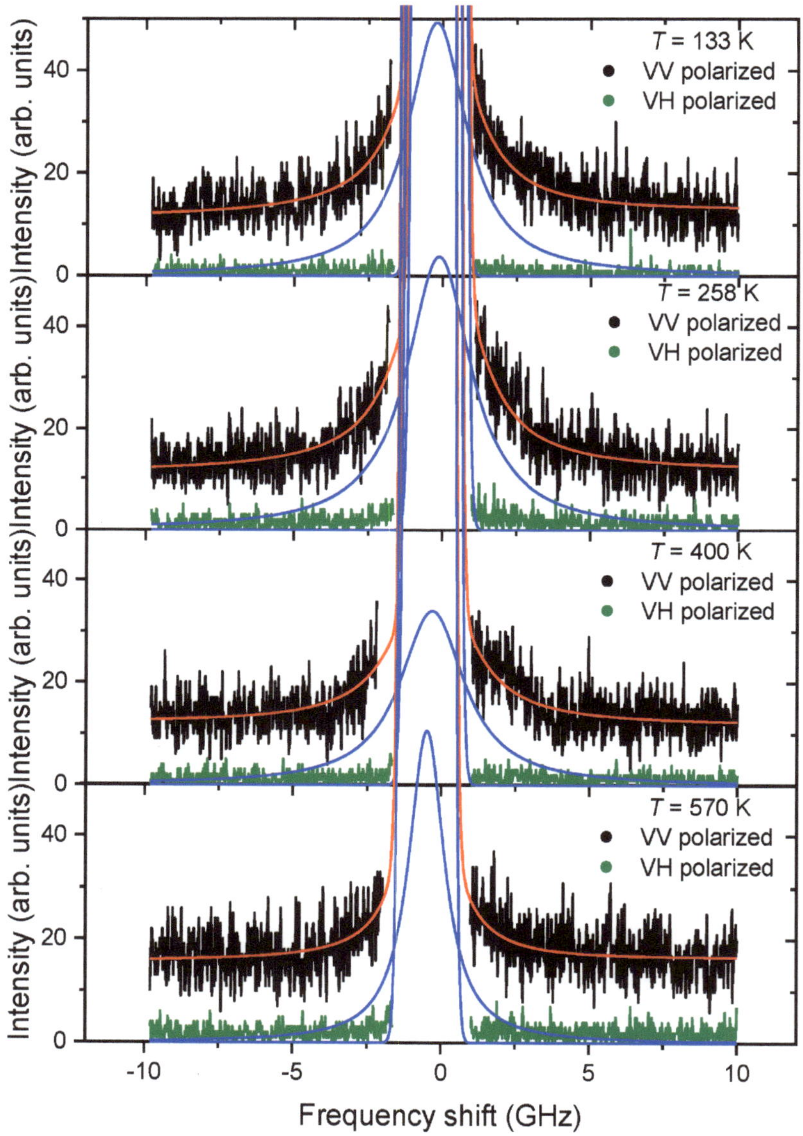

Figure 2. Polarized spectra of Brillouin light scattering in PMN (FSR = 12 GHz) and examples of their processing at different temperatures. The solid red line is the result of calculations using expression (4); fitting was carried out for the VV polarized spectra. Blue line is the contour of the QELS obtained as a result of data fitting (see text).

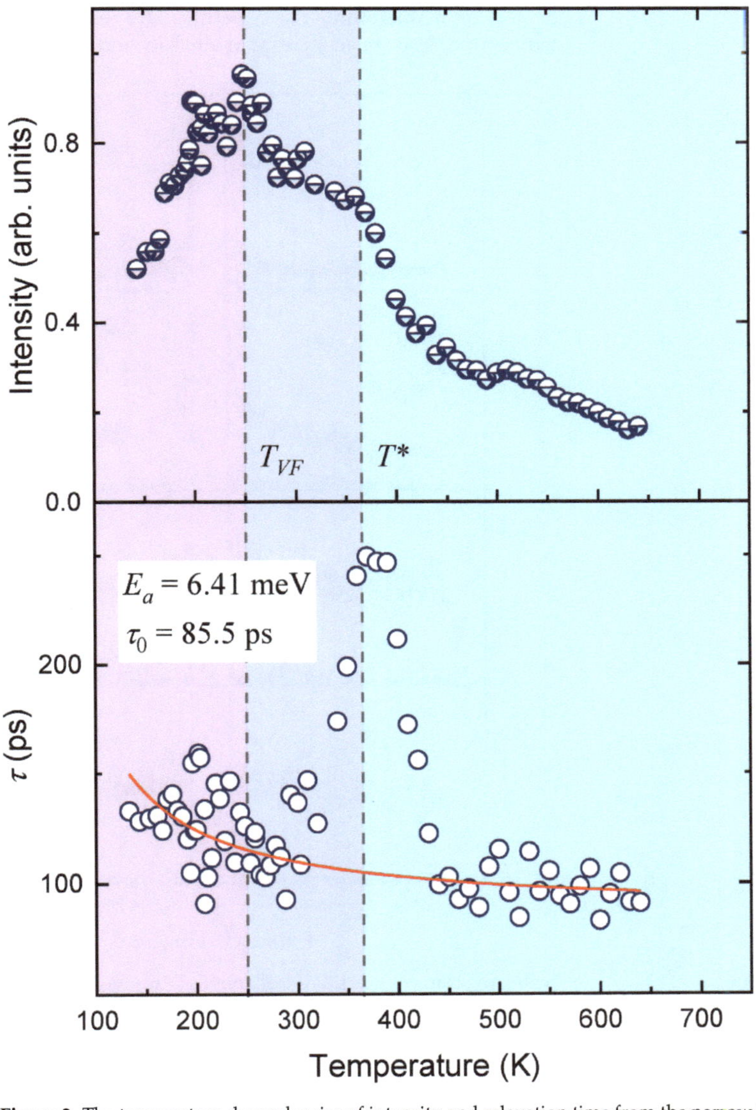

Figure 3. The temperature dependencies of intensity and relaxation time from the narrow QELS component in PMN. The circles are the results of calculations of experimental spectra with FSR = 12 GHz using Equations (1)–(4) (see text). The solid red line corresponds to the Arrhenius law. The colors in the figure highlight different temperature regions.

The behavior of Brillouin light scattering spectra in PMN with FSR = 800 GHz is shown in Figure 4. It reflects the temperature evolution of the broad QELS component. The scattering spectra, as can be seen in Figure 4, are polarized, and their changes in the VV polarization are clearly visible with decreasing temperature. The longitudinal acoustic phonon is not observed in the spectra because the width of the intense elastic scattering at a given FSR is comparable to the phonon frequency. The Brillouin spectra with FSR = 800 GHz (Figure 4) clearly show that, at high temperatures, the QELS is very wide, and its intensity is low. In Figure 1, it shows itself as an almost horizontal part of the high-frequency region of the combined spectrum, which increases slope angle with decreasing temperature. The experimental spectra in Figure 4 show that a decrease in temperature leads to an increase in intensity and a decrease in width.

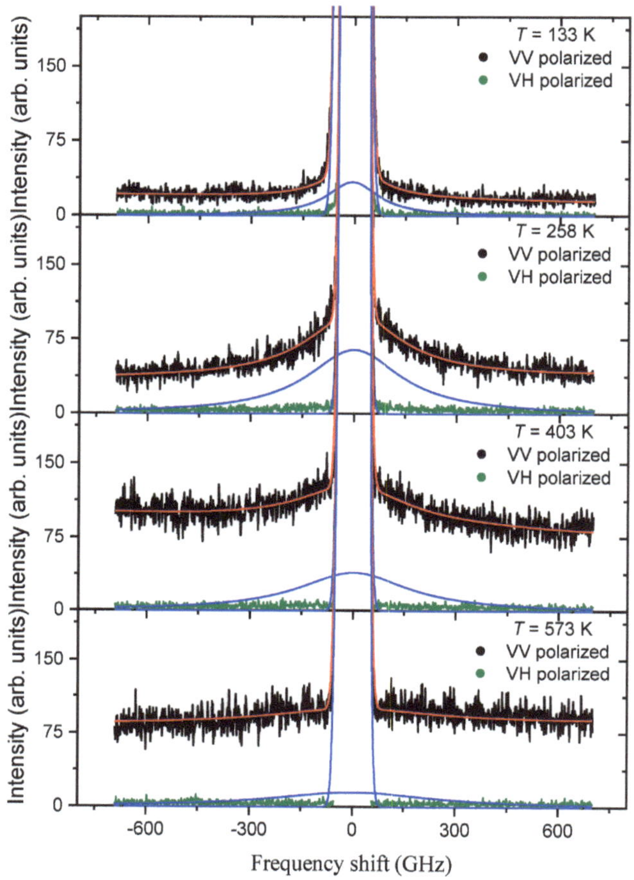

Figure 4. Polarized spectra of Brillouin light scattering in PMN (FSR = 800 GHz) and examples of their processing at different temperatures. The solid red line is the result of calculations using expression (4); fitting was carried out for the VV polarized spectra. Blue line is the contour of the QELS obtained as a result of data fitting (see text).

The results of fitting the experimental scattering spectra with FSR = 800 GHz are shown in Figure 5. It is clearly seen that the intensity of the QELS changes dramatically with decreasing temperature: a rather intense and broad component in the vicinity of 700 K decreases to minimum values in the region of 550 K and then begins to grow, increasing by more than an order of magnitude. The changes in intensity in the high-temperature region can be associated with the contribution of the soft optical mode. A broad anomaly in the temperature dependence of intensity, starting from 500 K, has a complex shape against the background of a maximum in the vicinity of 250 K, and a "shoulder" in the vicinity of 380 K can be distinguished. This intensity behavior correlates well with what we observe for the narrow QELS component (compare Figures 3 and 5). Below 250 K, the QELS intensity drops, changing its value by less than a quarter of the maximum intensity. Apparently, the wide and structured intensity maximum reflects the existence of several processes that contribute to its anomalous behavior. The relaxation time (τ_f) obtained from the broad QELS component of the Brillouin spectra measured with FSR = 800 GHz differs by two orders of magnitude from the relaxation time of the narrow component τ_s. In the vicinity of room temperature, $\tau_f \approx 1$ ps, and $\tau_s \approx 100$ ps. A change in temperature leads to an increase in τ_f, and this increase is not monotonous. It is necessary to highlight the temperature range between T^* and T_{VF}, where τ_f is practically independent of temperature. We tried to describe $\tau_f(T)$ in the temperature ranges $T > T^*$ and $T < T_{VF}$ using the Arrhenius law. At $T > T^*$, the temperature dependence of $\tau_f(T)$ is well described by the activation energy E_a = 35.2 meV and τ_0 = 0.3 ps, and at temperatures below T_{VF}, E_a = 11.3 meV and τ_0 = 0.74 ps. These values do not agree with the values of E_a and τ_0 known to us from the literature for the cluster-flipping mechanism in PMN [30]. Studies on the behavior of longitudinal and transverse polarization fluctuations of the PNRs in QELS in PMN-56% PT [13] made it possible to determine the relaxation time and activation energy in the paraelectric phase. The relaxation time and activation energy of the longitudinal polarization fluctuations in PMN-56PT in the paraphase (E_a = 26 meV and τ_0 = 0.55 ps) are in accordance with the values, which were obtained during the characterization of the thermally activated changes in the relaxation time τ_f of PMN at $T > T^*$.

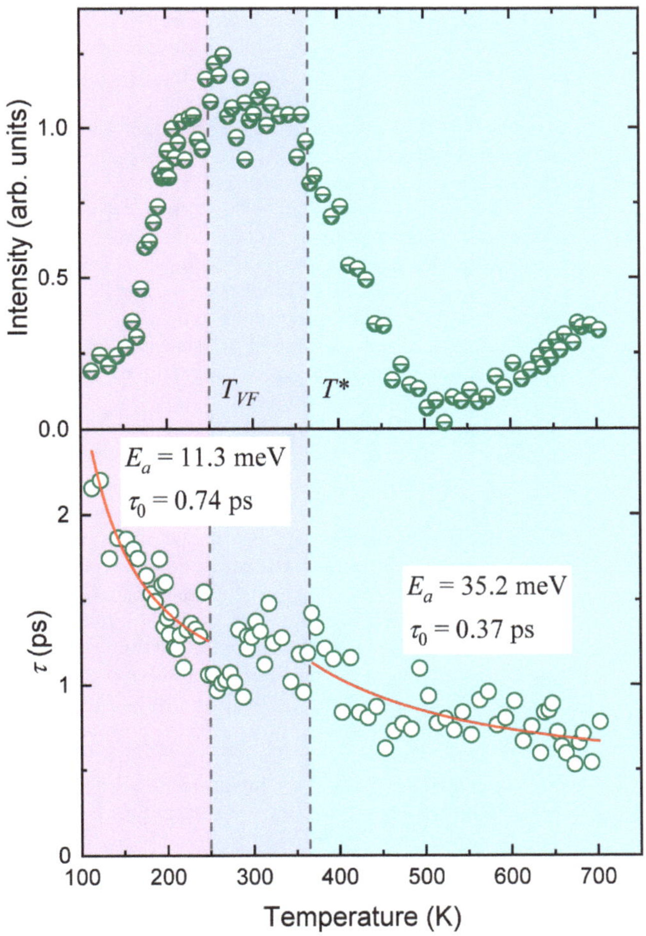

Figure 5. Temperature behavior of the intensity and relaxation time of the broad QELS component in PMN Brillouin spectra with FSR = 800 GHz. The circles are the results of fitting the experimental spectra by Equations (1)–(4); the line corresponds to the approximation of the data using the Arrhenius law. The colors in the figure highlight different temperature regions.

The dynamics of the PNRs at temperatures $T < T_{VF}$, according to the results of studies on the dielectric response, are mainly determined by the vibrations of the walls of these regions, and the contribution of the polarization fluctuations from the PNR reorientation motion disappears. If we assume that this model is correct, then the behavior of $\tau_f(T)$ in the considered temperature range is determined by the relaxation of the PNR walls. In the intermediate temperature range $T_{VF} < T < T^*$, the behavior of $\tau_f(T)$ is difficult to analyze due to the large scatter of points.

4. Conclusions

When summarizing the obtained results, it can be stated that in the Brillouin scattering of light in PMN within a frequency range from 0.7 GHz to 600 GHz, two-component quasielastic light scattering is observed. The narrow component is observed in the frequency range from 0.9 GHz to 11 GHz, and the broad component is observed in the frequency range from 80 GHz to 600 GHz. The intensity distribution in each of the components is described by a power law: $I \sim e\nu^{\alpha}$ with different exponents. These parts of the combined Brillouin spectra correspond to either the different distribution functions of relaxation time or a bimodal distribution function of relaxation time. In the intermediate frequency range between 11 GHz and 80 GHz, a LA phonon is visible, the line shape of which depends on temperature. At $T^* < T < T_B$, the shape of the LA phonon line is distorted, and a noticeable "dip" is also observed on the high-frequency side of the line profile. Below T^*, the line shape of the LA phonon is described by the Lorentz function. Similar distortions of the line shape are observed at the Fano resonance and are associated with the interaction of an acoustic phonon with a low-lying excitation.

The temperature behavior of the narrow and broad components of the QELS in PMN was analyzed based on the measured Brillouin spectra, with two specific frequency windows being selected, i.e., FSR = 12 GHz and 800 GHz, where different power-law behaviors for intensity were observed. Fitting the experimental spectra with the indicated free spectral intervals made it possible to plot the temperature dependencies of integral intensity and relaxation time. It was shown that the integrated intensity of both components behaves in a similar way, demonstrating a wide anomaly with two overlapping maxima in the vicinity of T^* and T_{VF}. The relaxation times of the broad and narrow components determined in the calculations differ by two orders of magnitude and are 1 ps and 100 ps, respectively (in the vicinity of room temperature). The behavior of the broad component with temperature change is different in different temperature ranges: at $T > T^*$, the relaxation time $\tau_f(T)$ is well described by the Arrhenius law and, apparently, is determined by longitudinal polarization fluctuations; at $T < T_{VF}$, the behavior of $\tau_f(T)$ is also thermally activated, with different activation energy. The temperature dependence of the relaxation time of the narrow component QELS $\tau_s(T)$ is determined by two mechanisms: a thermally activated process with deceleration at low temperatures and critical deceleration with a maximum in the vicinity of T^*. We cannot compare the activation energy and the characteristic attempted time τ_0, which determine the behavior of the relaxation time in the frequency range from 0.9 to 11 GHz, with the data available in the literature since they are absent.

Special attention should be paid to the anomalous behavior of the Brillouin spectra of PMN in the vicinity of T^*. Indeed, in the vicinity of 380 K, a "masked" maximum in the temperature behavior of the integrated intensity of the narrow and broad components of QELS can be observed, along with the critical behavior of the relaxation time $\tau_s(T)$. A possible explanation for these anomalies might be an additional narrow component in QELS, which appears in the vicinity of 400 K. This assumption is consistent with the results of studies strictly into the elastic scattering in inelastic neutron scattering [5,24,32], where an increase in the intensity of this scattering is observed in the considered temperature range. In the present experiments, the width of this central peak is beyond our resolution.

Another important result of this study can be mentioned here. In the Brillouin scattering studies of partially disordered crystals, it is necessary to analyze the combined scattering spectra with different frequency ranges. This makes it possible to focus on the areas of study which are going to be investigated further.

Author Contributions: Conceptualization, S.G.L. and S.K.; methodology, S.G.L., S.K. and J.-H.K.; software, N.K.D.; validation, S.G.L., S.K. and J.-H.K.; formal analysis, N.K.D. and A.I.F.; investigation, S.G.L. and A.I.F.; resources, S.G.L. and S.K.; data curation, S.G.L.; writing—original draft preparation, S.G.L.; writing—review and editing S.G.L., S.K. and J.-H.K.; visualization, N.K.D.; supervision, S.G.L.; project administration, S.G.L.; funding acquisition, S.G.L. and S.K. All authors have read and agreed to the published version of the manuscript.

Funding: This work was supported in part by the National Research Foundation of Korea (NRF) grant funded by the Korean Government (MSIT) (no. 2020R1A2C101083111).

Institutional Review Board Statement: Not applicable.

Informed Consent Statement: Not applicable.

Data Availability Statement: Not applicable.

Acknowledgments: One of the authors (S.G.L.) is thankful to the JSPS long-term Fellow Program for his stay at the University of Tsukuba.

Conflicts of Interest: The authors declare no conflict of interest.

References

1. Fleury, P.A.; Lyons, K.B. Central peaks near structural phase transitions. In *Modern Problems in Condensed Matter Sciences*; Cummins, H.Z., Levanyuk, A.P., Eds.; Elsevier: Amsterdam, The Netherlands, 1983; Volume 5, pp. 449–502.
2. Dasgupta, R.; Ballabh, T.K.; Tarafdar, S. Quasielastic incoherent scattering in fractal systems. *Phys. Rev. B* **1995**, *52*, 10021. [CrossRef]
3. Wiedersich, J.; Adichtchev, S.V.; Rössler, E. Spectral Shape of Relaxations in Silica Glass. *Phys. Rev. Lett.* **2000**, *84*, 2718. [CrossRef] [PubMed]
4. Kojima, S. 100th Anniversary of Brillouin Scattering: Impact on Materials Science. *Materials* **2022**, *15*, 3518. [CrossRef] [PubMed]
5. Cowley, R.A.; Gvasaliya, S.N.; Lushnikov, S.G.; Roessli, B.; Rotaru, G.M. Relaxing with relaxors: A review of relaxor ferroelectrics. *Adv. Phys.* **2011**, *60*, 229–327. [CrossRef]
6. Cros, L.E. Relaxor ferroelectrics. *Ferroelectrics* **1987**, *76*, 241–267. [CrossRef]
7. Siny, I.G.; Lushnikov, S.G.; Katiyar, R.S.; Rogacheva, E.A. Central peak in light scattering from the relaxor ferroelectric $PbMg_{1/3}Nb_{2/3}O_3$. *Phys. Rev. B* **1997**, *56*, 7962. [CrossRef]
8. Kojima, S.; Aharé, M.; Sivasubramanian, V.; Bokov, A.A.; Ye, Z.-G. Precursor dynamics of $Pb(B_{1/2}B'_{1/2})O_3$-type relaxor ferroelectrics studied by broadband micro-brillouin scattering. *J. Adv. Diel.* **2012**, *2*, 1241004. [CrossRef]
9. Kamba, S. Soft-mode spectroscopy of ferroelectrics and multiferroics: A review. *APL Mater.* **2021**, *9*, 020704. [CrossRef]
10. Lushnikov, S.G.; Jiang, F.M.; Kojima, S. Central peak in the vibrational spectrum of the relaxor ferroelectric lead scandotantalate. *Solid State Commun.* **2002**, *122*, 129–133. [CrossRef]
11. Koreeda, A.; Taniguchi, H.; Saikan, S.; Itoh, M. Fractal dynamics in a single crystal of a relaxor ferroelectric. *Phys. Rev. Lett.* **2012**, *109*, 197601. [CrossRef]
12. Tsukada, S.; Kojima, S. Broadband light scattering of two relaxation processes in relaxor ferroelectric $0.93Pb(Zn_{1/3}Nb_{2/3})O_3$-$0.07PbTiO_3$ single crystals. *Phys. Rev. B* **2008**, *78*, 144106. [CrossRef]
13. Helal, M.; Aftabuzzaman, M.; Tsukada, S.; Kojima, S. Role of polar nanoregions with weak random fields in Pb-based perovskite ferroelectrics. *Sci. Rep.* **2017**, *7*, 44448. [CrossRef] [PubMed]
14. Bokov, A.A.; Ye, Z.-G. Dielectric relaxation in relaxor ferroelectrics. *J. Adv. Diel.* **2012**, *2*, 1241010. [CrossRef]
15. Hlinka, J. Do we need the ether of polar nanoregions? *J. Adv. Diel.* **2012**, *2*, 1241006. [CrossRef]
16. Eremenko, M.; Krayzman, V.; Bosak APlayford, H.Y.; Chapman, K.W.; Woicik, J.C.; Ravel, B.; Levin, I. Local atomic order and hierarchical polar nanoregions in a classical relaxor ferroelectric. *Nat. Commun.* **2019**, *10*, 2728. [CrossRef]
17. Fedoseev, A.I.; Popova, E.A.; Syrnikov, P.P.; Kojima, S.; Lushnikov, S.G. Multi-component quasi-elastic light scattering in $Na_{1/2}Bi_{1/2}TiO_3$ as studied by broadband Brillouin scattering. *JETP Lett.* **2015**, *102*, 789–795. [CrossRef]
18. Gvasaliya, S.N.; Lushnikov, S.G.; Sashin, I.L.; Siny, I.G. Fractons in vibrational spectrum of the $PbMg_{1/3}Nb_{2/3}O_3$ relaxor ferroelectric. *Cryst. Rep.* **1999**, *44*, 250–254.
19. Gvasaliya, S.N.; Lushnikov, S.G.; Moriya, Y.; Kawaji, H.; Atake, T.; Smirnov, M.B.; Kazimirov, V.Y. Specific Heat of Cubic Relaxor Ferroelectrics. *J Phys. Condens. Matter.* **2004**, *16*, 8981–8990. [CrossRef]
20. Nuzhnyy, D.; Petzelt, J.; Bovtun, V.; Kempa, M.; Kamba, S.; Hlinka, J. Infrared, terahertz, and microwave spectroscopy of the soft and central modes in $Pb(Mg_{1/3}Nb_{2/3})O_3$. *Phys. Rev. B* **2017**, *96*, 174113. [CrossRef]
21. Mathan, N.; Husson, E.; Calvarn, G.; Gavarri, J.R.; Hewat, A.W.; Morell, A. A structural model for the relaxor $Pb(Mg_{1/3}Nb_{2/3})O_3$ at 5 K. *J. Phys. Cond. Matter.* **1991**, *3*, 8159–8171. [CrossRef]

22. Siny, I.G.; Smirnova, T.A. Preceding paraphases in "diffuse transition" ferroelectrics. *Ferroelectrics* **1989**, *90*, 191–194. [CrossRef]
23. Burns, G.; Scott, B.A. Index of refraction in 'dirty' displacive ferroelectrics. *Solid State Commun.* **1973**, *13*, 423–426. [CrossRef]
24. Gvasaliya, S.N.; Roessli, B.; Cowley, R.A.; Huber, P.; Lushnikov, S.G. Quasi-elastic scattering, random fields and phonon-coupling effects in $PbMg_{1/3}Nb_{2/3}O_3$. *J. Phys. Condens. Matter* **2005**, *17*, 4343–4359. [CrossRef]
25. Dkhil, B.; Gemeiner, P.; Al-Barakaty, A.; Bellaiche, L.; Dul'kin, E.; Mojaev, E.; Roth, M. Intermediate temperature scale T^* in lead-based relaxor systems. *Phys. Rev. B* **2009**, *80*, 064103. [CrossRef]
26. Lushnikov, S.G.; Fedoseev, A.I.; Gvasaliya, S.N.; Kojima, S. Anomalous dispersion of the elastic constants at the phase transformation of the $PbMg_{1/3}Nb_{2/3}O_3$ relaxor ferroelectrics. *Phys. Rev. B* **2008**, *77*, 104122. [CrossRef]
27. Grigalaitis, R.; Banys, J.; Kania, A.; Slodczyk, A. Distribution of relaxation times in PMN single crystal. *J. Phys. IV France* **2005**, *128*, 127–131. [CrossRef]
28. Struzhkin, V.V.; Goncharov, A.F.; Hemley, R.J.; Mao, H.-K. Cascading Fermi resonances and the soft mode in dense ice. *Phys. Rev. Lett.* **1997**, *78*, 4446–4449. [CrossRef]
29. Rahaman, M.M.; TImai, T.; Sakamoto, T.; Tsukada, S.; Kojima, S. Fano resonance of Li-doped $KTa_{1-x}Nb_xO_3$ single crystals studied by Raman scattering. *Sci. Rep.* **2016**, *6*, 23898. [CrossRef]
30. Bovtun, V.; Kamba, S.; Pashkin, A.; Savinov, M.; Samoukhina, P.; Petzelt, J.; Bykov, I.P.; Glinchuk, M.D. Central-peak components and polar soft mode in relaxor $PbMg_{1/3}Nb_{2/3}O_3$ crystals. *Ferroelectrics* **2004**, *298*, 23–30. [CrossRef]
31. Cowley, R.A. Soft modes and structural phase transitions. *Integr. Ferroelectr. Int. J.* **2012**, *133*, 109–117. [CrossRef]
32. Gehring, P.M.; Hiraka, H.; Stock, C.; Lee, S.-H.; Chen, W.; Ye, Z.-G.; Vakhrushev, S.B.; Chowdhuri, Z. Reassessment of the Burns temperature and its relationship to the diffuse scattering, lattice dynamics, and thermal expansion in relaxor $Pb(Mg_{1/3}Nb_{2/3})O_3$. *Phys. Rev. B* **2009**, *79*, 224109. [CrossRef]

materials

MDPI

Review

Brillouin Scattering Study of Electro-Optic $KTa_{1-x}Nb_xO_3$ Crystals

Md. Mijanur Rahaman [1,2,*] and Seiji Kojima [2]

1 Department of Materials Science and Engineering, University of Rajshahi, Rajshahi 6205, Bangladesh
2 Graduate School of Pure and Applied Sciences, University of Tsukuba, Tsukuba 305-8573, Japan
* Correspondence: mijan_mse@ru.ac.bd; Tel.: +88-017-2817-1446

Abstract: The functionality enhancement of ferroelectrics by local polar clusters called polar nanoregions (PNRs) is one of the current interests in materials science. $KTa_{1-x}Nb_xO_3$ (KTN) with perovskite structure is a well-known electro-optic crystal with a large Kerr effect. The existence of PNRs in relaxor-like ferroelectric Nb-rich KTN with homovalent B-site cations is controversial. This paper reviews recent progress in understanding precursor dynamics in Nb-rich KTN crystals studied using Brillouin scattering. The intense central peak (CP) and significant softening of sound velocity are observed above the Curie temperature (T_C) due to the polarization fluctuations in PNRs. The effects of Li-doping, defects, and electric fields on the growth and/or creation of PNRs are found using changes in acoustic properties. The electric-field-induced T_C, which is shifted to higher values with increases in applied voltage, including critical endpoint (CEP) and field gradient by trapped electrons, are discussed as well. This new knowledge may give new insight into advanced functionality in perovskite ferroelectrics.

Keywords: perovskite; ferroelectric; polar nanoregions; random fields; macro- and nanodomains; Brillouin scattering

1. Introduction

Lead-based ferroelectrics have been widely utilized in our daily lives owing to their immense piezoelectricity and giant dielectric response; however, lead has adverse environmental effects which limit the future applicability of these materials. Therefore, it is urgent to develop lead-free ferroelectric materials with excellent functionality for materials science and engineering. The role of local polar clusters called polar nanoregions (PNRs) is crucial to enhance the functionality of lead-free ferroelectric materials [1]. The oxide ferroelectrics with ABO_3-type perovskite structure are one of the most interesting research fields to explore the role of PNRs in relation to relaxor ferroelectrics (RFEs) and their phase transitions. Amid RFEs, the lead-free relaxor-like ferroelectric potassium tantalate niobate ($KTa_{1-x}Nb_xO_3$, KTN) with no charge disorder by homovalent occupancy of B-site cations can be a suitable candidate for investigating dynamic anomalies related to PNRs, because of its structural simplicity in comparison with lead-based RFEs such as $Pb(Mg_{1/3}Nb_{2/3})O_3$ (PMN) and $Pb(Zn_{1/3}Nb_{2/3})O_3$ (PZN) with charge disorder by heterovalent occupancy of B-site cations. KTN is the solid solution of quantum paraelectric potassium tantalate ($KTaO_3$, KTO) and classical ferroelectric potassium niobate ($KNbO_3$, KNO). The ferroelectric phase transition of one of the end member KTO at a very low temperature is suppressed by quantum fluctuations [2,3], whereas another end member KNO displays normal ferroelectric nature and undergoes successive phase transitions from the rhombohedral (R) to orthorhombic (O), tetragonal (T), and cubic (C) phases on heating from a low temperature [4,5]. Moreover, the ferroelectric phase transitions and the associated optical properties of KTN can be customized by changing the Nb content [6,7]. In addition, KTN has drawn much scientific attention because of its excellent quadratic electro-optic (EO) coefficient and large photorefractive effect in the vicinity of the Curie temperature (T_C) [8–11]. Different

Citation: Rahaman, M.M.; Kojima, S. Brillouin Scattering Study of Electro-Optic $KTa_{1-x}Nb_xO_3$ Crystals. *Materials* **2023**, *16*, 652. https://doi.org/10.3390/ma16020652

Academic Editor: Ioana Pintilie

Received: 28 October 2022
Revised: 28 November 2022
Accepted: 4 January 2023
Published: 9 January 2023

types of optical devices such as varifocal lenses, modulators, deflectors, and EO switches based on KTN crystals have been developed because of their giant EO effect [12–15].

RFEs are delineated by the random fields (RFs) that induce the remarkable frequency dispersion and the broad peak in the temperature dependence of the dielectric permittivity. It is well established that such peculiar properties are attributed to RFs related to PNRs [16]. The relaxor nature of ferroelectrics stems from the structural disorder [17,18], which attributed to the random occupation of various types of ions (e.g., Mg^{2+} and Nb^{5+} in PMN; Zn^{2+} and Nb^{5+} in PZN) on a crystallographically equivalent site. Their heterovalent random occupancy is a common source of structural disorder of RFEs [17,18], which can lead to the creation of chemical order regions (CORs) and RFs associated with PNRs. The most important feature of KTN is the off-center displacement of Nb ions that occurs at the B-site and its microscopic origin was rationalized by the pseudo-Jahn–Teller effect (PJTE) [19]. Therefore, the formation of PNRs in KTN stems from the off-center displacements of Nb ions at the B-site, which lead to the relaxor-like nature. On the other hand, the ordered regions of KTN are distinct from those of PMN and PZN, because there is no COR in KTN that induces the diffusive phase transition of a Pb-based RFE with the heterovalent cations at the B-site [20]. Using X-ray absorption fine-structure (XAFS) measurements, the off-center displacements of Nb ions at the B-site of KTN were precisely determined by Hanske-Petitpierre et al. [21]. The off-centering of Nb ions in KTN was also discussed with the eight-site model where Nb ions displace along the eight equivalent [111] directions [22]. The study of acoustic emission of a high-quality KTN (x = 0.32, KTN/0.32) single crystal revealed its relaxor-like nature by observing two characteristic temperatures, namely the Burns temperature, T_B~620 K, and the intermediate temperature, T^*~310 K [23]. Such a relaxor-like nature was also found in normal ferroelectric $BaTiO_3$ [24]. The distinct types of experimental studies such as Brillouin scattering [25–27], Raman scattering [28–32], dielectric spectroscopy [33], refractive index, linear birefringence [34], an ultrasonic pulse-echo technique [35], and infrared absorption [36] revealed the presence of PNRs in a paraelectric cubic phase of non-doped KTN and Li-doped KTN crystals. However, the dynamic behavior and microscopic origin of PNRs still have a long way to apprehend.

Nowadays, the study of Li-doping to KTO and KNO is one of the active research areas in materials science and condensed matter physics. The ferroelectricity appeared and quantum paraelectricity was suppressed with Li-doping to KTO. The substitutional Li ions in $K_{1-y}Li_yNbO_3$ (KLN) and $K_{1-z}Li_zTiO_3$ (KLT) take up one of six off-center sites along the [100] directions at the A-site [37,38], which can lead to the creation of RFs that enhance the appearance of PNRs. However, there is a phase boundary of KLT at about z = 0.022, below which KLT freezes into a dipole glass state at a glass transition temperature, T_g, and above which it undergoes a ferroelectric phase transition at the T_C [37]. Moreover, the KLT/0.026 (z = 0.026) showed a relatively broad phase transition, while KLT/0.063 (z = 0.063) exhibited a sharp phase transition under a moderate electric field along the [100] direction [39]. Zhou et al. [40] reported that the electromechanical coupling constant, k_{31}, and the piezoelectric coefficient, d_{33}, were enhanced up to 40.4% and 431 pC/N, respectively, by Li-doping to KTN. Li et al. [41] also studied the shear components of the piezoelectric response, d_{31}, and the pyroelectric coefficient of high-quality Li-doped KTN single crystals to apply these materials in pyroelectric and piezoelectric applications such as actuators, sensors, and transducers. Therefore, the studies of electric field and Li-doping effects on the precursor dynamics in the vicinity of the ferroelectric phase transition temperature are crucial for accelerating the understanding and application of these materials.

It is also significant that the structure and physical properties of non-doped KTN and Li-doped KTN are strongly dependent on their chemical compositions [6,7]. Among the various physical properties, the elastic stiffness constants, C_{ij} (i,j = 1~6), which are closely related to the phase velocities of acoustic phonons at the Brillouin zone center and inter-atomic potential, play a vital role in materials science. The ultrasonic pulse-echo technique revealed that the elastic constants gradually decrease with an increase in Nb content [35]. Similar results were also found in KTN ($0.003 \leq x \leq 0.057$) studied by Rytz et al. [42]. Using Brillouin scattering, the change of elastic constants was found in the case of $Pb[(Zn_{1/3}Nb_{2/3})_xTi_{1-x}]O_3$ ($x = 0.09, 0.045$) [43]. In the cubic phase of KTN, the investigation of temperature-dependent elastic constants denoted a strong coupling between PNRs and acoustic phonons [27,35,44], but the composition gradient of these physical constants is rather scarce. The value of T_C in the vicinity of room temperature (RT) is desirable for integrated optics; thus, it is important to investigate the composition gradient of KTN in the range where T_C is about RT. Additionally, defects significantly affect the elastic properties of materials. A theoretical study of the defect-dependent elastic properties of silicon revealed that both vacancy and interstitial defects, random and singly in combination, decrease the elastic constants C_{44} and C_{11} [45]. Burnett and Briggs also observed the decrease in elastic constants C_{11} and C_{44} of silicon with the increase in defects [46]. To the best of our knowledge, the effects of defects on the elastic properties of high-quality non-doped KTN and Li-doped KTN are rare.

Recently, space-charged controlled electro-optic (EO) devices such as optical beam deflectors, which are applied in light detection and ranging (LiDAR) systems and laser displays, are interesting topics in applied physics. The space charge in KTN crystals is formed due to the injection of electrons from the cathode with the application of a voltage to the electrode pairs [15,47]. The phenomenon of optical beam deflection by space charge was found in KTN crystals [15,47]. Thus, the space charge creates the distribution of electric fields at first; after that, the refractive index distribution occurs because of the EO effect. This refractive index distribution bends a light beam, and the deflector may achieve deflection angles exceeding several degrees, which is much higher than those observed from the conventional EO effect [48]. Interestingly, the injected electrons are stabilized in KTN crystals trapped by the localized states. Nakamura et al. [15] found the effect of the space charge on the optical beam deflection and observed that the electric fields induced by space charges are not uniform inside a KTN crystal. On the contrary, the injected electrons remarkably increase the capacitance and permittivity of a Li-doped KTN crystal [49]. It was reported that the critical endpoint (CEP) is the driving mechanism of polarization rotation and is associated with colossal electromechanical response in RFEs. The electric-field–temperature phase diagram including the CEP was reported for Li-doped KTN crystals without electron injection [49]. However, the investigation of the effects of injected and/or trapped electrons on the PNRs and CEP of non-doped KTN and Li-doped KTN crystals is deficient, which is crucial for application to electromechanical devices.

In the present study, we review the recent extensive studies on precursor dynamics in KTN crystals using inelastic light scattering spectroscopy. The effects of Li-doping, defects, composition gradient, and injected electrons on PNRs of KTN crystals are discussed as well. The acquired knowledge, which provides new insights into the various dynamics of PNRs, may help to enhance the functionality of ferroelectric materials.

2. Methodology: Micro-Brillouin Scattering

The elastic properties, including different types of elastic moduli and acoustic absorption coefficients, are one of the fundamental properties of a material. The elastic constants (C_{ij}) of a material, which are directly related to the interatomic forces in the crystal, can be determined using the velocity of the sound wave related to the acoustic phonons and density of the material. There are different techniques to estimate the value of elastic constants. Among them, the mechanical resonance and the ultrasonic pulse-echo techniques have been well established [35,50,51]. However, Brillouin spectroscopy, which covers the range of 1~1000 GHz frequency, is a powerful, non-destructive, non-contact tool to measure elastic constants under various external conditions such as temperature, pressure, and electric fields. Brillouin scattering has been utilized to study not only acoustic phonons but also the low-frequency optical soft mode [52], magnon [53], fracton [54], and soliton [55].

The optical arrangement of a typical Brillouin scattering experiment, which was developed by Sandercock [56], uses a combination of two Fabray–Pérot interferometers (FP1 and FP2) (JRS TEP-1, JRS, Zwillikon, Switzerland) with slightly different free spectral ranges as shown in Figure 1. Laser light is applied to the optical microscope with the help of a small mirror to focus the samples. The scattered light is gathered using the objective lens and entered into the FP1 through the aperture A1 and adjustable pinhole P1. The mirror M1 reflects the light toward the lens L1, which redirects it via mirror M2 to FP1 after the collimation. The deflected light from M2 is then passed through puncture 1 of the mask A2 and is directed to FP2 via mirror M3. The light hits the 90° prism PR1 after transmission through FP2, is reflected downwards, goes back parallel to itself towards FP2, and continues to FP1 through aperture 2 of A2. The light is passed through the lens L1 and underneath the mirror M1 after transmission through FP1 and is focused onto the mirror M4. Mirror M4 redirects the light to FP1 through lens L1 and mirror M2. The combination of mirror M4 and lens L1 lying at its focus is known as catseye and acts as a spatial filter that removes undesirable beams such as the beams reflected from the rear surfaces of the interferometer mirrors. The light is directed to the prism PR2 via the mirror M5 after the final pass through the interferometers. The combination of lens L2, prism PR2, and the output pinhole P2 forms a bandpass filter with a width estimated using the size of the pinhole. The mirror M6 sends the light onto the photomultiplier through the lens L2, aperture A3, and output pinhole P2. Finally, the scattered signal is collected using a conventional photon counting system.

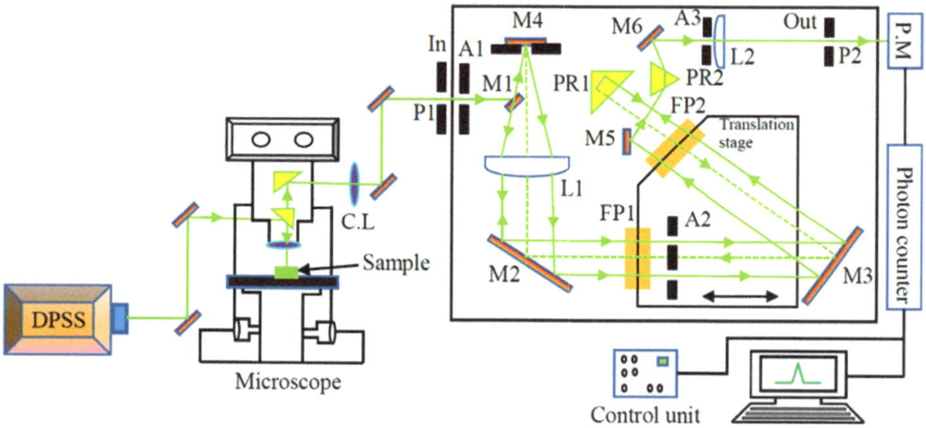

Figure 1. A schematic illustration of a Sandercock 3 + 3 pass tandem interferometer equipped with a photomultiplier and optical microscope [57,58].

The micro-Brillouin scattering spectra of KTN and Li-doped KTN single crystals were measured using a 3 + 3 pass Sandercock-type tandem Fabray–Pérot interferometer (JRS, TEP-1) equipped with a photon counting system. The KTN and Li-doped KTN single crystals grown using the top-seeded solution growth (TSSG) method at NTT Corporation (Tokyo, Japan) were cut into different dimensions. The largest (100) faces were polished to optical grade to obtain inelastic light scattering spectra. All the Brillouin scattering spectra were recorded at a backward scattering geometry [47]. The Brillouin scattering was excited with a wavelength of 532 nm and a power of about 100 mW from a single-frequency diode pumped solid state (DPSS) laser. The free spectral range (FSR) of the spectrometer was fixed at 300 GHz for the central peak (CP) and 75 GHz for the longitudinal acoustic (LA) and transverse acoustic (TA) mode measurements. The temperature of the sample was controlled with a heating and/or cooling stage (THMS600, Linkam, Salfords, UK,) with the temperature stability of ± 0.1 K overall temperature. For applying the electric field, the appropriate surfaces of the sample were coated with silver paste, and gold contact wires were attached to the electrodes.

3. Evolution of PNRs: Models

In a paraelectric cubic phase of RFEs, the PNRs are defined by two temperatures, namely the Burns temperature, T_B [59], below which the dynamic PNRs begin to appear, and the intermediate temperature, T^*, at which the dynamic PNRs start to transform into static and/or rapid growth of these PNRs up to the percolation limit in which the reorientation is completely suppressed [25–27,30,60,61]. It is believed that the formation of dynamic PNRs begins at T_B owing to a short-lived correlation among off-center ions, whereas the dynamic-to-static transition of these PNRs begins at T^* with the sudden growth in size due to the long-lived correlation among the atomic displacements of off-center ions. However, the physical origin and mechanism of the formation PNRs and their dynamic behavior are still a long way from our understanding.

There are various approaches to understanding the formation of PNRs [25–27,60–69]. All of them can be separated into two models suggested by A.A. Bokov and Z.-G. Ye [16]. One model considers that PNRs are a result of local phase fluctuations or local phase transitions, so that the crystal consists of nanosized polar clusters embedded in a nonpolar paraelectric cubic phase in which the symmetry of the lattice remains unchanged [62–65]. The second model considers the local transition to occur in all the crystal regions containing low-symmetry nanodomains detached not by the regions of cubic symmetry but by the domain walls [66,67]. They also reported that the symmetry of local clusters in the cubic matrix is not expected to be cubic and the domain wall thickness is comparable with the size of nanodomains [16]. However, they did not provide information about the local symmetry of PNRs. Without defining the local symmetry of polar clusters in RFEs, J. Toulouse [68] suggested that the static PNRs' so-called polar nanodomains (PNDs) appear at T^* upon cooling. He also suggested that PNDs (static PNRs) represent the permanent polarization appearing with local strain fields as PNDs do not mean that they are "frozen" or inert, but their reorientation may contribute to the relaxor nature below T^* in RFEs. The reorientation and vibrational-field-driven dynamics of PNRs in the vicinity of the T_C of relaxor-like KTN crystal were studied by Tan et al. [70]. The neutron pair distribution function analysis revealed local rhombohedral *R3m* symmetry of the PNRs of a PMN single crystal [71]. Taniguchi et al. [72] reported the local cubic $Fm\bar{3}m$ symmetry for the COR and rhombohedral *R3m* symmetry for the PNRs in the paraelectric cubic phase of PMN single crystal using Raman scattering. They also reported that the local symmetry breaking due to the existence of PNRs in the cubic phase of the PMN with heterovalent B-site cations stems from the $E(x,y)$ mode of rhombohedral *R3m* symmetry [72]. Recently, Helal et al. [69] reported that the PNRs of PMN-0.56PT ($x = 0.56$) with weak RFs belong to the local tetragonal symmetry with the point group *4mm*, which is distinct from local rhombohedral *R3m* symmetry of PMN with strong RFs.

Recently, Dey et al. [73] visualized randomly oriented nanodomains within macro-domain states in the ferroelectric phase of the $(1-x)Ba(Zr_{0.2}Ti_{0.8})O_3 - x(Ba_{0.7}Ca_{0.3})TiO_3$ (BZT-*x*BCT, $x = 0.60$) ceramics with homovalent cations at both the A- and B-sites. In the ferroelectric phases of the BZT-*x*BCT ceramics, the randomly oriented nanodomains within macro-domain states were also found by observing the prominent CP in Brillouin scattering spectra, which is related to the precursor's dynamics [74]. Such a type of prominent broad CP, which was attributed to random nanodomain states, was observed in Co-doped BTO ceramics as well [75]. It is well established that the dynamic-to-static transition of PNRs starts at the intermediate temperature, T^*, and most of the PNRs become static in the vicinity of T_C. Thus, it is believed that the static PNRs become randomly oriented nanodomain states due to the freezing out of local polarization in a ferroelectric phase [74,75]. The temperature evolution of PNRs in RFEs with heterovalent cations at the B-site is well explained, but extensive studies on PNRs in relaxor-like ferroelectrics with homovalent cations at B-sites such as KTN are rather scarce. The microscopic scenario of PNRs in KTN may differ from Pb-based RFEs because the PNRs in KTN arise only for the off-centering of Nb ions at the B-site. In the present investigation, the microscopic scenario of PNRs with local structural symmetry including vibrational symmetry in relaxor-like KTN crystals is discussed using a model, as shown in Figure 2, based on the results of inelastic light scattering experiments.

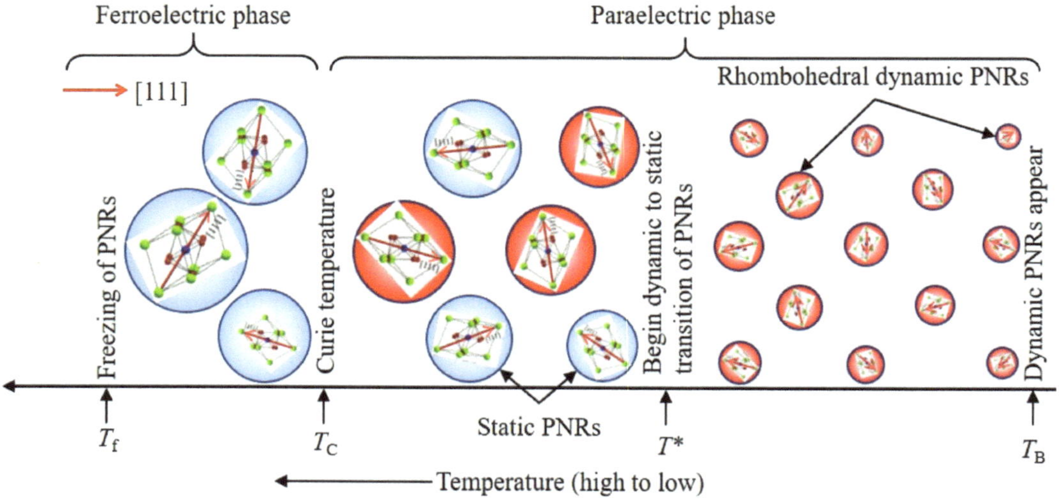

Figure 2. A schematic illustration of the temperature evolution of PNRs in relaxor-like KTN crystals based on the inelastic light scattering results. Red denotes the dynamic PNRs and blue indicates static PNRs, where the arrow denotes the orientation of PNRs. The dynamic PNRs appear at the so-called Burns temperature, T_B. Upon cooling, the size of the dynamic PNRs increases slowly and dynamic PNRs become more sluggish due to the cooperative interaction with the neighboring PNRs. Eventually, the dynamic PNRs start to become static PNRs at the intermediate temperature, T^*. For further cooling, the size of PNRs becomes intense due to their growth in addition to the merger of neighboring PNRs.

3.1. Experimental Evidence of PNRs

The analysis of temperature dependencies of the refractive index, n, which is linear at $T > T_B$ but starts to deviate from linearity at $T < T_B$, was the first experimental evidence for PNRs [59]. The deviation of n from linearity below T_B was attributed to the variation in n induced by local polarization inside the PNRs via the quadratic EO effect. In RFE PMN, the presence of PNRs was confirmed using synchrotron X-ray and neutron diffuse scattering about the reciprocal lattice points [76,77]. High-resolution neutron elastic diffusion scattering revealed the size of emerging PNRs at about 1.5 nm, which was almost temperature-independent at high temperatures; smaller PNRs reach around 7 nm on cooling [78]. The size of the emergence of PNRs in PMN below T_B was also observed using transmission electron microscopy (TEM) [79], but their size was larger than that determined from neutron diffuse scattering. The temperature-dependent dielectric properties of PMN were extensively studied to characterize the dielectric relaxation related to PNRs as well [80,81]. However, it is very difficult to record the broadband spectra between 1 to 1000 GHz using conventional dielectric spectroscopic measurements. On the other hand, Brillouin spectroscopy enables the acquisition of such broad spectra with a single scan. Brillouin scattering spectroscopy is a powerful tool to observe the dynamics of local polarization fluctuations in ferroelectrics because it is very sensitive to local heterogeneities associated with compositional and structural disorder among a few techniques that can be used to directly probe the local symmetry and structure. In addition, Brillouin scattering has a spatial resolution comparable to the wavelength of an exciting laser, and it is possible to measure the optic soft mode and acoustic phonons in different types of perovskite ferroelectrics.

3.2. PNRs in Relaxor-like Ferroelectric KTN

KTN crystals undergo a successive phase transition of a rhombohedral–orthorhombic–tetragonal–cubic sequence the same as that of $KNbO_3$ (x = 1.00) for solid solutions of $x > 0.05$ [6,7] on heating, and the phase transition's temperature can be customized with the Nb content. However, KTN crystals with x = 0.008–0.05 undergo only a cubic–rhombohedral phase. Samara [82] reported relaxational glass-like behavior rather than a ferroelectric phase transition for KTN with a small Nb content ($x < 0.02$). Samara also found that, unless subject to high pressure, KTN crystal with x = 0.02 did not show any frequency dispersion [82]. On the other hand, refractive index and linear birefringence measurements revealed the diffuse ferroelectric phase transition of KTN/0.02 crystal [34]. The structural phase transition and the existence of PNRs well above T_C of KTN/0.012 crystal were studied using dielectric, polarization, and Raman measurements [28], which led to the relaxor-like nature of the KTN crystals. The presence of PNRs in KTN/0.09 crystal was also investigated using Raman scattering as well [83]. Using the ultrasonic pulse-echo technique, Knauss et al. [35] found a reduced softening of elastic constants in the vicinity of T_C in KTN crystals with x = 0.012, 0.03, 0.09, and 0.157, which was accompanied by local strain fields related to PNRs that could couple to the sound waves, resulting in a reduced softening of the elastic constants. The relaxor-like nature of a high-quality KTN/0.32 single crystal was seen by observing Burns and intermediate temperatures associated with PNRs at T_B~620 K and T^*~310 K, respectively, using acoustic emission measurement [23]. The relaxor-like dynamics near T_C of the KTN/0.32 crystal were also investigated using Brillouin scattering [26]. Hence, there is noticeable controversy on the optimum composition at which KTN crystals show a relaxor-to-normal crossover ferroelectric phase transition. In this study, the existence of PNRs and their dynamic nature in Li-doped and non-doped Nb-rich KTN were investigated using Brillouin scattering to clarify the elastic anomaly in the GHz range caused by the coupling of acoustic phonons with the PNRs and their slowing down.

3.2.1. Elastic Anomaly

The measurement of temperature-dependent Brillouin scattering spectra below and above T_C was carried out to clarify the elastic anomaly in the vicinity of the paraelectric to a ferroelectric phase transition. The Brillouin scattering spectra of high-quality Nb-rich KTN/0.40 single crystal at some selected temperatures measured with FSR = 75 GHz are shown in Figure 3a. As can be seen in Figure 3a, the two doublets and a CP at zero frequency shift appeared in each spectrum in a wide range of temperatures. The low- and high-frequency doublets correspond to the transverse acoustic (TA) and longitudinal acoustic (LA) modes, respectively. It is apparent from Figure 3a that the TA mode persists in the paraelectric cubic phase of the KTN/0.40 crystal, which is forbidden at a backscattering geometry by the Brillouin selection rule [84]. Therefore, the presence of the prominent TA mode in the paraelectric phase may denote the local breaking of cubic symmetry. In the cubic phase of KTN/0.40, the local symmetry breaking owing to the A_1(z) mode of PNRs with rhombohedral $R3m$ symmetry was confirmed by the intense first-order Raman mode in the cubic phase that is inactive for the cubic $Pm\bar{3}m$ symmetry by the Raman selection rule [25,85].

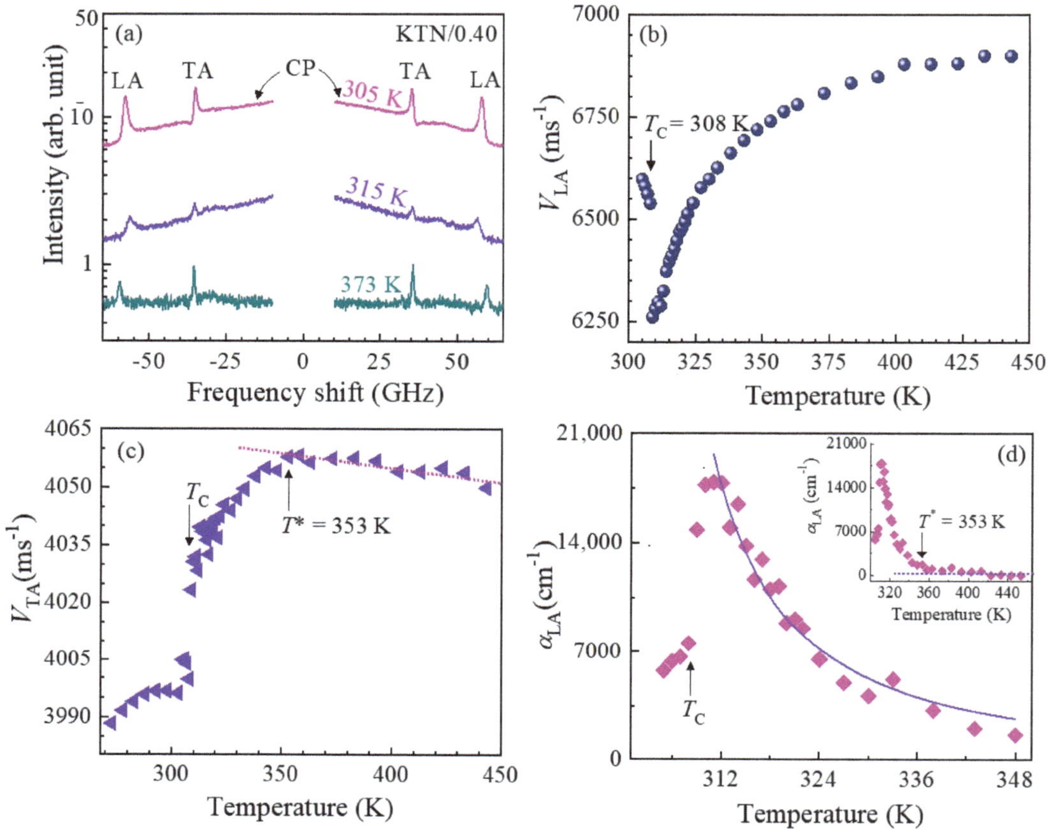

Figure 3. (**a**) The observed Brillouin spectra of non-doped KTN/0.40 single crystal at some selected temperatures measured with FSR = 75 GHz. (**b**) The longitudinal velocity of the LA mode, and (**c**) the transverse velocity of the TA mode in a KTN/0.40 single crystal as a function of temperature. (**d**) The temperature dependence of the longitudinal sound attenuation in a KTN/0.40 single crystal. The short dashed and solid lines in (**c**) and (**d**) are the guide to eyes and the best fit result obtained using Equation (7), respectively. The inset in figure (**d**) exhibits the plot on a large scale to denote the T^*. The data were adapted from Refs. [25,58].

The doublets were fitted using the Voigt function to determine the frequency shift and damping of the acoustic phonon modes. The sound velocity (V) and sound attenuation coefficient (α), which are related to the Brillouin frequency shift (v_B) and full width at half maximum (FWHM), (Γ_B), were calculated with the following equations:

$$\Delta v_B = \pm \frac{2mV}{\lambda_0} sin\left(\frac{\theta}{2}\right) \quad (1)$$

$$\alpha = \frac{\pi \Gamma_B}{V} \quad (2)$$

where m, λ_0, and θ are the refractive index of the observed sample, the wavelength of the incident light, and the scattering angle, respectively. The temperature dependencies of LA and TA velocities of the KTN/0.40 crystal were determined and plotted as shown in

Figure 3b,c, respectively. In the cubic phase, the observed LA and TA sound velocities are related to the elastic constants C_{11} and C_{44}, respectively, by the following equations [83]:

$$C_{11} = \rho V_{LA}^2 \quad (3)$$

$$C_{44} = \rho V_{TA}^2 \quad (4)$$

Here, ρ is the density of the crystal. The LA velocity exhibits a significant softening near the ferroelectric tetragonal to paraelectric cubic phase transition temperature, T_C = 308 K, while the TA velocity displays only a modest softening and a small discontinuity at T_C, denoting the behavior of a slightly first-order phase transition [25].

The anomalous changes in the acoustic properties of relaxor-like ferroelectrics can be explained using the Landau theory considering the coupling between strain and order parameters in the free energy expansion. In ferroelectrics, the Landau expression for the free energy, F, generally consists of three parts: the first part is related to the order parameter, i.e., polarization, P, in the present case; the second part is pertinent to the elastic energy; and the third part is the coupling term between strain and the order parameter. The effective elastic constant C_{ij}, owing to the variation in polarization, driven by the modification in strain, can be expressed as follows [86]:

$$C_{ij} = C_{ij}^{\infty} - \sum_{k,l} \left(\frac{\partial^2 F}{\partial x_i \partial P_k} \right) \left(\frac{\partial^2 F}{\partial P_k \partial P_l} \right)^{-1} \left(\frac{\partial^2 F}{\partial x_j \partial P_l} \right) \quad (5)$$

where C_{ij}^{∞} is the polarization-independent elastic constant, namely, the bare elastic constant in which the temperature dependence is determined by the lattice anharmonicity. There is no macroscopic spontaneous polarization, i.e., $\langle P \rangle = 0$ in the paraelectric cubic phase of relaxor-like ferroelectric KTN due to the random orientations of PNRs. Hence, we envisage the quadratic piezoelectric coupling between strain and the order parameter would dominate the coupling terms in the free energy below T_B. The variation in elastic constant due to the quadratic piezoelectric coupling can be modelled with the following expression [87]:

$$\Delta C = C - C^{\infty} = \gamma < P^2 > \chi \quad (6)$$

where γ is the piezoelectric coupling coefficient, $< P^2 >$ is the mean value of the squared local polarization of PNRs, and χ is the susceptibility for the order parameter. For the second-order phase transition, the classical Landau theory anticipates only a step-like anomaly for $C(T)$ at the phase-transition temperature. However, in relaxor-like ferroelectrics the local polarizations emerge at T_B and they steadily develop with decreasing temperature down to T_C. The appearance of local polarization, although the average polarization is zero, contributes to the decrease in the elastic constants due to the quadratic behavior of the piezoelectric interaction. The decrease in elastic constants evolves steadily without any abrupt change until the local polarizations of PNRs is saturated upon cooling. Hence, the significant softening and the clear anomaly in both the V_{TA} (related to C_{44}) and V_{LA} (related to C_{11}) towards the T_C is attributed to the coupling of the TA and LA modes with fluctuation of local polarizations in PNRs. Moreover, a clear anomaly and the remarkable increase in the LA mode damping towards T_C was seen on cooling from high temperature, as shown in the inset of Figure 3d. It is apparent from the inset of Figure 3d that the almost constant temperature-dependent sound attenuation at high temperatures starts to deviate at around 353 K, which was assigned as the intermediate temperature (T^*) at which the rapid growth of PNRs begins, which induces the significant softening of the TA mode as well (Figure 3b). Landau and Khalatnikov [88] performed a mean-field theory of the sound attenuation anomaly and found that this anomaly arises owing to the coupling between the strains and order parameters. Levanyuk et al. [89] observed the temperature-dependent sound attenuation in the following form with the critical exponent of n = 1.5

for the order—disorder nature of ferroelectric phase transition based on the mean-field approximation.

$$\alpha = \alpha_1 (T - T_0)^{-n} \quad (7)$$

where α_1 is the constant. T_0 is the characteristic temperature, in which $T_0 = T_C$ describes the second-order while $T_0 < T_C$ explains the first-order nature of the ferroelectric phase transition. The sound attenuation was fitted between T_C and T_C + 40 K using Equation (7) (Figure 3d) to explain the anomalous nature and order—disorder behavior of the phase transition. The experimental results were well reproduced by the theory with the critical exponent of 1.5, indicating the order—disorder nature of the ferroelectric phase transition of the high quality KTN/0.40 single crystal. The parameters obtained by fitting the sound attenuation are given in Ref. [23]. Moreover, the good consistency between the observed sound attenuation and the theoretical line indicates that the anomaly arises due to coupling between the LA phonons and polarization fluctuations in the PNRs. These characteristics near T_C were observed in various lead-based perovskite relaxors such as 0.65Pb($Mg_{1/3}Nb_{2/3}$)O_3–0.35PbTiO_3 (PMN–0.35PT), 0.70Pb($Sc_{1/2}Nb_{1/2}$)O_3–0.30PbTiO_3 (PSN–0.30PT), 0.85Pb($Zn_{1/3}Nb_{2/3}$)O_3–0.07PbTiO_3 (PZN–0.07PT), 0.65Pb($In_{1/2}Nb_{1/2}$)O_3–0.35PbTiO_3 (PIN–0.35PT), etc. [84–90].

3.2.2. Critical Slowing Down

The temperature-dependent broad central peak (CP) measured with FSR = 300 GHz, which is associated with the relaxation process of polarization fluctuations in PNRs, was investigated to clarify the precursor dynamics in KTN crystals. The broad CP of the KTN/0.40 crystal at some selected temperatures is shown in Figure 4a. As can be seen in Figure 4, the CP develops upon cooling and its intensity becomes a maximum at T_C. It is significant that the intensity of the CP drastically increases towards T_C below T^* as shown in Figure 4b. The increase in the CP intensity upon cooling reflects the increase in the correlation among PNRs. The anomalous changes in the CP intensity and optical Raman modes of PZN crystals revealed that T^* exists between T_B and T_C [60]. They also reported that the temperature region between T^* and T_B is characterized by the formation of short-lived correlations between the off-centered ions resulting in the formation of small dynamic PNRs, while the region between T_C and T^* is characterized by coupling between adjacent PNRs and their aggregation to form larger PNRs [60]. Such anomalous variation of optical Raman modes and CP intensity was also observed in a KTN/0.40 crystal using Raman scattering [85]. Hence, the significant increase in damping of the LA mode below T^* indicates the rapid growth of PNRs, which also induces a remarkable increase in the CP intensity and a significant softening of V_{TA} [25,26,91–94].

The phenomenon of critical slowing down is known as the order–disorder nature of a structural phase transition. The relaxation time of the polarization fluctuations related to ferroelectric phase transitions was found using broadband Brillouin scattering spectroscopy [92,93]. The relaxation time was determined using the width Γ_{CP} of the CP using the following equation [93].

$$\pi \Gamma_{CP} \tau_{CP} = 1 \quad (8)$$

where Γ_{CP} and τ_{CP} are the width of the CP and relaxation time, respectively. The inverse relaxation time as a function of temperature (Figure 4c) exhibits a linear dependence between T_C and T_C + 49 K, which denotes the fact that a paraelectric cubic to ferroelectric tetragonal phase transition is the order–disorder type. The critical slowing down near T_C is given by

$$\frac{1}{\tau} = \frac{1}{\tau_0} + \frac{1}{\tau_1}\left(\frac{T - T_C}{T_C}\right) \quad (9)$$

τ_0 and τ_1 are the relaxation time and the characteristic time, respectively [25,26,95]. The observed values of τ_0 and τ_1 are 1.61 ps 14.8 ps, respectively, in the temperature range between T_C and T_C + 49 K [25]. It is significant to compare the value of τ_0 with a typical relaxor system that showed the order–disorder nature of the ferroelectric phase transition near T_C. The observed value of τ_0 of the KTN/0.40 crystal is comparable to that of the PZN–0.15PT crystal [96]. Hence, one can conclude from these results that an order–disorder mechanism in the KTN/0.40 single crystal contributes to slowing down the dynamic PNRs.

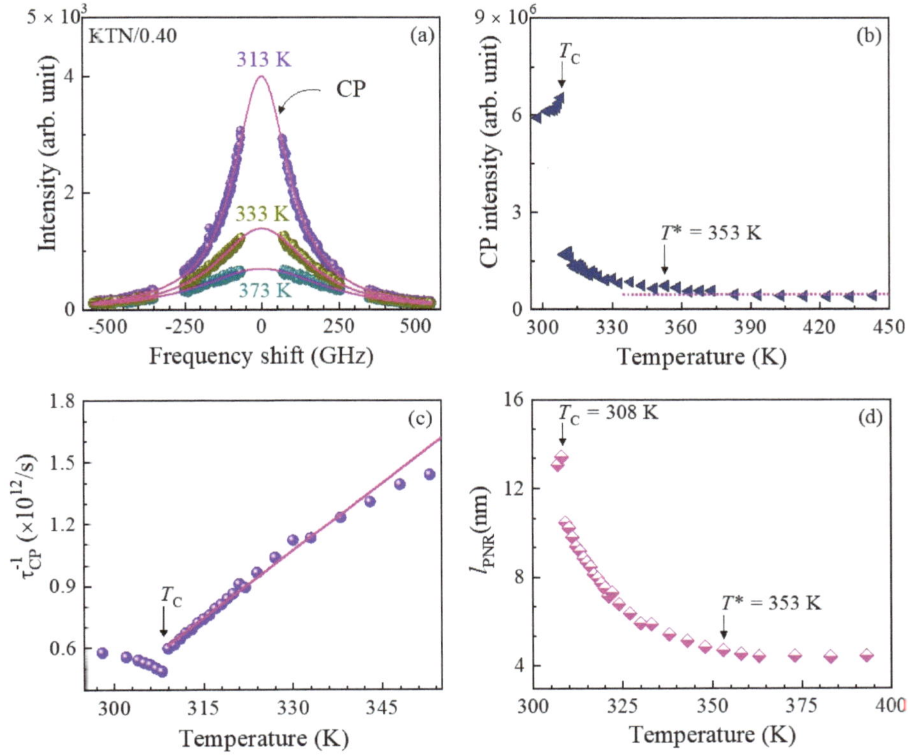

Figure 4. (**a**) Brillouin scattering spectra the KTN/0.40 crystal at selected temperatures measured with FSR = 300 GHz, where the solid lines denote the curves fitted using the Voigt function. (**b**) The CP intensity of the KTN/0.40 crystal as a function of temperature. (**c**) The temperature dependence of the inverse of the relaxation time determined from the CP width, where the solid line is the result of best fit using Equation (9). (**d**) The mean size of the dynamic PNRs of KTN/0.40 single crystal as a function of temperature. The data were adapted from Refs. [25,58].

For a better understanding of the temperature evolution of the dynamic PNRs, the average size of the dynamic PNRs, l_{PNRs}, was determined using the equation $l_{PNR} = V_{LA} \times \tau_{CP}$, because the maximum of the characteristic length of the strain fluctuations is given by the propagation length of strain within the relaxation time [97]. It is apparent that the average size of the dynamic PNRs increases towards T_C upon cooling, as shown in Figure 4d, indicating the growth of the PNRs up to the percolation limit in which the reorientation is completely suppressed [61]. The average size of PNRs is about 4.5 nm at an intermediate temperature, T^*, which is in agreement with the size of a dynamic polar cluster of 4–6 nm [26].

3.3. Li-Doping Effects on PNRs in KTN

The lattice dynamic properties related to a relaxor ferroelectric phase transition of high quality 2.5%Li-doped KTN (KLTN/0.025/0.38) and 5%Li-doped KTN (KLTN/0.05/0.27) single crystals were investigated and compared to the results of the non-doped KTN/0.40 crystals to understand the effects of Li-doping on PNRs. The observed Brillouin scattering spectra of the KLTN/0.05/0.27 single crystal at some selected temperatures are shown in Figure 5a. The Brillouin scattering spectra of the KLTN/0.05/0.27 crystal are qualitatively similar to spectra of a high-quality KTN/0.40 single crystal with no Li content. The elastic constant C_{11} was calculated from the sound velocities with Equation (3) using a density of ρ = 6630 kg/m^3 for Li-doped KTN [40] and ρ = 6078.65 kg/m^3 for non-doped KTN/0.40 [98]. The estimated elastic constant C_{11} of non-doped and 5%Li-doped KTN crystals as a function of temperature is displayed in Figure 5b. It is seen that the elastic constants of non-doped KTN are smaller than those of the Li-doped KTN crystals in the studied temperature range. However, it is found from Figure 5b that Li-doping broadens the temperature region of the elastic anomaly near the T_C. The broadening of the temperature region may be caused by the growth of PNRs induced by Li ions. Hence, the elastic anomaly broadening near T_C of Li-doped KTN crystals in comparison to that of the non-doped KTN can be an indication of the expansion of the temperature region of the elastic anomaly [27].

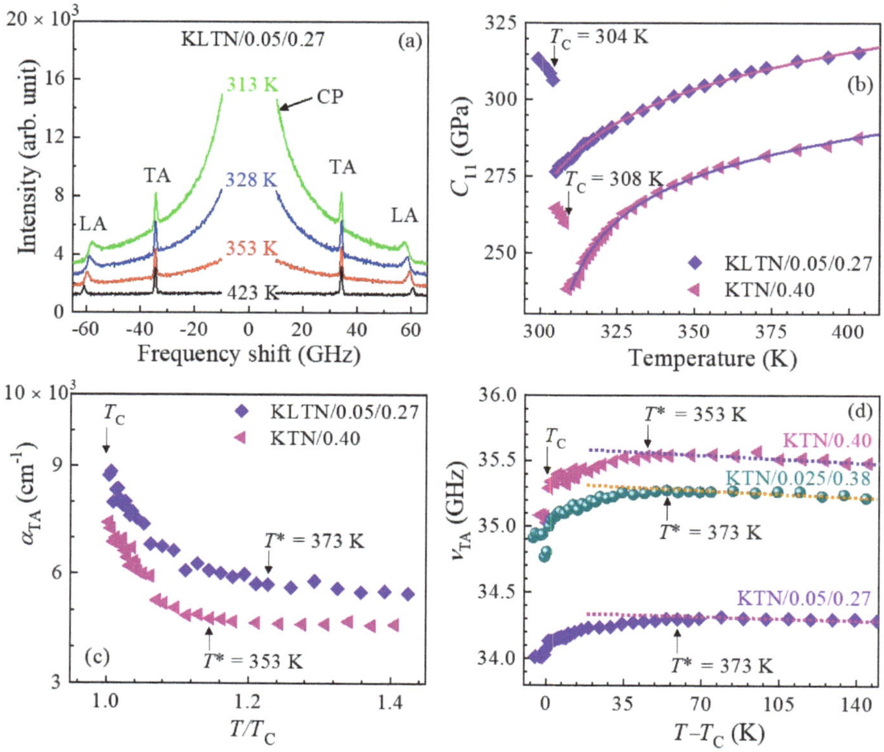

Figure 5. (**a**) Brillouin scattering spectra of the KLTN/0.05/0.27 single crystal at some selected temperatures recorded with FSR = 75 GHz. (**b**) Temperature-dependent elastic constant C_{11} of both KTN/0.40 and KLTN/0.05/0.27 single crystals. The solid lines are the best fit result obtained from Equation (11) (**c**) The attenuation of the TA mode in the KTN/0.40 and KLTN/0.05/0.27 crystals as a function of T/T_C. (**d**) The observed frequency shift of the TA mode of KTN/0.40, KLTN/0.025/0.38, and KLTN/0.05/0.27 crystals as a function of normalized temperature. The short dashed lines are guides to the eyes. The data were adapted from Refs. [27,58].

We also studied the frequency and damping of the acoustic phonons (Figure 5c,d to clarify the extension of the temperature region by Li-doping. It was challenging to make decisive data of the LA mode damping on fitting spectra using Lorentzian functions because of its asymmetric nature (Figure 5a). Thus, we concentrated on the frequency and damping of the TA mode. It is apparent from Figure 5c,d that the temperature interval between T^* and T_C was enhanced by the Li-doping. As can be seen in Figure 5c, the attenuation of the TA mode remarkably increases with Li-doping, which means the TA phonons were markedly scattered by the Li-doping, resulting in an increase in the damping and decrease in the frequency of the TA phonons. These results are a clear indication of enhanced growth of PNRs by Li-doping. Hence, the expansion of the temperature interval between T_C and T^* can be attributed to Li-ion-induced enhanced stability of PNRs. With Li-doping, the enhanced stability of PNRs persuades the elastic anomaly broadening, which can be an indication of the enhanced relaxor nature of KTN crystals with Li-doping. A comparative study of the dielectric properties among non-doped, 2.5%Li-doped, and 5%Li-doped KTN crystals was performed to confirm the enhanced relaxor nature of KTN with Li-doping. The dielectric constant follows the well-known Curie–Weiss law $1/E = (T - T_C)/C$ above the T_C in the case of normal ferroelectrics, where C is the Curie constant, whereas this law is obeyed only at temperatures higher than the T_B in the case of RFEs [79,80]. The dielectric constant below T_B in the paraelectric phase of the relaxor-like ferroelectrics can be discussed with the following equations:

$$\frac{1}{\varepsilon} = \frac{1}{\varepsilon_m}\left[1 + \frac{(T - T_m)^\gamma}{2\delta^2}\right], (T > T_m); 1 < \gamma \le 2 \tag{10}$$

where ε_m denotes the magnitude of the dielectric peak at the T_m. δ and γ are fitting parameters denoting the degree of the diffuseness of the phase transition. When $\gamma = 2$ it describes a typical relaxor property, while at $\gamma = 1$, Equation (10) is the same as the Curie–Weiss relation. In RFEs, such a peculiar property is attributed to RFs related to PNRs. It was found that the inverse of the dielectric constant of non-doped, 2.5%Li-doped, and 5%Li-doped KTN crystals is well fit by Equation (10) with the critical exponent γ of 1.12 ± 0.006, 1.19 ± 0.003, and 1.31 ± 0.005, respectively, indicating a relaxor-like nature of the non-doped KTN and the Li-doped KTN crystals [58,99]. It is significant that the values of fitting parameters gradually increase with the Li content, which confirms the Li ions induced enhanced relaxor natures in KTN crystals.

To explain the Li-doping effects on elastic anomaly caused by PNRs, the elastic constant C_{11} curves of the KTN/0.40 and KLTN/0.05/0.27 single crystals (Figure 5b) were fit with the following equation [35,100,101]:

$$C_{11} = A_1 + A_2T - A_3(T - T_0)^{-q} \tag{11}$$

The second term describes the anharmonic behavior. The critical softening owing to polarization fluctuations in PNRs is described by the third term, where A_3 denotes the elastic anomaly. The value of the critical exponent q was predicted to be 0.5 for three-dimensional fluctuations, 1.0 for two-dimensional fluctuations, and 1.5 for one-dimensional fluctuations [102]. A good consistency between the fitting and experimental results was obtained with an exponent of $q = 0.5$ reflecting the three-dimensional fluctuations of polarization in a cubic phase of non-doped and Li-doped KTN crystals [27]. It is significant to compare the value of the anomalous part, i.e., A_3, of the non-doped KTN ($A_3 = 242.55 \pm 8.49$) and 5%Li-doped KTN ($A_3 = 358.92 \pm 10.50$) crystals [27]. It is clear that the magnitude of the anomalous part was significantly increased by the Li-doping. The remarkable increase in the amplitude of the elastic anomaly in Li-doped KTN is clear evidence of the expansion of the temperature region of the elastic anomaly attributed to PNRs in comparison to that of the non-doped KTN.

In order to have a better understanding Li-doping's effect on PNRs, the comparative study of the CP behavior as a function of temperature was performed, which is related to the relaxation process in fluctuating PNRs. The CP spectra of the KTN/0.40 and KLTN/0.05/0.27 single crystals measured at 353 K with FSR = 300 GHz are shown in Figure 6a. It was seen that the intensity of the CP significantly increases with Li-doping, as shown in Figure 6a,b. The remarkable increase in intensity of the CP of the Li-doped KTN is owing to the growth of PNRs, which is induced by Li ions. A comparative study of the mean relaxation time, which was estimated by the CP width, between non-doped KTN and Li-doped KTN crystals was also performed to show the Li-doping effects of the precursor dynamics. The temperature-dependent inverse of the relaxation time is displayed in Figure 6c. It was found that the dynamic susceptibility, which was determined from the CP intensity, of the 2.5%Li-doped KTN crystal follows the extended Curie–Weiss law [99]. Since the dynamic susceptibility of the Li-doped KTN crystal follows the extended Curie–Weiss law, the generalized Lyddane–Sachs–Teller (LST) relation anticipates that the inverse of the relaxation time can be explained by the stretched-type slowing near T_C. Under the influence of RFs related to PNRs, such a stretched-type slowing-down behavior of the relaxation time might be described by the following empirical relation [97].

$$\frac{1}{\tau} = \frac{1}{\tau_0} + \frac{1}{\tau_1}\left(\frac{T - T_C}{T_C}\right)^{\beta}, (\beta \geq 1)\ for\ T > T_C$$

where β denotes the stretched index associated with the RFs. β = 1.0 indicates the normal critical slowing down with almost negligible RFs and it assumes values >1.0 depending on the strength of the RFs. The observed inverse of the relaxation time of 2.5%Li-doped and 5%Li-doped KTN crystals with the critical exponent β of 1.10 [99] and 1.18 [58], respectively, indicates the order–disorder behavior of the ferroelectric phase transition with weak RFs.

The average size of the dynamic PNR of the Li-doped KTN crystals was compared with that of the non-doped one to clarify the effects of Li-doping on PNRs. The average size of the dynamic PNR of the non-doped and Li-doped KTN crystals as a function of T/T_C is shown in Figure 6d. It is clear from Figure 6d that Li-doping significantly increases the size of the dynamic PNR. Hence, it is concluded that the intense CP, the broadening of elastic anomaly, and the enhanced relaxor nature of the Li-doped KTN in comparison to that of the non-doped KTN can be attributed to the enhanced growth of PNRs induced by Li ions.

One of the most important effects of Li-doping is the change in the physical origin of the CP of the KTN in the paraelectric cubic phase. It was observed that the CP in the cubic phase of non-doped and 2.5%Li-doped KTN crystals comes from the $A_1(z)$ mode of PNRs [25,85,99], while the CP of the 5%Li-doped KTN stems from the $E(x,y)$ mode of PNRs with a rhombohedral *R3m* symmetry [28] based on angular dependence of Raman scattering spectra with different scattering geometries. Hence, the intense CP of the 5%Li-doped KTN in comparison to the non-doped KTN could be owing to the transformation of the local mode of PNRs in addition to the growth of PNRs. Sokoloff et al. [22] found that the CP intensity caused by A-mode fluctuations is less intense than the E-mode fluctuations in the tetragonal phase with *P4mm* symmetry of the KTN/0.28 crystal. It is important to explain the microscopic origin of the change in the local mode of PNRs in 5%Li-doped KTN crystals. The change in the local mode of PNRs in 5%Li-doped KTN can be explained by two possibilities:

i. In non-doped KTN, the PNRs stem only from the off-center displacements of Nb ions at the B-site along the equivalent [111] directions corresponding to one set of atomic displacements related to 180° fluctuations, i.e., longitudinal polarization fluctuations in PNRs, as shown in Figure 7a [69,103]. Hence, the $A_1(z)$ mode of PNRs is reasonable in the case of non-doped KTN crystals. Moreover, it is believed that off-center displacements of Li ions in Li-doped KTN crystals occur at the A-site along the equivalent [100] directions [37,38]. However, the local [111] displacements are correlated in the medium range with the displacements along the [100] direction [71]. The measurement of X-ray absorption fine-structure (XAFS) revealed the off-centering of Nb ions in KTN at the B-site along the equivalent [111] direction at about 0.15 Å [21], while the off-centering of Li ions at the A-site along the [100] direction is expected to be 1.2 Å [38,104]. Since the off-center displacement of Li ions is much larger than that of Nb ions, the Li ion has a relatively larger dipole in comparison with the Nb ion. As a result, it is predicted that the correlations of displacements may lead along the equivalent [100] direction in Li-ion-rich KTN crystals. In the case of KLTN/0.025/0.38, the correlations of displacements are dominated by the Nb ions because of small concentrations of Li ions, and on average Li ions for the [100], [010], and [001] displacements are along an equivalent [111] direction [71,103], resulting in $A_1(z)$ mode of PNRs in 2.5%Li-doped KTN crystals. The invariance of the local mode of PNRs of the KLTN/2.5/38 crystal in comparison with that of the non-doped KTN indicates that the Li ions occupy only the A-site up to 2.5%Li-doping. Because of the off-center displacements of Li ions at the A-site, the doping of small amounts of Li ions in KTN may induce additional polarizations. These additional polarizations are cooperative with the neighboring PNRs, resulting in larger PNRs [27], and give rise to a weak RF due to the homovalent cations at both A- and B-sites [99]. On the other hand, the correlation of displacements between Li and Nb ions becomes stronger due to the relatively large Li concentrations in 5%Li-doped KTN; therefore, their displacements are predominantly along the [100], [010], and [001] directions. The switching of directions of polarization among [100], [010], and [001] may enhance the transverse polarization as shown in Figure 7c [69,103], resulting in the $E(x,y)$ mode of PNRs in 5%Li-doped KTN.

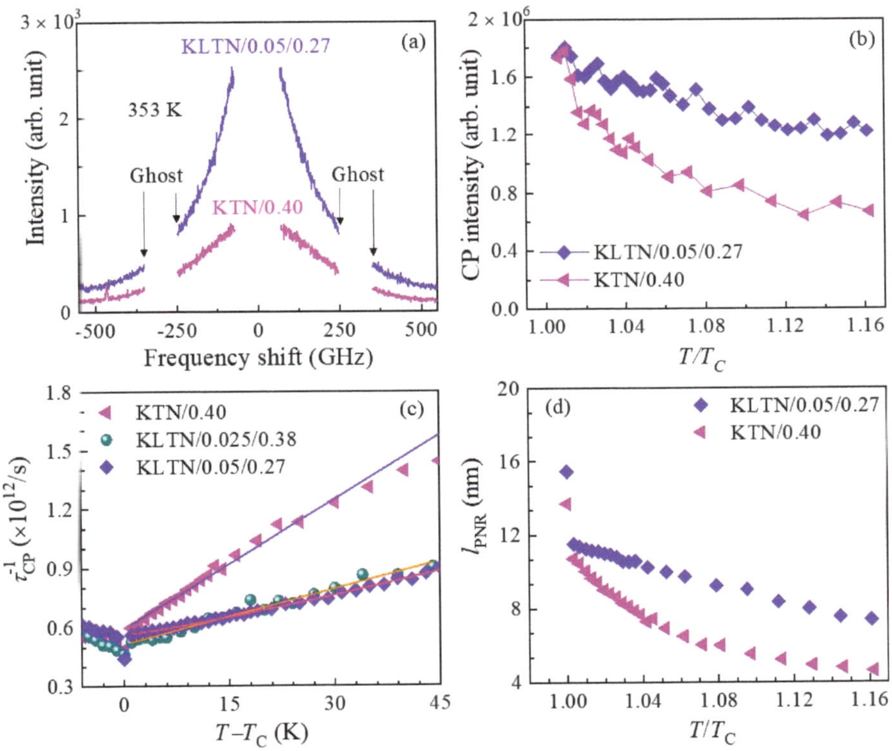

Figure 6. (**a**) The CP spectra of KTN/0.40 and the KLTN/0.05/0.27 single crystals measured with the FSR = 300 GHz at 353 K. (**b**) The CP intensity of KTN/0.40 (solid diamonds) and KLTN/0.05/0.27 (solid triangles) single crystals as a function of T/T_C. (**c**) The inverse of the relaxation time estimated from the width of the CP of non-doped KTN, 2.5%Li-doped KTN, and 5%Li-doped KTN single crystals as a function of normalized temperature. The solid lines are the lines fit with Equation (12) (**d**) The size of a dynamic PNR of non-doped KTN and 5%Li-doped KTN crystals as a function of T/T_C. The data were adapted from Refs. [27,58].

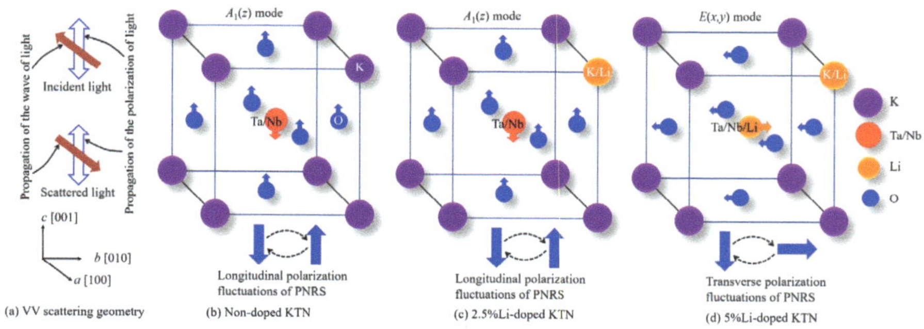

Figure 7. (**a**) A typical depiction of the VV scattering geometry. A schematic interpretation of the polarization fluctuations of PNRs caused by $A_1(z)$ mode in (**b**) non-doped KTN, (**c**) 2.5%Li-doped KTN, and $E(x,y)$ mode in (**d**) 5%Li-doped KTN crystals observed in the VV scattering under the condition of fixed A-site cations [103].

ii. In 5%Li-doped KTN, another possibility for the change in the local $A_1(z)$ to the $E(x,y)$ mode of PNRs may be the partial occupancy of a small percentage of the Li ions at the B-site. Wakimoto et al. [104] reported that the Li^+–Li^+ dipolar pairs in Ca-doped KLT/0.05 were reoriented with the reorientation of two nearest-neighbor Li^+ ions. In 5%Li-doped KTN, it is predicted that the dipolar pairs may be present, and these dipolar pairs can be reoriented with the reorientation of the nearest neighbor Li^+ ions. However, the interaction strength of dipole–dipole pairs depends on the magnitude of both dipoles, their orientations, and their proximity. In general, the proximity effect intensifies with an increase in the concentrations of dopants. Hence, the noticeable repulsive interaction between neighboring Li^+ and dipolar pairs may exist in 5%Li-doped KTN owing to the proximity effect. The random orientations at high temperatures above T^* may cause the occupancy of a small percentage of the Li ions at the B-site through hopping from the A-site, as shown in Figure 8c. Thus, there is a noticeable interaction between Li and Nb ions in fluctuating PNRs corresponding to two sets of atomic displacements, giving rise to the $E(x,y)$ mode of PNRs in 5%Li-doped KTN crystals. The $E(x,y)$ mode of PNRs was also found in a relaxor PMN single crystal, in which off-center displacements of cations occurred at both A- and B-sites and the charge disorder existed only at the B-site [72].

In order to clarify the partial occupancy of Li ions at the B-site, we calculated the total energy, which is the key parameter for an equilibrium structure, of a unit cell using the density functional theory (DFT)-based CASTEP code [105]. Ye et al. [106] reported the occupancy of Sn ions at A- and B-sites from the energy point of view using first-principles calculations. XAFS measurements also revealed the occupancy of Mn ions at both A- and B-sites in Mn-doped $SrTiO_3$ single crystals [107]. From the structural point of view, in non-doped KTN with a perovskite structure, the B-site is occupied by smaller Ta^{5+} (ionic radius 0.64 Å) and Nb^{5+} (ionic radius 0.64 Å) ions and the A-site is occupied by the larger K^{1+} ions (ionic radius 1.38 Å) [108], as shown in Figure 8a. Therefore, the RFs in non-doped KTN are too weak to show a relaxor-like nature [109], because B-site cations are homovalent and comparable in size and KTN possesses no charge disorder. Owing to the similar chemistry between K^{1+} and Li^{1+}, the Li^{1+} (ionic radius 0.76 Å) would be expected to substitute for K^{1+} at the A-site in the Li-doped KTN crystal. However, the occupation of Li ions is slightly unfavorable due to the size consideration of the perovskite structure [110], and $LiNbO_3$ is not a perovskite structure. In Ca-doped KTN, G. A. Samara and L. A. Boatner reported that the substitution of Ca occurs at the B-site by considering the ionic size [110]. Thus, there is a probability of partial occupancy of Li ions at the B-site in Li-doped KTN. For a better understanding of the partial occupancy of Li ions, the total energy per unit cell of the Li-doped KTN/0.27 was evaluated with first-principles calculations, as shown in Figure 8d–f.

At first, the calculation of the total energy of the unit cell of KTN/0.27 was carried out by the Li-doping at the A-site, as shown in Figure 8d. As can be seen in Figure 8d, the total energy debases with Li-doping and shows a minimum when the Li content is 4.9% at the A-site. This result is a clear indication of the occupancy of Li ions at the A-site only in 2.5%Li-doped KTN crystal. Thus, the invariance of the local mode of PNRs in non-doped and 2.5%Li-doped KTN may be due to homovalent cations at both A- and B-sites. It is worth noting that the total energy starts to increase above the 4.9%Li-doping at the A-site (Figure 8e), which stimulates to study of the partial occupancy of Li ions at the B-site in 5%Li-doped KTN crystals. We calculated the energy of the Li occupancy at both A- and B-sites to clarify the partial occupancy of Li ions at the B-site. It is important to note that the total energy shows a more minimum value when the substitution level was 0.8% at the B-site and 4.2% at the A-site, as shown in Figure 8f. It is concluded that 0.8% of Li ions occupy the B-site. In 5%Li-doped KTN single crystals, the change in the local $A_1(z)$ to $E(x,y)$ mode of PNRs is caused by the partial occupancy of Li ions at the B-site. The partial occupancy of Li ions at the B-site may enhance the RFs due to the charge disorder at the B-site as well. The enhanced RFs in addition to the growth of PNRs may induce the broadening of the elastic anomaly of 5%Li-doped KTN crystal in comparison with that of the non-doped KTN crystal.

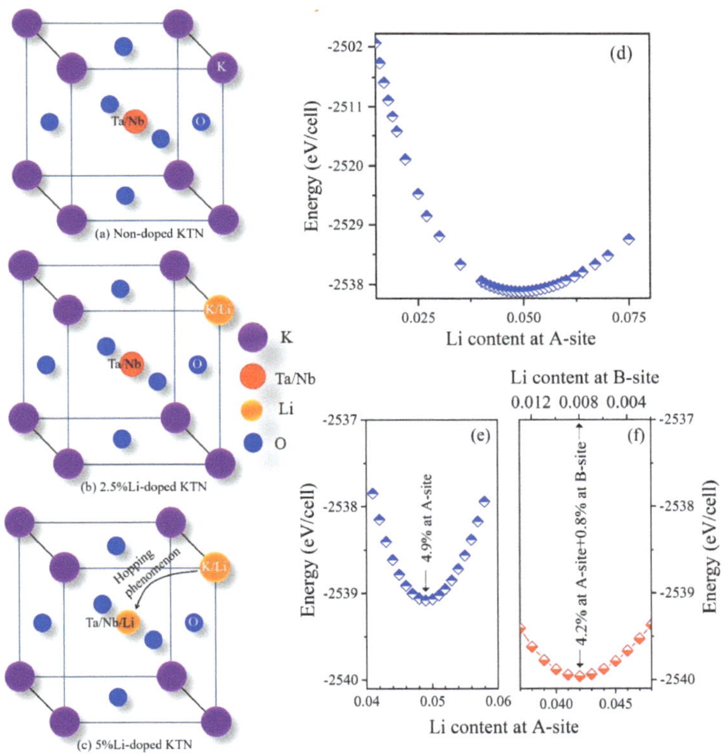

Figure 8. A schematic illustration of the lattice showing the location of substituents in the cubic perovskite, (**a**) non-doped KTN, (**b**) 2.5%Li-doped KTN, and (**c**) 5%Li-doped KTN single crystals, where a small percentage of Li ions occupy the B-site in 5%Li-doped KTN and the Li ions only occupy the A-site in 2.5%Li-doped KTN. The total energy per formula unit obtained using the density functional theory (DFT) calculation as a function of the Li content (**d**) at the A-site (**e**) near the 5%Li-doping at the A-site and (**f**) at A- and B-sites of the Li-doped KTN/0.27 single crystal. The data were adapted from Ref. [103].

3.4. Effects of Defects on PNRs in Li-Doped KTN

A schematic illustration of the composition gradient Li-doped KTN crystal wafer consists of dislocations that develop at the time of crystal growth along the [100] direction from a seed and the spatial distribution of T_C denoting the composition gradient are schematically shown in Figure 9a,b, respectively [58,111]. The dislocation density was high near the center (light yellow color) of the sample, that is, from the bottom to the top in Figure 9a. Therefore, the wafer was divided into two parts to observe the effects of defects on the composition gradient of Li-doped KTN. The "high defects", represented with closed diamonds (blue), denote the region where the dislocation density is high, whereas "low defects", marked with closed circles (red), indicate the region where the number density of dislocations is low. The defects' origins might be essentially screw dislocations, and additionally protons or oxygen vacancies [112].

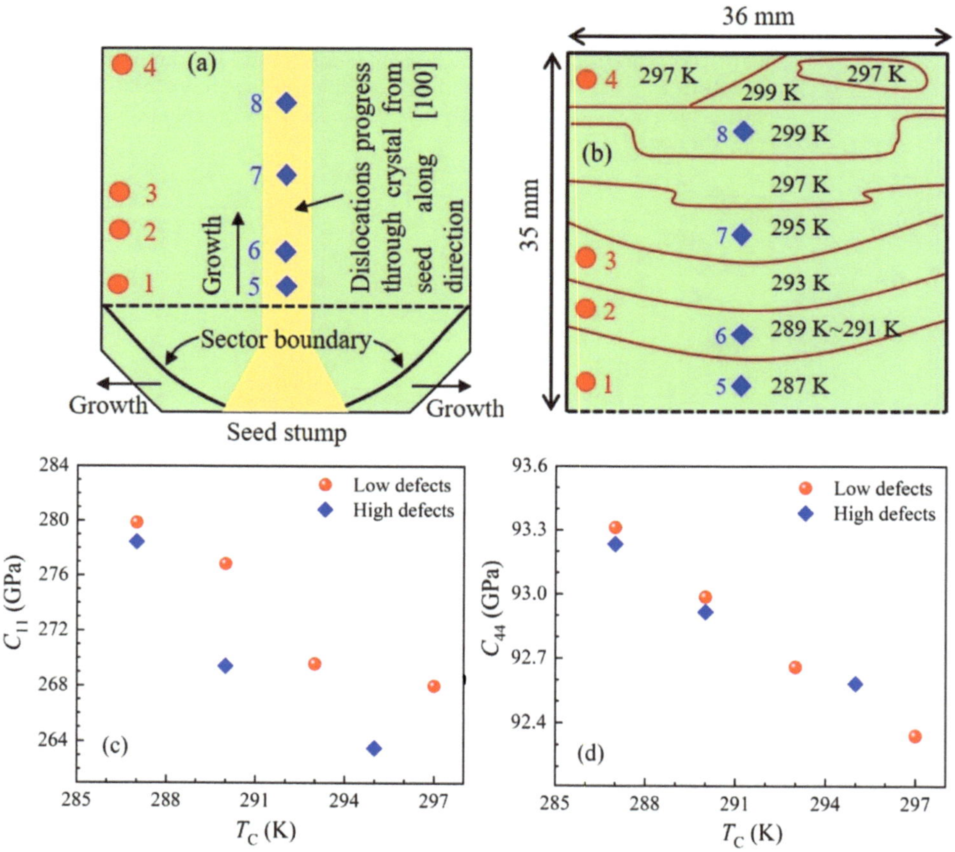

Figure 9. Schematic illustration of (**a**) spatial heterogeneity and (**b**) spatial dependence of T_C in a composition gradient on Li-doped KTN wafer. Diamonds (blue) and closed circles (red) denote the regions with high and low defect density, respectively. The elastic constants (**c**) C_{11} and (**d**) C_{44} of a composition gradient Li-doped KTN wafer as a function of position. The data were adapted from Refs. [58,111].

The positional dependence of elastic constants C_{11} (associated with LA phonon) and C_{44} (associated with TA phonon) of the Li-doped KTN wafer measured at room temperature (RT) is shown in Figure 9c,d. The variation in elastic constants with position reflects the composition gradient. These results are similar to those found in KTN single crystals with the ultrasonic pulse echo technique [35]. In the ferroelectric tetragonal phase at RT, the change in elastic constants in PZN–0.045PT and PZN–0.09PT crystals was studied by Ko et al. [43], and they hypothesized that the distinct behavior of the elastic constants was owing to the difference in the contribution from the domain walls and domain motion [43]. However, the change in elastic constants in Li-doped KTN was observed in the paraelectric cubic phase (Figure 9c,d), where there was no extrinsic contribution from the domain configurations. This fact suggests that the distinction of C_{11} and C_{44} with the position of the Li-doped KTN wafer can be caused by their different composition and the number density of defects. Another possibility is that there might be frequency dispersion within the Li-doped KTN wafer owing to the distinct number densities of PNRs and their correlation length depending mainly on the Nb content. In the high-defect regions, the acoustic phonons were considerably scattered by defects, so the sound attenuation increased, while velocity slightly decreased with the increase in defect densities, as shown in Figure 9c,d. The decrease in phonons frequency with the increase in the number density of defects suggests that defects may enhance the creation of PNRs. A similar change in elastic constants C_{11} and C_{44} with the number density of defects was reported in the case of non-ferroelectric materials such as silicon [45,46].

3.5. *Effects of Electric Fields on PNRs in Li-Doped KTN*

To understand the effects of the electric field on precursor dynamics, we investigated the 5%Li-doped KTN with $x = 0.26$ under zero-field cooling (ZFC) and an externally applied dc field using Brillouin scattering. The Brillouin spectra of the KLTN/0.05/0.26 single crystal at some selected temperatures recorded in a narrow frequency range under the ZFC process are shown in Figure 10a [113]. As can be seen in Figure 10a, the spectra are qualitatively similar to those observed in non-doped KTN in the paraelectric cubic phase. However, one interesting result observed in the ferroelectric phase is that the splitting of the TA mode occurs just below T_C.

The frequency shift of the TA mode as a function of temperature is displayed in Figure 10b. The splitting of the TA mode indicates that the ferroelectric tetragonal phase of the Li-doped KTN crystals consists of two domain states. The high-frequency TA (HTA) mode stands for ferroelectric macro-domain states, whereas the low-frequency TA (LTA) denotes the random nanodomain states within macro-domain states, which were attributed to the freezing up of the polarization fluctuations of PNRs where the growth into macro-domains was blocked by the RFs [113]. Recently, Dey et al. [73] visualized randomly oriented nanodomains within macro-domain states in the ferroelectric phase of $(1-x)Ba(Zr_{0.2}Ti_{0.8})O_3-x(Ba_{0.7}Ca_{0.3})TiO_3$ (BZT–0.60BCT, $x = 0.60$) ceramics using piezo response force microscopy (PRFM). In the ferroelectric phase, the splitting of the acoustic modes due to the existence of multi-domain states was also found in typical RFEs [114].

The effect of the electric field in the ferroelectric phase of the KLTN/0.05/0.26 crystal was studied to discuss the microscopic origin of the splitting of the TA mode. The Brillouin scattering spectra of the KLTN/0.05/0.26 crystal measured under a constant temperature of 291 K at some selected electric fields applied along the [100] direction are shown in Figure 10c. The TA frequency shift as a function of electric fields under a constant temperature of 291 K is displayed in Figure 10d. It is important to note that the LTA and HTA modes become closer to each other with an increasing electric field along the [100] direction. At the critical electric field of $E_C = 0.9$ kV/cm, the split TA mode merges into a single peak, indicating the switching of the ferroelectric macro- and nanodomains along the direction of the applied field at E_C. This result is a clear indication of the change of crystal from multi-domain to single domain state close to HTA by the suppression of pinning by RFs. Such a switching of multi-domain states into a single domain state under a moderate electric field was also observed in typical Refs [114].

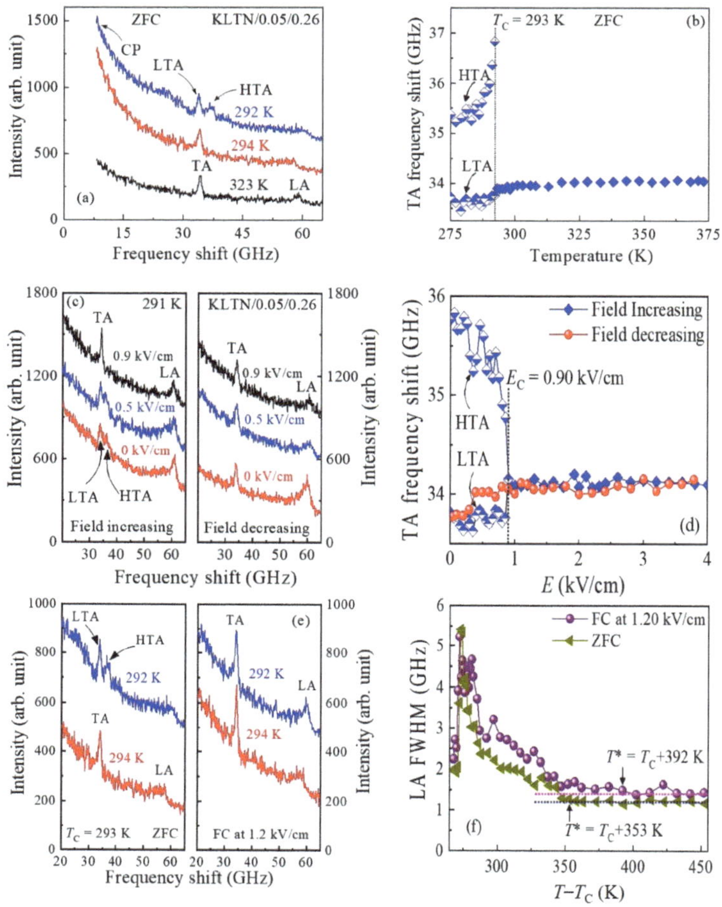

Figure 10. (**a**) Brillouin scattering spectra of the KLTN/0.05/0.26 single crystal at some selected temperatures measured at zero field cooling (ZFC). (**b**) The temperature dependence of the frequency shift of the TA mode of the KLTN/0.05/0.26 single crystal. (**c**) The Brillouin scattering spectra of the KLTN/0.05/0.26 crystal at some selected electric fields measured in the ferroelectric phase under a constant temperature of 291 K, where the electric field was applied along the [100] direction. (**d**) The frequency shift of the TA mode observed in the ferroelectric phase as a function of electric fields. (**e**) The typical Brillouin spectra of the KLTN/0.05/0.26 crystal at some selected temperatures measured under ZFC and field cooling (FC) at 1.20 kV/cm. (**f**) The temperature dependence of the FWHM of the LA mode observed under ZFC and FC process [113].

A comparative study of the acoustic properties was performed between the field cooling (FC) and ZFC at 1.2 kV/cm to clarify the effect of electric fields on PNRs. The Brillouin scattering spectra of the KLTN/0.05/0.26 crystal observed under FC and ZFC at some selected temperatures are displayed in Figure 10e. As can be seen in Figure 10e, the TA mode splitting disappears under the FC at 1.2 kV/cm. Since the critical field of E_C = 0.9 is much smaller than the applied electric field of 1.2 kV/cm, the KLTN/0.05/0.26 crystal at 1.2 kV/cm mainly comprises a single domain state due to the complete switching of the random nanodomain states to the macro-domain state or single domain state resulting in the disappearance of splitting of the acoustic modes. Notably, the FWHM of the LA mode increases under FC in comparison with that of the ZFC process (Figure 10f). This means that the TA phonon under an electric field of 1.2 kV/cm was markedly scattered by PNRs, and the sound velocity may also decrease with the scattering. This result suggests that the electric field may enhance the creation and growth of PNRs. A significant decrease in the sound velocities with the growth and the number densities of PNRs was also observed in Li-doped KTN single crystal and wafer, respectively [27,111]. Under a moderate electric field of 1.2 kV/cm, the temperature interval between T^* and T_C was markedly increased (Figure 10f). The enhancement of the temperature interval between T^* and T_C can be attributed to the enhanced stability of PNRs [27], induced by the electric field along the [100] direction.

3.6. Critical End Point (CEP) and Injected Electrons in Li-Doped KTN

The study of temperature-field phase diagrams, including the critical end point (CEP), is one of the current topics in advanced materials such as ferroelectric materials owing to the giant electromechanical response near the CEP [115]. The CEP is the driving mechanism for the rotation of polarization and the associated giant electromechanical response of RFEs studied by M. Aftabuzzaman and S. Kojima [114]. Without injected electrons in 5%Li-doped KTN with $x = 0.27$, the temperature-field phase diagram concluding the CEP $(E, T) = (3.3$ kV/cm, $T_C + 6$ K) was investigated by Imai et al. [49]. In addition, a significant increase in the capacitance and permittivity of a Li-doped KTN crystal was observed using injected electrons [49]. However, injected electrons are trapped by the localized states in the crystal and stabilized. Therefore, a space-charge-induced electric field gradient is expected which significantly affects the CEP of the Li-doped KTN crystals. A study of the temperature dependence of the CP behavior was carried out to show the effect of injected electrons on the CEP. In this study, titanium electrode pairs were used to inject electrons into the KLTN/0.05/0.26 crystals by the application of voltages along the [100] direction [116]. The electric voltage dependence of the CP intensity at some selected temperatures of the KLTN/0.05/0.26 crystal in the paraelectric cubic phase is shown in Figure 11a [116]. Significant variation in the intensity of the CP was found at the critical voltages, reflecting the paraelectric cubic to the ferroelectric tetragonal phase transition. It is believed that most of the dynamic PNRs become static near T_C in RFEs. Therefore, the remarkable increase in the intensity of the CP at the critical voltage may be due to the alignment of randomly oriented dynamic and static PNRs along the direction of the applied voltage resulting in a [100] time-averaged polarization, and the crystal transforming into a lower tetragonal *P4mm* symmetry. A small hysteresis behavior of the CP intensity was observed in the paraelectric to ferroelectric phase transition at V_{C1} on heating and the ferroelectric to paraelectric phase transition at V_{C2} on cooling (Figure 11b). This fact indicates the slightly first-order nature of the ferroelectric phase transition. As can be seen in Figure 11b, the paraelectric to ferroelectric phase transition voltages, i.e., the critical electric voltages, were seen to shift to higher values with increasing temperature. However, the shift of V_{C1} is smaller than the shift of V_{C2}; hence, V_{C1} and V_{C2} merge at around $(V,T) = (190$ V, $T_C + 3.4$ K), equivalent to $(E,T) = (1.6$ kV/cm, $T_C + 3.4$ K). This point in which the hysteresis behavior gradually diminished to zero was identified as the critical endpoint (CEP) of the KLTN/0.05/0.26 crystal. This point also denotes the change of the first-order to second-order nature of the ferroelectric phase transition [49]. Without electrons injected into the KLTN/0.05/0.27 crystal, the value of the CEP was found at $(E,T) = (3.3$ kV/cm, $T_C + 6$ K) [49]. The value of the critical electric field observed at the CEP of the KLTN/0.05/0.26 crystal with injected electrons is lower in comparison to that of the value found in KLTN/0.05/0.27 crystal with no injected electrons. Hence, the lowering of the average electric field of $E_{av} = 1.7$ kV/cm of the KLTN/0.05/0.26 crystal is due to the space-charge-induced electric field, which is generated by the injected and/or trapped electrons.

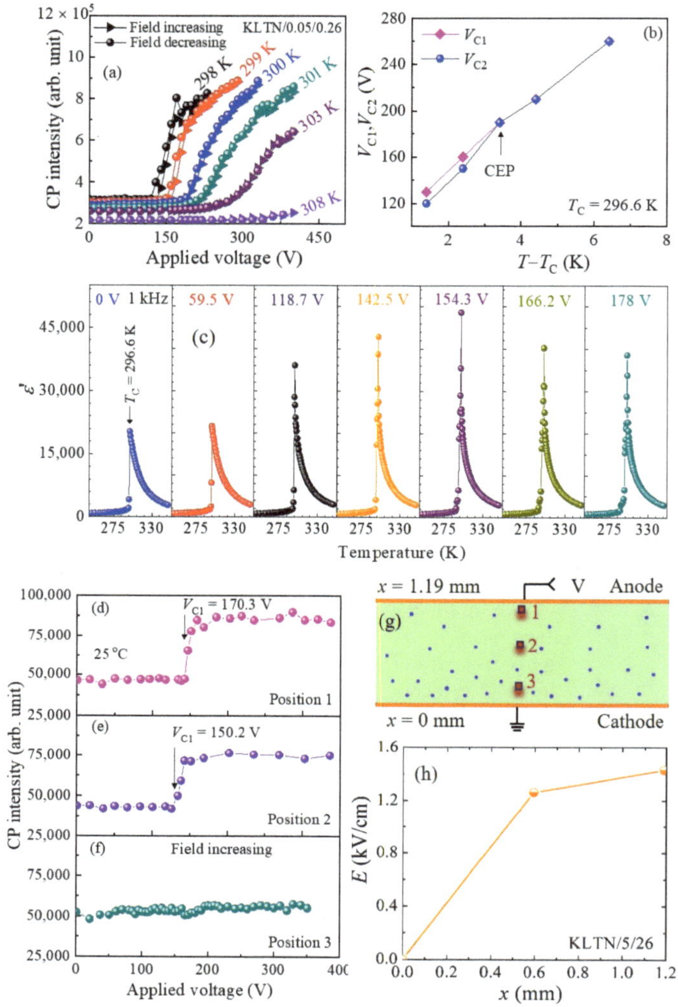

Figure 11. (**a**) The CP intensity of the KLTN/0.05/0.26 crystal at some selected temperatures as a function of applied voltage. (**b**) Temperature dependence of critical electric voltages of a phase transition, where V_{C2} denotes the transition from ferroelectric to paraelectric and V_{C1} indicates the transition from paraelectric to ferroelectric phase. (**c**) The dielectric constants of the KLTN/0.05/0.26 crystals measured at 1 kHz under the DC biasing voltages as a function of temperature. The electric voltage dependence of the observed CP intensity at (**d**) close to the anode (position 1), (**e**) close to the center (position 2), and (**f**) close to the cathode (position 3). The three positions in the KLTN/0.05/0.26 crystal are schematically drawn in (**g**). (**h**) The critical electric field of the KLTN/0.05/0.26 crystal as a function of position. The data were adapted from Ref. [116].

The temperature dependence of the dielectric properties of the KLTN/0.05/0.26 crystal was examined to investigate the effect of injected and/or trapped electrons on the CEP as well. The temperature-dependent real part of the dielectric constant (E') under selected applied voltages is shown in Figure 11c. The dielectric anomaly was clearly seen at the paraelectric to ferroelectric phase transition temperature, T_C (Figure 11c). It is important to note that the slightly broad peak exhibited at about T_C at V = 0 V changes into a very sharp peak when approaching the CEP at V = 154.3V or E = 1.3 kV/cm. A sharp phase transition in the vicinity of the CEP was also observed in typical RFEs such as PMN-xPT [114,115]. It is also significant that the value of T_C is shifted to higher values in the range of 296.6–299.5 K under the application of applied voltage up to 178.0 V. It is possible to measure a very small area of a specimen to clarify the local heterogeneity using a combination of Brillouin scattering spectroscopy and optical microscopy [57]. However, the average value of the dielectric constant of the investigated specimen was obtained using conventional dielectric spectroscopy. Hence, the observed discrepancy of the CEP between dielectric and Brillouin spectroscopy measurements may be due to the field gradient induced by the injected and/or trapped electrons.

We also studied the position dependence of the Brillouin scattering spectra of the KLTN/0.05/0.26 crystal to clarify the effect of trapped electrons on the critical voltages. Figure 11g shows a schematic illustration of three positions in the crystal [116]. The electric voltage dependence of the CP intensity at three positions is shown in Figure 11d–f. The electron density is not uniform inside the crystal, and the electron density tends to be large near the cathode compared with the part near the anode when the electron injecting voltage is low. Hence, the absolute value of the injected-electron-induced electric field is almost zero at the cathode and it increases with the distance from the cathode and becomes the maximum at the anode. However, the crystal symmetry becomes tetragonal by an external electric field along the [100] direction because of the electrostriction. Hence, under a moderate electric field, the crystal part near the cathode tends to remain paraelectric (Figure 11g), whereas the part near the anode tends to become ferroelectric (Figure 11e) because the electric field is strong. It is worth noting that the value of the critical field near the center is somewhat smaller than that of the value near the anode (Figure 11d,e).

We also investigated the position-dependent critical electric field plot of the KLTN/0.05/0.26 crystal to understand the charge distribution inside the crystal. It is challenging to rationalize the actual behavior of the electric field inside the crystal because of the limited number of observation positions. For simplicity, we consider the positions $x = d$ (close to anode), $x = d/2$ (close to center), and position $x = 0$ (close to cathode), where x is the distance from the cathode and d is the thickness of the sample. The detailed calculation procedure of the position-dependent critical electric field was reported in Ref. [116]. Figure 11h shows the position dependence of the critical field of the KLTN/0.05/0.26 crystal. The position-dependent electric field exhibits a slight deviation from linear behavior based on the results of the position dependence of the intensity of the CP (Figure 11h). This result suggests that the deviation of the electric field from linear behavior could be due to an inhomogeneous distribution of charges injected into the crystal with applied voltages. A difference between the calculated and observed capacitance owing to the inhomogeneous distribution of charges was also reported in Ref. [49]. It is also important to note that the CP intensity near the center was slightly smaller than that of the CP intensity near the anode (Figure 11d,e). As the electric field is strong enough near the anode in comparison with the other parts of the crystal, the randomly oriented dynamic and static PNRs may start to align along the direction of the applied field near the anode, and bring about an intense CP near the anode.

4. Conclusions

The understanding of precursor dynamics associated with polar clusters called PNRs is crucial for the enhancement of the functionality of ferroelectric materials. The precursor dynamics in relaxor-like ferroelectric non-doped KTN and Li-doped KTN crystals near the paraelectric to ferroelectric phase transition temperate, T_C, were investigated using Brillouin scattering. A significant softening of sound velocity and a remarkable increase in sound attenuation in the vicinity of T_C were observed in both non-doped KTN and Li-doped KTN single crystals. A significant increase in the intensity of the central peak (CP) and sound attenuation as well as a remarkable softening of the transverse sound velocity, V_{TA}, were clearly found below the intermediate temperature, T^*, denoting the beginning of the rapid growth of PNRs. In both crystals, the temperature evolution of the average size of dynamic PNRs is evaluated, and below T^* it markedly increases towards the T_C. The relaxation time, which was estimated from the width of a broad central peak (CP) of non-doped KTN and Li-doped KTN crystals, exhibits a critical slowing down towards T_C, indicating the order–disorder behavior of a ferroelectric phase transition.

The effects of Li-doping on PNRs in KTN crystals were clearly observed by the broadening of the elastic anomaly and the stretching of the slowing down with weak random fields (RFs) related to PNRs in the Li-doped KTN crystals in comparison with the non-doped KTN one. A significant increase in the average size of the PNRs towards T_C was also found with Li-doping, which indicates a broadening and extension of the temperature region of the elastic anomaly. The expansion of the temperature region and broadening of the elastic anomaly could indicate the enhanced relaxor nature of Li-doped KTN single crystals. The analysis of dielectric properties confirms the enhanced relaxor nature of Li-doped KTN crystals. A change of local $A_1(z)$ to $E(x,y)$ modes of PNRs was observed in 5%Li-doped KTN crystal. A first-principles calculation based on density functional theory (DFT) revealed that 4.2% of Li ions occupy the A-site and 0.8% of Li ions occupy the B-site in 5%Li-doped KTN. The heterovalence charge disorder at the B-site induced by the partial occupancy of Li ions could be the microscopic origin of the change in the local mode of PNRs.

The co-existance of macro-domain and random nanodomains associated with PNRs in the ferroelectric phase of Li-doped KTN was observed by the splitting of the TA mode. The switching of nano- and macro-domain states into a single-domain state was found at a critical field of E_C = 0.9 kV/cm. An increase in number density and growth of PNRs under a moderate electric field was suggested by a significant increase in the damping of the acoustic mode in comparison with that of the ZFC process. The defect induces different number densities of PNRs found in Li-doped KTN wafer as well.

The temperature-field phase diagram including the critical end point (CEP) of Li-doped KTN crystal was clarified. The trapped electrons' effects were evidenced by the lower value of the CEP in comparison with the Li-doped KTN crystal with no trapped electrons in the literature. The difference in CEP between the dielectric result averaged over all areas of a sample and the Brillouin result at a fixed small area revealed the inhomogeneous distribution of injected and/or trapped electrons in Li-doped KTN single crystal. The inhomogeneous distribution of injected and/or trapped electrons was confirmed by the observation of position dependence of the CP intensity.

Author Contributions: M.M.R. carried out the experiment and theoretical calculations and wrote the manuscript. S.K. analyzed the results and reviewed the manuscript. All authors have read and agreed to the published version of the manuscript.

Funding: This research was supported in part by the Japan Society for the Promotion of Science KAKENHI Grant Number JP17K05030 and the Faculty of Engineering, University of Rajshahi, Rajshahi 6205, Bangladesh, Fund Number 106/5/52/R.U./Eng.

Institutional Review Board Statement: Not applicable.

Informed Consent Statement: Not applicable.

Data Availability Statement: Not applicable.

Acknowledgments: The authors are thankful to Tatsuya Mori, Md. Al-Helal, and Md. Aftabuzzaman for their fruitful discussion.

Conflicts of Interest: The authors declare no conflict of interest.

References

1. Keen, D.A. Perovskites take the lead in local structure analysis. *IUCrJ* **2016**, *3*, 8–9. [CrossRef] [PubMed]
2. Barrett, J.H. Dielectric constant in perovskite type crystals. *Phys. Rev.* **1952**, *86*, 118. [CrossRef]
3. Müller, K.A.; Burkard, H. SrTiO3: An intrinsic quantum paraelectric below 4 K. *Phys. Rev. B* **1979**, *19*, 3593. [CrossRef]
4. Fontana, M.D.; Metrat, G.; Servoin, J.L.; Gervais, F. Infrared spectroscopy in $KNbO_3$ through the successive ferroelectric phase transitions. *J. Phys. C Solid State Phys.* **1984**, *16*, 483–514. [CrossRef]

5. Toumanari, A.; Khatib, D.; Kinase, W. Theory of the successive phase transitions in $KNbO_3$. *Phase Transit.* **2006**, *71*, 13–37. [CrossRef]
6. Triebwasser, S. Study of ferroelectric transitions of solid-solution single crystals of $KNbO_3$-$KTaO_3$. *Phys. Rev.* **1959**, *114*, 63–71. [CrossRef]
7. Rytz, D.; Scheel, H.J. Crystal growth of $KTa_{1-x}Nb_xO_3$ ($0 < x \leq 0.04$) solid solution by a slow-cooling method. *J. Cryst. Growth* **1982**, *59*, 468–484. [CrossRef]
8. Chang, Y.C.; Wang, C.; Yin, S.; Hoffman, R.C.; Mott, A.G. Giant electro-optic effect in nanodisordered KTN crystals. *Opt. Lett.* **2013**, *38*, 4574–4577. [CrossRef]
9. Agranat, A.; Hofmeister, R.; Yariv, A. Characterization of a new photorefractive material: $K_{1-y}L_yT_{1-x}N_x$. *Opt. Lett.* **1992**, *17*, 713–715. [CrossRef]
10. Shang, A., Jr.; Chen, C.-J.; Liu, R.; Lee, Y.G.; Chao, J.-H.; Yina, S. Giant electrostrictive coefficient in rapidly cooled nanodisordered $KTa_{1-x}Nb_xO_3$ lead-free single crystals. *AIP Adv.* **2021**, *11*, 035020. [CrossRef]
11. Shang, A., Jr.; Liu, R.; Chen, C.-J.; Lee, Y.-G.; Chao, J.-H.; Zhang, W.; Wetherington, M.; Yin, S. Domain engineering enabled giant linear electro-optic effect and high transparency in ferroelectric $KTa_{1-x}Nb_xO_3$ single crystals. *Phys. Status Solidi. RRL* **2022**, *16*, 2200005. [CrossRef]
12. Imai, T.; Yagi, S.; Toyoda, S.; Miyazu, J.; Naganuma, K.; Sasaura, M.; Fujiura, K. Fast response variable focal-length lenses using $KTa_{1-x}Nb_xO_3$ crystals. *Appl. Phys. Express* **2014**, *4*, 022501. [CrossRef]
13. Sasaki, Y.; Okabe, Y.; Ueno, M.; Toyoda, S.; Kobayashi, J.; Yagi, S.; Naganuma, K. Resolution enhancement of $KTa_{1-x}Nb_xO_3$ electro-optic deflector by optical beam shaping. *Appl. Phys. Express* **2013**, *6*, 102201. [CrossRef]
14. Wang, X.; Liu, B.; Yang, Y.; Zhang, Y.; Lv, X.; Hong, G.; Shu, R.; Yu, H.; Wang, J. Anomalous laser deflection phenomenon based on the interaction of electro-optic and graded refractivity effects in Cu-doped $KTa_{1-x}Nb_xO_3$ crystal. *Appl. Phys. Lett.* **2014**, *105*, 051910. [CrossRef]
15. Nakamura, K.; Miyazu, J.; Sasaki, Y.; Imai, T.; Sasaura, M.; Fujiura, K. Space-charge-controlled electro-optic effect: Optical beam deflection by electro-optic effect and space-charge controlled electrical conduction. *J. Appl. Phys.* **2008**, *104*, 013105. [CrossRef]
16. Bokov, A.A.; Ye, Z.-G. Recent progress in relaxor ferroelectrics with perovskite structure. *J. Mater. Sci.* **2006**, *41*, 31–52. [CrossRef]
17. Cross, L.E. Relaxor ferroelectrics. *Ferroelectrics* **1987**, *76*, 241–267. [CrossRef]
18. Phelan, D.; Stock, C.; Rodriguez-Rivera, J.A.; Chi, S.; Leão, J.; Long, X.; Xie, Y.; Bokov, A.A.; Ye, Z.-G.; Ganesh, P.; et al. Role of random electric fields in relaxors. *Proc. Natl. Acad. Sci. USA* **2014**, *111*, 1754–1759. [CrossRef]
19. Polinger, V. Off-center instability of Nb^{5+} in $KNbO_3$ under ambient pressure. *Chem. Phys.* **2015**, *459*, 72–80. [CrossRef]
20. Ko, J.-H.; Kim, T.H.; Kojima, S.; Long, X.; Bokov, A.A.; Ye, Z.-G. Effect of chemically ordered regions on the acoustic behaviors in $P(Mg_{1/3}Nb_{2/3})O_3$ studied by Brillouin scattering. *J. Appl. Phys.* **2010**, *107*, 054108. [CrossRef]
21. Hanske-Petitpierre, O.; Yakoby, Y.; Mustre De Leon, J.; Stern, E.A.; Rehr, J.J. Off-center displacement of the Nb ions below and above the ferroelectric phase transition of $KTa_{0.91}Nb_{0.09}O_3$. *Phys. Rev. B* **1991**, *44*, 6700–6707. [CrossRef]
22. Sokoloff, J.P.; Chase, L.L.; Boatner, L.A. Low-frequency relaxation modes and structural disorder in $KTa_{1-x}Nb_xO_3$. *Phys. Rev. B* **1990**, *41*, 2398–2408. [CrossRef]
23. Dul'kin, E.; Kojima, S.; Roth, M. Characteristic temperatures and field effect in $KTa_{1-x}Nb_xO_3$ relaxor crystals seen via acoustic emission. *Euro. Phys. Lett.* **2012**, *97*, 57004. [CrossRef]
24. Dul'kin, E.; Petzelt, J.; Kamba, S.; Majaev, E.; Roth, M. Relaxor-like behavior of $BaTiO_3$ crystals from acoustic emission study. *Appl. Phys. Lett.* **2010**, *97*, 032903. [CrossRef]
25. Rahaman, M.M.; Imai, T.; Miyazu, J.; Kobayashi, J.; Tsukada, S.; Helal, M.A.; Kojima, S. Relaxor-like dynamics of ferroelectric $K(Ta_{1-x}Nb_x)O_3$ crystals probed by inelastic light scattering. *J. Appl. Phys.* **2014**, *116*, 074110. [CrossRef]
26. Ohta, R.; Zushi, J.; Ariizumi, T.; Kojima, S. Order–disorder behavior of ferroelectric phase transition of $KTa_{1-x}Nb_xO_3$ probed by Brillouin scattering. *Appl. Phys. Lett.* **2011**, *98*, 092909. [CrossRef]
27. Rahaman, M.M.; Imai, T.; Kobayashi, J.; Kojima, S. Effect of Li-doping on polar-nanoregions in $K(Ta_{1-x}Nb_x)O_3$ single crystals. *Jpn. J. Appl. Phys.* **2015**, *54*, 10NB01. [CrossRef]
28. Toulouse, J.; DiAntonio, P.; Vugmeister, B.E.; Wang, X.M.; Knauss, L.A. Precursor effects and ferroelectric macroregions in $KTa_{1-x}Nb_xO_3$ and $K_{1-y}Li_yTaO_3$. *Phys. Rev. Lett.* **1992**, *68*, 232–235. [CrossRef] [PubMed]
29. Rahaman, M.M.; Tsukada, S.; Svirskas, S.; Banys, J.; Kojima, S. Vibrational dynamics of ferroelectric $K(Ta_{1-x}Nb_x)O_3$ studied by inelastic light scattering. *Ferroelectrics* **2019**, *538*, 96–104. [CrossRef]
30. Rahaman, M.M.; Imai, T.; Sakamoto, T.; Tsukada, S.; Kojima, S. Fano resonance of Li-doped $KTa_{1-x}Nb_xO_3$ single crystals studied by Raman scattering. *Sci. Rep.* **2016**, *6*, 23898. [CrossRef]
31. Kojima, S.; Rahaman, M.M.; Sase, R.; Hoshina, T.; Tsurumi, T. Vibrational dynamics of ferroelectric $K(Ta_{1-x}Nb_x)O_3$ studied by far-infrared spectroscopic ellipsometry and Raman scattering. *Jpn. J. Appl. Phys.* **2018**, *57*, 11UB05. [CrossRef]
32. Wu, Y.; Liang, F.; Wang, X.; Wang, J.; Yu, H.; Zhang, H. Temperature dependent Raman spectroscopic study of Fano resonance in perovskite ferroelectric $KTa_{1-x}Nb_xO_3$ single crystal. *Opt. Mater. Express* **2022**, *12*, 247–255. [CrossRef]
33. Parravicini, J.; DelRe, E.; Perego, S.; Acciarri, M.; Binetti, S.; Garcia, Y.; Parapelitsa, G.; Agranat, A.J.; Parravicini, G.-B. Key role of polar nanoregions in the cubic-to-tetragonal phase transition of potassium-based perovskites. *Phys. Rev. B* **2022**, *106*, 064107. [CrossRef]

34. Kleemann, W.; Schäfer, F.J.; Rytz, D. Diffuse ferroelectric phase transition and long-range order of dilute $KTa_{1-x}Nb_xO_3$. *Phys. Rev. Lett.* **1985**, *54*, 2038–2041. [CrossRef]
35. Knauss, L.A.; Wang, X.M.; Toulouse, J. Polarization-strain coupling in the mixed ferroelectric $KTa_{1-x}Nb_xO_3$. *Phys. Rev. B* **1995**, *52*, 13261–13268. [CrossRef]
36. Pashkin, A.; Železný, V.; Petzelt, J. Infrared spectroscopy of $KTa_{1-x}Nb_xO_3$ crystals. *J. Phys. Condens. Matter* **2005**, *17*, L265. [CrossRef]
37. Cai, L.; Toulouse, J.; Harrier, L.; Downing, R.G.; Boatner, L.A. Origin of the crossover between a freezing and a structural transition at low concentration in the relaxor ferroelectric $K_{1-x}Li_xTaO_3$. *Phys. Rev. B* **2015**, *91*, 134106. [CrossRef]
38. Machado, R.; Sepliarsky, M.; Stachiotti, M.G. Off-center impurities in a robust ferroelectric material: Case of Li in $KNbO_3$. *Phys. Rev. B* **2012**, *86*, 094118. [CrossRef]
39. Schremmer, H.; Kleemann, W.; Rytz, D. Field-induced sharp ferroelectric phase transition in $K_{0.937}Li_{0.063}TaO_3$. *Phys. Rev. Lett.* **1989**, *62*, 1896–1899. [CrossRef]
40. Zhou, Z.; Li, J.; Tian, H.; Wang, Z.; Li, Y.; Zhang, R. Piezoelectric properties of the lead-free $K_{0.95}Li_{0.05}Ta_{0.61}Nb_{0.39}O_3$ single crystal. *J. Phys. D Appl. Phys.* **2009**, *42*, 125405. [CrossRef]
41. Li, J.; Li, Y.; Zhou, Z.; Guo, R.; Bhalla, A. Electrical Properties of Lead-free Niobium Rich Piezoelectric $(K_{0.95}Li_{0.05})(Ta_{1-x}Nb_x)O_3$ Single Crystals. *Ferroelectrics* **2011**, *130*, 65. [CrossRef]
42. Rytz, D.; Châtelain, A.; Höchli, U.T. Elastic properties in quantum ferroelectric $KTa_{1-x}Nb_xO_3$. *Phys. Rev. B* **1983**, *27*, 6830–6840. [CrossRef]
43. Ko, J.-H.; Kim, D.H.; Kojima, S. Central peaks, acoustic modes, and the dynamics of polar nanoregions in $Pb[(Zn_{1/3}Nb_{2/3})_xTi_{1-x}]O_3$ single crystals studied by Brillouin spectroscopy. *Phys. Rev. B* **2008**, *77*, 104110. [CrossRef]
44. Svitelskiy, O.; Suslov, A.V.; Betts, J.B.; Migliori, A.; Yang, G.; Boatner, L.A. Resonant ultrasound spectroscopy of $KTa_{1-x}Nb_xO_3$ ferroelectric relaxor crystals. *Phys. Rev. B* **2008**, *78*, 064113. [CrossRef]
45. Clark, L.A.; Xianglong, Y.; Martin, Z.B.; Linn, W.H. Elastic constants of defected and amorphous silicon with the environment-dependent interatomic potential. *Phys. Rev. B* **2004**, *70*, 134113. [CrossRef]
46. Burnett, P.J.; Briggs, G.A.D. The elastic properties of ion-implanted silicon. *J. Mater. Sci.* **1986**, *21*, 1828–1836. [CrossRef]
47. Imai, T.; Miyazu, J.; Kobayashi, J. Charge distributions in $KTa_{1-x}Nb_xO_3$ optical beam deflectors formed by voltage application. *Opt. Express* **2014**, *22*, 14114–14126. [CrossRef]
48. Chen, F.S.; Geusic, J.E.; Kurtz, S.K.; Skinner, J.G.; Wemple, S.H. Light modulation and beam deflection with potassium tantalate-niobate crystals. *J. Appl. Phys.* **1966**, *37*, 388. [CrossRef]
49. Imai, T.; Toyoda, S.; Miyazu, J.; Kobayashi, J.; Kojima, S. Permittivity changes induced by injected electrons and field-induced phase transition in $KTa_{1-x}Nb_xO_3$ optical beam deflectors. *Jpn. J. Appl. Phys.* **2014**, *53*, 09PB02. [CrossRef]
50. Eros, S.; Reitz, J.R. Elastic Constants by the Ultrasonic Pulse Echo Method. *J. Appl. Phys.* **1958**, *29*, 683. [CrossRef]
51. Sumino, Y.; Ohno, I.; Goto, T.; Kumazawa, M. Measurement of the elastic constant and internal frictions on single-crystal MgO by rectangular parallelepiped resonance. *J. Phys. Earth* **1976**, *24*, 263. [CrossRef]
52. Maczka, M.; Hanuza, J.; Ko, J.-H.; Kojima, S. High-resolution broadband Brillouin scattering study of antiferroelectric phase transition in $CsBi(MoO_4)_2$. *Phys. Rev. B* **2003**, *68*, 174101. [CrossRef]
53. Grunberg, P. Brillouin scattering from spin waves in thin ferromagnetic films. *J. Magn. Mater.* **1980**, *766*, 15–18. [CrossRef]
54. Courtens, E.; Pelous, J.; Phalippou, J.; Vacher, R.; Woignier, T. Brillouin-scattering measurements of phonon-fracton crossover in silica aerogels. *Phys. Rev. Lett.* **1987**, *58*, 128. [CrossRef]
55. Damzen, M.J.; Vlad, V.I.; Babin, V.; Mocofanescu, A. *Stimulated Brillouin Scattering*; Institute of Physics Publishing: London, UK, 2003.
56. Mock, R.; Hillebrands, B.; Sandercock, R. Construction and performance of a Brillouin scattering set-up using a triple-pass tandem Fabry-Perot interferometer. *J. Phys. E Sci. Instrum.* **1987**, *20*, 656–659. [CrossRef]
57. Kojima, S. Gigahertz acoustic spectroscopy by Micro-Brillouin scattering. *Jpn. J. Appl. Phys.* **2010**, *49*, 07HA01. [CrossRef]
58. Rahaman, M.M. Ferroelectric Phase Transition and Dynamical Behavior of PNRs of Potassium Tantalate Niobate Crystals: Broadband Inelastic Light Scattering Study. Ph.D. Thesis, University of Tsukuba, Tsukuba, Japan, 2016.
59. Burns, G.; Dacol, F.H. Crystalline ferroelectric with glassy polarization behavior. *Phys. Rev. B* **1983**, *28*, 2527–2530. [CrossRef]
60. Toulouse, J.; Jiang, F.; Svitelskiy, O.; Chen, W.; Ye, Z.-G. Temperature evolution of the relaxor dynamics in $Pb(Zn_{1/3}Nb_{2/3})O_3$: A critical Raman analysis. *Phys. Rev. B* **2005**, *72*, 184106. [CrossRef]
61. Pirc, R.; Blinc, R. Vogel-Fulcher freezing in relaxor ferroelectrics. *Phys. Rev. B* **2007**, *76*, 020101. [CrossRef]
62. Glinchuk, M.D.; Farhi, R. A random field theory based model for ferroelectric relaxors. *J. Phys. Condens. Matter* **1996**, *8*, 6985. [CrossRef]
63. Bokov, A.A. Influence of disorder in crystal structure on ferroelectric phase transitions. *J. Exp. Theor. Phys.* **1997**, *84*, 994. [CrossRef]
64. Timonin, P.N. Griffiths' phase in dilute ferroelectrics. *Ferroelectrics* **1997**, *199*, 69–81. [CrossRef]
65. Bokov, A.A. Ferroelectric domains and polar clusters in disordered structures. *Ferroelectrics* **1997**, *190*, 197. [CrossRef]
66. Westphal, V.; Kleemann, W.; Glinchuk, M.D. Diffuse phase transitions and random-field-induced domain states of the relaxor ferroelectric $PbMg_{1/3}Nb_{2/3}O_3$. *Phys. Rev. Lett.* **1992**, *68*, 847. [CrossRef] [PubMed]

67. Ishchuk, V.M. Was it necessary to introduce the notion relaxor ferroelectrics?—The problem of phase transitions in (Pb,$Li_{1/2}$-$La_{1/2}$)(Zr,Ti)O_3, (Pb,La)(Zr,Ti)O_3, Pb($Mg_{1/3}Nb_{2/3}$)O_3, Pb($In_{1/2}Nb_{1/2}$)O_3. and related materials. 1. model conceptions. *Ferroelectrics* **2001**, *255*, 73. [CrossRef]
68. Toulouse, J. The three characteristic temperatures of relaxor dynamics and their Meaning. *Ferroelectrics* **2008**, *369*, 203–213. [CrossRef]
69. Helal, M.A.; Aftabuzzaman, M.; Tsukada, S.; Kojima, S. Role of polar nanoregions with weak random fields in Pb-based perovskite ferroelectrics. *Sci. Rep.* **2017**, *7*, 44448. [CrossRef]
70. Tan, P.; Tian, H.; Mao, C.; Hu, C.; Meng, X.; Li, L.; Shi, G.; Zhou, Z. Field-driven electro-optic dynamics of polar nanoregions in nanodisordered $KTa_{1-x}Nb_xO_3$ crystal. *Appl. Phys. Lett.* **2017**, *111*, 012903. [CrossRef]
71. Jeong, I.-K.; Darling, T.W.; Lee, J.K.; Proffen, T.; Heffner, R.H.; Park, J.S.; Hong, K.S.; Dmowski, W.; Egami, T. Direct observation of the formation of polar nanoregions in Pb($Mg_{1/3}Nb_{2/3}$)O_3 using neutron pair distribution function analysis. *Phys. Rev. Lett.* **2005**, *94*, 147602. [CrossRef]
72. Taniguchi, H.; Itoh, M.; Fu, D. Raman scattering study of the soft mode in Pb($Mg_{1/3}Nb_{2/3}$)O_3. *J. Raman Spectrosc.* **2011**, *42*, 706–714. [CrossRef]
73. Dey, K.; Ahad, A.; Gautam, K.; Tripathy, A.; Majid, S.S.; Francoual, S.; Richter, C.; Singh, M.N.; Sagdeo, A.; Welter, E.; et al. Coexistence of local structural heterogeneities and long-range ferroelectricity in Pb-free (1-*x*)Ba($Zr_{0.2}Ti_{0.8}$)O_3-*x*($Ba_{0.7}Ca_{0.3}$)TiO_3 ceramics. *Phys. Rev. B* **2021**, *103*, L100205. [CrossRef]
74. Dash, S.; Pradhan Dhiren, K.; Kumari, S.; Ravikant; Rahaman, M.M.; Cazorla, C.; Brajesh, K.; Kumar, A.; Thomas, R.; Rack Philip, D.; et al. Enhanced ferroelectric and piezoelectric properties of BCT-BZT at the morphotropic phase boundary driven by the coexistence of phases with different symmetries. *Phys. Rev. B* **2021**, *104*, 224105. [CrossRef]
75. Pradhan Dhiren, K.; Mohanty, H.S.; Kumari, S.; Bhoi, K.; Tang, N.; Ravikant; Rahaman, M.M.; Pradhan Dillip, K.; Kumar, A.; Gilbertah Dustin, A. Ferroic phase transitions and magnetoelectric coupling in cobalt doped $BaTiO_3$. *J. Mater. Chem. C* **2021**, *9*, 12694–12711. [CrossRef]
76. Vakhrushev, S.; Nabereznov, A.; Sinha, S.K.; Feng, Y.P.; Egami, T. Synchrotron X-ray scattering study of lead magnoniobate relaxor ferroelectric crystals. *J. Phys. Chem. Solids* **1996**, *57*, 1517–1523. [CrossRef]
77. Hirota, K.; Ye, Z.-G.; Wakimoto, S.; Gehring, P.M.; Shirane, G. Neutron diffuse scattering from polar nanoregions in the relaxor Pb($Mg_{1/3}Nb_{2/3}$)O_3. *Phys. Rev. B* **2022**, *65*, 104105. [CrossRef]
78. Xu, G.; Shirane, G.; Copley, J.R.D.; Gehring, P.M. Neutron elastic diffuse scattering study of Pb($Mg_{1/3}Nb_{2/3}$)O_3. *Phys. Rev. B* **2004**, *69*, 064112. [CrossRef]
79. Yoshida, M.; Mori, S.; Yamamoto, N.; Uesu, Y. TEM Observation of polar domains in relaxor ferroelectric Pb($Mg_{1/3}Nb_{2/3}$)0_3. *Ferroelecrrics* **1998**, *217*, 327–333. [CrossRef]
80. Bokov, A.A.; Ye, Z.-G. Phenomenological description of dielectric permittivity peak in relaxor ferroelectrics. *Solid State Commun.* **2000**, *116*, 105–108. [CrossRef]
81. Bokov, A.A.; Ye, Z.-G. Dielectric relaxation in relaxor ferroelectrics. *J. Adv. Dielectr.* **2012**, *2*, 1241010. [CrossRef]
82. Samara, G.A. Glasslike behavior and novel pressure effects in $KTa_{1-x}Nb_xO_3$. *Phys. Rev. Lett.* **1984**, *53*, 298. [CrossRef]
83. Yacoby, Y. Defect induced fluctuations in the paraelectric phase of $KTa_{0.94}Nb_{0.06}O_3$. *Z. Physik B* **1978**, *31*, 275. [CrossRef]
84. Vacher, R.; Boyer, L. Brillouin scattering: A tool for the measurement of elastic and photoelastic constants. *Phys. Rev. B* **1972**, *6*, 639–673. [CrossRef]
85. Rahaman, M.M.; Imai, T.; Sakamoto, T.; Tsukada, S.; Kojima, S. The role of polar-nanoregions in $KTa_{1-x}Nb_xO_3$ single crystals studied by Raman scattering. *Ferroelectrics* **2016**, *503*, 85–93. [CrossRef]
86. Slonczewski, J.C.; Thomas, H. Interaction of elastic strain with the structural transition of strontium titanate. *Phys. Rev. B* **1970**, *1*, 3599. [CrossRef]
87. Smolenskiĭ, G.A.; Yushin, N.K.; Smirnov, S.I. Acoustic properties of a lead magnoniobate crystal, a ferroelectric with a broad phase transition. *Sov. Phys. Solid State* **1985**, *27*, 492–495.
88. Landau, L.D.; Lifshitz, E.M. *Physical Kinetics*; English translation of Fizicheskaya Kinetika (Nauka, Moscow); Pergamon Press: Oxford, UK, 1979.
89. Levanyuk, A.P.; Minyukov, S.A.; Vallade, M. Elastic anomalies at structural phase transitions: A consistent perturbation theory. I. One component order parameter. *J. Phys. I Fr.* **1992**, *2*, 1949. [CrossRef]
90. Jiang, M.; Kojima, S. Relaxation mode in 0.65Pb($Mg_{1/3}Nb_{2/3}$)O_3-0.35$PbTiO_3$ relaxor single crystals studied by micro-Brillouin scattering. *Phys. Rev. B* **2000**, *62*, 8572. [CrossRef]
91. Kojima, S.; Ko, J.-H. Broadband micro-Brillouin scattering spectroscopy of Pb($B_{1/3}B'_{2/3}$)O_3-based relaxor ferroelectrics. *Curr. Appl. Phys.* **2011**, *11*, S22. [CrossRef]
92. Sivasubramanian, V.; Tsukada, S.; Kojima, S. Anomalous behavior of acoustic phonon and central peak in [$(In_{1/2}Nb_{1/2})_{0.65}Ti_{0.35}$]$O_3$ single crystal. *J. Appl. Phys.* **2009**, *105*, 014108. [CrossRef]
93. Tsukada, S.; Kojima, S. Broadband light scattering of two relaxation processes in relaxor ferroelectric 0.93Pb($Zn_{1/3}Nb_{2/3}$)O_3-0.07$PbTiO_3$ single crystals. *Phys. Rev. B* **2008**, *78*, 144106. [CrossRef]
94. Kojima, S.; Tsukada, S.; Hidaka, Y.; Bokov, A.A.; Ye, Z.-G. Broadband gigahertz dynamics of relaxor ferroelectric Pb($Sc_{1/2}Nb_{1/2}$)O_3-*x*$PbTiO_3$ single crystal probed by Brillouin scattering. *J. Appl. Phys.* **2011**, *109*, 084114. [CrossRef]

95. Hushur, A.; Gvasaliya, S.; Roessli, B.; Lushnikov, L.; Kojima, S. Ferroelectric phase transition of stoichiometric lithium tantalate studied by Raman, Brillouin, and neutron scattering. *Phys. Rev. B* **2007**, *76*, 064104. [CrossRef]
96. Mishra, K.K.; Sivasubramanian, V.; Arora, A.K.; Pradhan, D. Anomalous behavior of acoustic phonon mode and central peak in $Pb(Zn_{1/3}Nb_{2/3})_{0.85}Ti_{0.15}O_3$ single crystal studied using Brillouin scattering. *J. Appl. Phys.* **2012**, *112*, 114109. [CrossRef]
97. Kojima, S.; Tsukada, S. Micro-Brillouin scattering of relaxor ferroelectrics with perovskite structure. *Ferroelectrics* **2010**, *405*, 32. [CrossRef]
98. Reisman, A.; Banks, E. Reactions of the Group VB Pentoxides. VIII. Thermal, Density and X-Ray Studies of the Systems $KNbO_3$-$NaNbO_3$ and $KTaO_3$-$KNbO_3$. *J. Am. Chem. Soc.* **1958**, *80*, 1877. [CrossRef]
99. Rahaman, M.M.; Imai, T.; Sakamoto, T.; Helal, M.A.; Tsukada, S.; Kojima, S. Ferroelectric phase transition of Li-doped $KTa_{1-x}Nb_xO_3$ single crystals with weak random fields: Inelastic light scattering study. *J. Alloys Compd.* **2018**, *735*, 1063–1070. [CrossRef]
100. Bell, R.O.; Rupprecht, G. Elastic constants of strontium titanate. *Phys. Rev.* **1963**, *129*, 90. [CrossRef]
101. Fossum, J.O.; Fossheim, K. Measurements of ultrasonic attenuation and velocity in Verneuil-grown and flux-grown $SrTiO_3$. *J. Phys. C Solid State Phys.* **1985**, *18*, 5549. [CrossRef]
102. Courdile, J.M.; Dumas, J.; Ziolkiewicz, S.; Joffrin, J. Soft mode characteristics in the KTN ferroelectric transition. *J. Phys. Fr.* **1977**, *38*, 1519–1525. [CrossRef]
103. Rahaman, M.M.; Kojima, S. Site occupancy of Li ions in Li-doped $K(Ta_{1-x}Nb_x)O_3$ single crystals: Reanalysis of inelastic light scattering data and first-principles calculations. *Ferroelectrics* **2022**, *586*, 10–20. [CrossRef]
104. Wakimoto, S.; Samara, G.A.; Grubbs, R.K.; Venturini, E.L.; Boatner, L.A.; Xu, G.; Shirane, G.; Lee, S.-H. Dielectric properties and lattice dynamics of Ca-doped $K_{0.95}Li_{0.05}TaO_3$. *Phys. Rev. B* **2006**, *74*, 054101. [CrossRef]
105. Clark, S.J.; Segall, M.D.; Probert, M.J.; Pickard, C.J.; Hasnip, P.J.; Payne, M.C. First principles methods using CASTEP. *Z. Kristallogr.* **2005**, *220*, 567. [CrossRef]
106. Ye, H.; Zhang, R.; Wang, D.; Cui, Y.; Wei, J.; Wang, C.; Xu, Z.; Qu, S.; Wei, X. First-principles calculation of lead-free perovskite $SnTiO_3$. *Int. J. Mod. Phys. B* **2013**, *27*, 1350144. [CrossRef]
107. Mancini, G.F.; Ghigna, P.; Mozzati, M.C.; Galinetto, P.; Makarova, M.; Syrnikov, P.; Jastrabik, L.; Trepakov, V.A. Structural investigation of manganese doped $SrTiO_3$ single crystal and ceramic. *Ferroelectrics* **2014**, *463*, 31. [CrossRef]
108. Shannon, R.D. Revised effective ionic radii and systematic studies of interatomic distances in halides and chalcogenides. *Acta Cryst.* **1976**, *32*, 751. [CrossRef]
109. Samara, G. The relaxational properties of compositionally disordered ABO_3 perovskites. *J. Phys. Condens. Matter* **2003**, *15*, R367. [CrossRef]
110. Samara, G.A.; Boatner, L.A. Ferroelectric-to-relaxor crossover and oxygen vacancy hopping in the compositionally disordered perovskites $KTa_{1-x}Nb_xO_3$:Ca. *Phys. Rev. B* **2000**, *61*, 3889. [CrossRef]
111. Rahaman, M.M.; Imai, T.; Miyazu, J.; Kobayashi, J.; Kojima, S. Micro-Brillouin scattering study on composition gradient Li doped $KTa_{1-x}Nb_xO_3$ wafer. *Ferroelectrics* **2015**, *487*, 47–54. [CrossRef]
112. Imai, T.; Yagi, S.; Yamazaki, H. Thermal fixing of photorefractive holograms in $KTa_{1-x}Nb_xO_3$ and its relation to proton concentration. *J. Opt. Soc. Am. B* **1996**, *13*, 2524. [CrossRef]
113. Rahaman, M.M.; Imai, T.; Sakamoto, T.; Kojima, S. Electric field effects on polar-nanoregions in Li- doped $KTa_{1-x}Nb_xO_3$ single crystals probed by micro-Brillouin scattering. *Integr. Ferroelectr.* **2017**, *185*, 22–30. [CrossRef]
114. Aftabuzzaman, M.; Kojima, S. Electric field effect of relaxor ferroelectric (1-x)$Pb(Mg_{1/3}Nb_{2/3})O_3$-$x$$PbTiO_3$ crystals near morphotropic phase boundary composition probed by Brillouin scattering. *Jpn. J. Appl. Phys.* **2016**, *55*, 70KB03. [CrossRef]
115. Kutnjak, Z.; Petzelt, J.; Blinc, R. The giant electromechanical response in ferroelectric relaxors as a critical phenomenon. *Nature* **2006**, *441*, 956. [CrossRef]
116. Rahaman, M.M.; Imai, T.; Sakamoto, T.; Kojima, S. Electric field induced critical points and field gradient by trapped electrons in Li-doped $KTa_{1-x}Nb_xO_3$ single crystals. *Appl. Phys. Lett.* **2017**, *111*, 032904. [CrossRef]

materials MDPI

Review

Anomalies of Brillouin Light Scattering in Selected Perovskite Relaxor Ferroelectric Crystals

Venkatasubramanian Sivasubramanian [1,2,*], Sarveswaran Ganesamoorthy [1,2] and Seiji Kojima [3]

1 Condensed Matter Physics Division, Indira Gandhi Centre for Atomic Research, Kalpakkam 603102, India
2 Homi Bhabha National Institute, Mumbai 400094, India
3 Division of Materials Science, University of Tsukuba, Tsukuba 305-8573, Japan
* Correspondence: shiva@igcar.gov.in

Abstract: Compositionally disordered perovskite compounds have been one of the exotic topics of research during the past several years. Colossal piezoelectric and electrostrictive effects have been observed in disordered perovskite ferroelectric materials. The key ingredient in the physical behavior of disordered perovskites is the nucleation and growth of the local dipolar regions called polar nanoregions (PNRs). PNRs begin to nucleate far above the temperature of the dielectric maximum T_m and exhibit varied relaxation behavior with temperature. The evidence for the existence of various stages in the relaxation dynamics of PNRs was revealed through the study of the temperature evolution of optical phonons by Raman scattering. The quasi-static regime of PNRs is characterized by the strong coupling between the local polarization and strain with the local structural phase transition and the critical slowing of the relaxation time. Strong anomalies in the frequency and the width of the acoustic phonons, and emergence of the central peak in the quasi-static region of the relaxation dynamics of PNRs have been observed through Brillouin scattering studies. In this review, we discuss the anomalies observed in Brillouin scattering in selected disordered perovskite ferroelectrics crystals such as $Pb(Mg_{1/3}Ta_{2/3})O_3$, $Pb(Sc_{1/2}Ta_{1/2})O_3$, 0.65PIN-0.35PT and $Sr_{0.97}Ca_{0.03}TiO_3$ to understand dynamical behavior of PNRs.

Keywords: disordered ferroelectrics; perovskites; polar nanoregions; Brillouin scattering; acoustic phonon; relaxation

Citation: Sivasubramanian, V.; Ganesamoorthy, S.; Kojima, S. Anomalies of Brillouin Light Scattering in Selected Perovskite Relaxor Ferroelectric Crystals. *Materials* **2023**, *16*, 605. https://doi.org/10.3390/ma16020605

Academic Editor: Efrat Lifshitz

Received: 23 November 2022
Revised: 3 January 2023
Accepted: 5 January 2023
Published: 8 January 2023

1. Introduction

The ABO_3 perovskite-type ferroelectric oxides have been extensively studied over several decades both from the fundamental and applications point of view. The prototype ferroelectrics such as $BaTiO_3$ and $KNbO_3$undergo a sequential structural phase transition from initial cubic phase to final rhombohedral phase, involving softening of zone-center optical phonon modes in each phase. On the other hand, $PbTiO_3$exhibits a single-phase transition from cubic to tetragonal ferroelectric phase and is regarded as a text book example of the soft-mode-driven phase transition. A ferroelastic transition at T_{tr} = 110 K driven by the softening of R_{25} zone boundary phonon mode was observed in $SrTiO_3$, while $KTaO_3$ remains paraelectric down to the lowest temperatures. In both $SrTiO_3$ and $KTaO_3$, the low temperature dielectric constant increases upon cooling but becomes temperature independent (<10 K for $SrTiO_3$and $KTaO_3$) due to quantum fluctuations. In both these systems, the quantum fluctuations suppress the ferroelectric instability and they enter a phase called quantum paraelectrics [1,2].

The substitution of ions of different valences and polarizabilities at both A and B sites in the parent perovskite compound results in the compositional disorder. This structural disorder gives rise to the formation of local dipolar impurities and defects that profoundly affect the static and dynamic properties of these materials [2]. The off centering of the A and/or B ion in the high polarizable perovskite host results in the development of net dipole moment in the unit cell. Each dipole polarizes the surrounding region, thus forming the local polar region called the polar nanoregion (PNR). The spatial correlation length r_c determines the size of the PNRs. The flipping of polarization between the equivalent directions within the PNR gives rise to the relaxation behavior. In the dilute limit of substitution, r_c is small, and each PNR behaves as an independent, noninteracting dipole with a single relaxation time. As substitution increases, so does r_c, and the PNRs begin to interact with one another. The interaction between the PNRs results in complex relaxation behavior with a distribution of relaxation times. The average relaxation time then becomes strongly temperature dependent and exhibits divergent behavior with a decrease in temperature [2]. The dielectric constant (ε_r) shows remarkable frequency dependence and broadening of the peak at T_m, the temperature of the dielectric maximum. It is for this reason that the compositionally disordered ferroelectric materials are generally termed "relaxor ferroelectrics" or simply "relaxors." In what follows, we discuss the important physical features exhibited by perovskite-type relaxor ferroelectric materials.

The relaxor behavior can be induced in a paraelectric host by the substitution of suitable ions at A and B sites, as in $K_{1-x}Li_xTaO_3$, $KTa_{1-x}Nb_xO_3$ and $Sr_{1-x}Ca_xTiO_3$. Relaxorbehavior may also be intrinsic due to the compositional disorder at the B site as in Pb-based compounds. Pb-based relaxor ferroelectrics are $Pb(Mg_{1/3}Nb_{2/3})O_3$ (PMN), $Pb(Mg_{1/3}Ta_{2/3})O_3$ (PMT), $Pb(Zn_{1/3}Nb_{2/3})O_3$ (PZN), $Pb(Sc_{1/2}Nb_{1/2})O_3$ (PSN), $Pb(Sc_{1/2}Ta_{1/2})O_3$ (PST) and $Pb(In_{1/2}Nb_{1/2})O_3$ (PIN). Numerous studies have been carried out on the relaxation dynamics of PNRs in these disordered perovskite compounds. The temperature evolution of relaxation behavior of PNRs has been studied through dielectric, X-ray and neutron diffraction, thermal expansion, ultrasonic spectroscopy, inelastic neutron, and Raman and Brillouin scattering [3–11]. The characteristic feature that is related to the dynamical behavior of PNRs in the disordered perovskites is manifested in the temperature and the frequency dependence of the dielectric constant. The maximum dielectric constant ε_r^{max} and its corresponding maximum temperature $T_m$varies strongly with the frequency of the applied electric field. It is observed that ε_r^{max} decreases while T_m increases with the increase in frequency [6].

$K_{1-x}Li_xTaO_3$ (KLT) exhibits relaxor behavior for the Li concentration up to 4%. Similar behavior is observed in $KTa_{1-x}Nb_xO_3$ (KTN) for Nb substitution up to 2% [2]. In KLT, the large ionic radius difference between K (1.51 Å) and Li (0.92 Å) causes Li^+ to be off-centered about 1.2 Å from ideal cubic position along the [100] direction. This results in the large dipole moment of Li^+ ion and forms PNRs through the interaction with neighboring unit cells. Similarly, off-center displacement of Nb^{5+} ion in the highly polarizable $KTaO_3$ host results in the formation of PNRs. Relaxor behavior has been observed for 1.5 and 3.5% substitution of Li in KLT through temperature and frequency-dependent dielectric studies [12]. Depending on the frequency of the applied field, both KLT and KTN show a broad dielectric maximum, with T_m varying from 40 to 80 K. A strong deviation from Curie–Weiss behavior was observed in the temperature dependence of dielectric permittivity at the temperature T^*, which is 20 K above T_m. T^* was identified as the temperature where PNRs enter into quasi-static regime. The symmetry-forbidden nonpolar Raman modes associated with PNRs appear for both KLT and KTN at T^* [13,14]. The intensity of these nonpolar modes was found to increase with a decrease in temperature below T^*. For both KLT and KTN, the average size of PNRs was estimated from the analysis of the Raman data [13,14]. The coupling between local polarization and strain in KTN was investigated through ultrasonic studies [9]. The elastic constants C_{11} and C_{44} exhibited a strong shift from the mean-field behavior at about T^*. The deviation was attributed to the electrostrictive coupling between the local polarization and strain in PNRs. Recently, the ferroelectric phase transition in KLT crystals has been studied by broadband dielectric spectroscopy. Several different types of dipolar relaxations covering a broad temperature range have been observed [15].

Relaxor behavior has been observed in $Sr_{1-x}Ca_xTiO_3$ (SCT-x) for small x. As the ionic radius of Ca^{2+} ion (1.12 Å) is considerably smaller than that of Sr^{2+} (1.26 Å), off-centering of

Ca^{2+} ion at the Sr^{2+} site gives rise to the formation of PNRs. The evidence for the existence of Ca-induced PNRs was ascertained through dielectric, Raman scattering and nonlinear optical studies [16,17]. For small x, SCT show a glassy dipolar behavior and above the critical concentration of $x = 0.0018$, a long-range ferroelectric order was observed [18]. Based on dielectric and X-ray diffraction studies, a temperature–composition phase diagram was proposed for whole range of SCT solid solutions [19]. According to the phase diagram, SCT-x exhibits a ferroelastic-type phase transition from cubic to tetragonal structure at high temperature, and ferroelastic to ferroelectric relaxor transition accompanied by the change in structure from tetragonal to orthorhombic at low temperature [19]. At $x = 0.04$, the ferroelastic phase transition from cubic to tetragonal phase with the transition temperature of 210 K and the ferroelastic to ferroelectric relaxor transition at 30 K was observed through dielectric, X-ray and neutron diffraction studies [20]. Recent studies reveal the intriguing complexity of the physical behavior in oxygen deficient SCT. SCT-x becomes metal and shows superconducting behavior with the coexistence of both ferroelectric and superconducting phases for extremely small oxygen vacancies [21]. As in the case of $SrTiO_3$, critical elastic anomalies are expected to occur in SCT-x at the ferroelastic phase transition temperature.

Pb-based perovskites $Pb(B'_{1/3}B''_{2/3})O_3$ and $Pb(B'_{1/2}B''_{1/2})O_3$ are a special group of relaxor ferroelectrics that exhibit complex structural and dynamical behaviors. Both types of relaxors have fixed chemical compositions and are intrinsically disordered at the B site. The antiparallel short-range displacement of Pb and B′/B″against oxygen octahedra along <111> direction leads to the formation of the PNRs [7]. In Pb-based relaxors, the nucleation of PNRs takes place at T_B, the Burns temperature, which is much higher than T_m. Non-Arrhenius behavior in the temperature dependence of relaxation time of PNRs was observed and the frequency dependence of the temperature of dielectric maximum T_m was explained through the phenomenological Vogel–Fulcher relation [22]. The nucleation and further growth of PNRs below T_B leads to the change in the thermal expansion, the deviation of Curie–Weiss law, specific heat, optical and acoustic phonon anomalies in Pb-based relaxors also [8,23–25]. In Pb-based relaxors, the temperature evolution of PNR dynamics was initially assumed to be characterized by two temperatures, namely T_B, and T_f, the freezing temperature. However, the detailed analysis of the Raman spectra of PMN, PZN and related relaxors revealed the existence of another intermediate percolation temperature, T^* between T_B and T_m [10,26,27]. The observed anomalous variations in the frequency and intensity of Raman active phonon modes and the appearance of central peaks confirmed the existence of T^*. T^* is proposed as the onset temperature for the development of strong correlations between the initial quasi-dynamic PNRs. The temperature evolution of relaxation dynamics of PNRs consists of four different temperature stages: (i) a fully dynamic without any correlations above T_B, (ii) quasi-dynamic phase between the temperature T_B and T^* with weak correlations, (iii) strongly correlated quasi-static phase between T^* and T_f and (iv) static below T_f [26]. The existence of the intermediate percolation temperature T^* was also confirmed through acoustic emission studies in PZN and related systems [28,29]. While the Burns temperature T_B widely varies, T^* was found to be almost same for all Pb-based relaxor ferroelectrics within the narrow temperature interval of 500 ± 30 K [30]. From the application point of view, the single crystals of relaxor-$PbTiO_3$ (PT) solid solutions such as PMN-xPT, PZN-xPT and PIN-xPT with the composition at the morphotropic phase boundary (MPB) show giant piezoelectric effect, making them the ideal choice for ultrasonic transducers, actuators and micropositioners [31–33].

In contrast to $Pb(B'_{1/3}B''_{2/3})O_3$-type RFEs, the $Pb(B'_{1/2}B''_{1/2})O_3$ relaxor group exhibits varied ferroelectric behavior depending on the degree of B-site ordering [34,35]. The degree of B-site ordering can be altered by thermal treatment. The disordered (thermally quenched) PSN, PST and PIN show relaxor-type ferroelectric behavior [36–38]. While the ordered (thermally annealed) PSN and PST exhibit normal ferroelectric-like phase transition [36,37], the ordered PIN exhibit an antiferroelectric phase transition [38]. Both disordered PSN and PST undergo a spontaneous relaxor to the ferroelectric phase transition accompanied by the change in structure from cubic to rhombohedral at the temperature T_c below T_m [35–37]. The relaxor behavior becomes suppressed and the transition temperature decreases with increase in the degree of B-site ordering [37]. The nuclear magnetic resonance (NMR) studies revealed a substantial difference in the polarization dynamics between the disordered and the ordered regions of $Pb(B'_{1/2}B''_{1/2})O_3$ relaxor group [39,40]. NMR studies also revealed that the long-range polar order occurs only in fully B-site ordered region of the crystal [39,40].

In RFEs, a significant change in the dynamical behaviour of PNRs takes place in the quasi-static regime between T^* and T_f. This temperature range gives rise to the formation of static PNRs due to the strongly correlated shifts of A and/or B ions and associated quenched random dipolar fields.The local dipolar field results in the variation in the local strain field and both become coupled through quadratic electrostrictive effect. Brillouin light scattering spectroscopy is a valuable technique to study the coupling between the strain and polarization and relaxation dynamics of PNRs in relaxor ferroelectrics. The characteristic frequency range associated with the relaxation dynamics of PNRs is rightly located in the low frequency part of the Brillouin scattering, which is normally inaccessible through conventional Raman and thermal neutron scattering. The anomalous behavior of elastic constants in the quasi-static region in PMN and PMT was studied through Brillouin light scattering spectroscopy. Below T^*, a broad central peak associated with the polarization fluctuations of PNRs was observed in PMN and PMT [41]. Both in PMN and PMT, the relaxation time of PNRs showed a characteristic slowing behavior with a marked anomaly below T_f. The anomalous behavior of acoustic phonons and the central peak has been studied extensively in various Pb-based relaxor systems by Brillouin light scattering spectroscopy [42–54]. A significant softening of LA and TA phonon modes accompanied by a broad central peak was observed below T^* in PZN-0.045PT and PZN-0.09PT [44]. In the Brillouin light scattering studies of PZN-xPT with $x = 0.07$, 0.1 and 0.12, a minimum was observed in the temperature derivative of the width of the LA phonon, $d\Gamma/dT$ at T^* [45]. Brillouin light scattering from acoustic phonons was investigated for a wide range of compositions in PMN-xPT solid solution [47]. As in the case of PZN-xPT, below T^*, a significant softening of frequency of the LA phonon followed by increase in the width was observed. The activation energy of flipping of PNRs calculated using a modified superparaelectric model showed a dramatic increase across the composition of MPB. In 0.65PIN-0.35PT, anomalous behavior of the LA phonon mode at the cubic–tetragonal structural phase transition and two-component central peaks were studied using Brillouin light scattering spectroscopy [50]. The width of the LA phonon showed Landau–Khalatnikov-type behavior in the vicinity of the structural phase transition in a highly ordered PSN [52]. This review focuses on the Brillouin light scattering studies of dynamical aspects of PNRs in some selected RFEs carried out by the authors in the past. In the following, the results obtained in the Brillouin light scattering studies on RFEs such as PMT, PST, 0.65PIN-0.35PT and SCT-0.03 are reviewed in detail.

2. Brillouin Scattering

In Brillouin scattering, the incident photon undergoes inelastic scattering by the near-Brillouin zone-center acoustic phonons ($q \approx 0$). The shift in the frequency of the scattered photon $\Delta\nu_B$, known as the Brillouin shift is given by [53]:

$$\Delta v_B = \pm\frac{2nV}{\lambda_0} sin\left(\frac{\theta}{2}\right) \quad (1)$$

where $\pm$ in the above Equation (1) represents the Stokes and anti-Stokes Brillouin scattering. λ_0, n, V and θ correspond to the wavelength of the incident photon, refractive index, phase velocity of the acoustic phonon and the scattering angle between the incident and scattered photons, respectively. The intensity of the scattered light $I(q,\omega)$ at the given wave vector q

and the frequency ω is related to the imaginary part of the complex dynamic susceptibility through the fluctuation–dissipation theorem, as given by [55]:

$$I(q,\omega) = \frac{\hbar}{\pi}[n(\omega)+1]\ Im\chi^*(q,\omega\ ) \quad (2)$$

where $n(\omega)$, the Bose–Einstein thermal population factor, is given by:

$$n(\omega) = \frac{1}{\exp\left(\frac{\hbar\omega}{k_B\ T}\right)-1} \quad (3)$$

where $\hbar$ is the Planck's constant, k_B the Boltzmann constant and T the temperature. For zone-center acoustic phonons, the dynamic susceptibility $\chi^*(q,\omega)$ is given by the Lorentzian response function:

$$\chi^*(0,\omega) = \frac{f_0}{\omega^2-\omega_0^2-i\gamma_0\omega} \quad (4)$$

where f_0 is the oscillator strength, and ω_o and γ_o are the frequency and damping factor of the acoustic phonon.

The high-contrast multi-pass Fabry–Perot interferometer (FPI) developed by Sandercock [56] is well suited to study Brillouin scattering in solids. A 3 + 3 pass tandem FPI provides adequate contrast and resolution to measure the frequency and the width of the acoustic phonons in solids in various scattering geometries. Since 3 + 3 pass FPI covers a large frequency range, 0.03–30 cm^{-1}, not only the acoustic phonons but also the features of the central peak can be studied.

3. Materials and Methods

Single crystals of PMT, 0.65PIN-0.35PT and PST were grown by the flux method as described elsewhere [50,57,58]. Single crystals of SCT-0.03 were grown by flame fusion technique [59]. The degree of B-site ordering, S parameter, for PST was determined using powder X-ray diffraction. Brillouin light scattering measurements for all the crystals were performed using a high-contrast 3 + 3 pass Sandercock tandem Fabry–Perot Interferometer in backscattering geometry. Brillouin spectra were excited using a single longitudinal mode diode-pumped solid-state laser operating at 532 nm. Measurements were carried out in different free spectral ranges (FSR) in the temperature range 870–100 K using a Linkam (THMS 600) stage. The frequency and the width of acoustic phonons were deduced by fitting the experimental spectra with the Voigt function. During fitting, the Gaussian part of the Voigt function was kept constant to account for the instrumental broadening. The instrumental broadening was obtained by fitting the transmitted peak of the interferometer with the Gaussian function. The full width at half maximum of the Gaussian fit is the measure of the instrumental broadening. The central peak was fitted using a Lorentzian function with $\omega_o = 0$.

4. Results and Discussion

4.1. Elastic Anomalies and Polarization Dynamics in PMT

The Brillouin spectra of PMT measured in the FSR range of 75 GHz at different temperatures are shown in Figure 1. A single LA phonon mode with a frequency around 42 GHz can be seen. Since the macroscopic structure of PMT is cubic, the selection rule permits only the excitation of LA phonon mode in the backscattering geometry. The corresponding elastic constant C_{11} is given by the following expressions [60]:

$$V_{11} = \frac{\lambda_o v_{LA}}{2n},\ C_{11} = \rho(V_{11})^2,\ \alpha = \frac{\pi\Gamma}{V_{11}} \quad (5)$$

where ν_{LA}, Γ, V_{11} and α in Equation (5) are the frequency, width, velocity and attenuation of the LA phonon, and ρ and n are the density and the refractive index of PMT and λ_o is the wavelength of the incident photon. Using the parameters of the LA phonon, C_{11} and α were calculated using the known values of $\rho = 9.65$ gm/cm^3 and the refractive index $n = 2.475$ for PMT [60]. The temperature dependence of the refractive index was not considered in the calculation of C_{11}, as it does not significantly affect its magnitude. Figure 2 shows the temperature dependence of C_{11} and α for PMT.

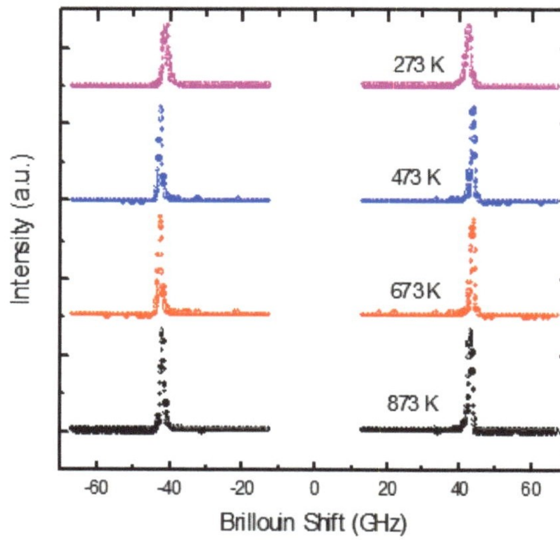

Figure 1. Brillouin spectra of PMT at different temperatures(Reprinted with kind permission from Ref. [41]. 2015, the *European Physical Journal* (EPJ)).

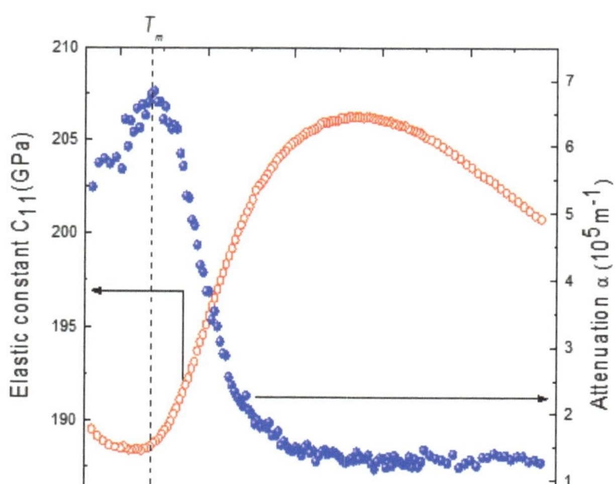

Figure 2. Temperature dependence of the elastic constant C_{11} (open circles) and attenuation α (closed circles) of PMT (Reprinted with kind permission from Ref. [41]. 2015, the *European Physical Journal* (EPJ)).

The elastic constant C_{11} shows a broad minimum and α shows a broad maximum in the region of the dielectric anomaly at T_m. Our data are in good agreement with those of the earlier studies on the elastic anomalies observed in PMT [60]. The minima of C_{11} agree well with the corresponding dielectric anomaly of the PMT [42]. In the present review, a mean field model that was previously used to study the coupling between the polarization and the strain in KTN is employed [9]. The temperature dependence of C_{11} according to this model is given by the following expression:

$$C_{11} = A_1 - A_2 T - A_3 (T - T_o)^{-p} \tag{6}$$

The first two terms of Equation (6) describe the normal anharmonic process, and the third term accounts for the critical softening of C_{11} due to the critical fluctuations. For three-dimensional (3D) fluctuations, the critical exponent $p = 0.5$, for 2D and 1D fluctuations; it is 1 and 1.5 [61]. Since PNRs begin to appear around T_B, the temperature dependence of C_{11} was fitted with this equation for temperatures higher than T_m. As the structure of PMT is cubic, it is reasonable to assume p to be 0.5. $C_{11}$was therefore fitted with the critical exponent p fixed at 0.5 and all the other parameters in Equation (6) are allowed to vary freely [9]. The parameters obtained from the fitting are: A_1 = 223.45 GPa, A_2 = 0.024 GPa/m, A_3 = 36.5 GPa and T_o = 463 K. Figure 3 shows the result of the fitting for PMT.

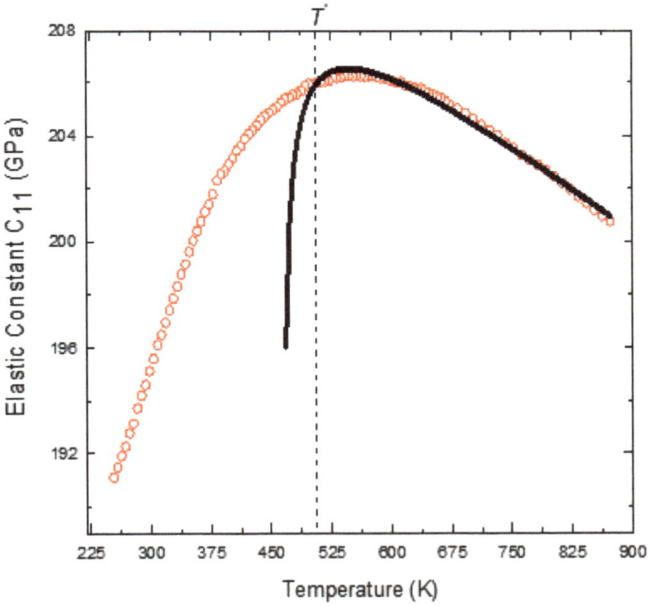

Figure 3. Fit of the experimental C_{11} (open circles) with Equation (6) (solid line) of PMT (Reprinted with kind permission from Ref. [41]. 2015, the *European Physical Journal* (EPJ)).

The experimental C_{11} shows good agreement with the mean-field model in the temperature above T_o + 40 K. C_{11} shows a marked deviation from the mean-field model below this temperature and exhibits reduced softening. The attenuation α also begins to rise precisely at the temperature where C_{11} deviates from the mean field model. The temperature at which C_{11} begins to show reduced softening can be identified with the characteristic percolation temperature T^* [26]. T^* obtained for PMT is 503 K, which is well in agreement with that of the X-ray diffraction data [30]. The temperature dependence of C_{11} between T_B and T^* is determined by the third term of Equation (6), which accounts for the quasi-dynamic fluctuations inpolarization.

The reduced softening of C_{11} and increased α below T^* indicates the appearance of excess strain in the crystal. This additional strain ΔS arises from the electrostrictive coupling between the polarization and strain and is expressed as $\Delta S = (\gamma_{11} + 2\gamma_{12})\langle P^2 \rangle$, where

γ_{11} and γ_{12} are the electrostriction coefficients [2]. Equation (6) does not account for the excess contribution from the electrostriction γ to the elastic constant C_{11}. The reduced softening of C_{11} below T^* can be qualitatively explained using Kubo's formalism [62]. According to this formalism, the polarization fluctuations δP contain two components with very distinct dynamics:

$$\delta P = \delta P_d + \delta P_s \quad (7)$$

Here, δP_d is the quasi-dynamic part, i.e., the third term of Equation (6), and δP_s corresponds to much slower fluctuations (quasi-static) with the development of long-range correlations between the PNRs. The quasi-static part δP_s is expressed in terms of the Edwards–Anderson order parameter $q^{EA} = V^{-1} \int \langle \delta P_s \rangle^2 dV$, where V is the volume over which long-range correlations exist [62,63]. The change in the elastic constant ΔC in the quasi-static region can be written as [63]:

$$\Delta C_s = -4(R_{11} + R_{12})^2 q^{EA} \chi' \chi'' \quad (8)$$

where χ' and χ'' are the real and imaginary parts of the low-frequency dielectric susceptibility. In the disordered $Rb_{1-x}(NH_4)_xH_2PO_4$, the quasi-static term of Equation (8) makes a dominant contribution to the anomalous region of the frequency and width of LA phonon mode [62,63]. A detailed study of the contribution ofthe quasi-static term ΔC_s on the temperature dependence of C_{11} for Pb-based relaxors is yet to be carried out. However, a qualitative idea of the temperature dependence of ΔC_s can be obtained from the study of q^{EA}. The direct determination of q^{EA} by ^{93}Nb NMR for PMN shows a strong increase with a decrease in temperature below T^* [64]. The deviation of the elastic constant C_{11} from the mean field behavior below T^* therefore evidently arises from the additional term ΔC_s of the quasi-static polarization within PNRs. In fact, in PMN, the reduced softening of C_{11} almost coincides with the rise inthe local polarization P_d below T^* [65]. Therefore, in PMT, the reduced softening of C_{11} and the increased attenuation in the quasi-static region are evidently associated with the appearance of additional strain due to the strong coupling between the polarization and the strain.

In PMT, a broad central peak with marked temperature dependence appears below T^*. The typical central peak spectra measured in the FSR range of 500 GHz at different temperatures are displayed in Figure 4a–d. The intensity of the central peak can be described as a Debye-type relaxation according to the following equation [66,67]:

$$I(\omega) \propto \frac{\omega\tau}{(1 + \omega^2\tau^2)} \quad (9)$$

where τ is the relaxation time.In the strong damping limit, ($\gamma_o >> \omega_o$), the Lorentzian function given by Equation (4) reduces to the Debye-type function with $\omega_o = 0$ [55]. The width of the central peak obtained from the fit of the Lorentzian function is then proportional to $1/\tau$. The width of the central peak obtained from the fit is shown in Figure 5. With the decrease in temperature below T^*, the width decreases with a change in slope between T_m and 160 K. An anomalous decrease in the width of the central peak can be observed below 160 K. This temperature corresponds to the freezing temperature T_f for PMT [68]. The relaxation time of the central peak corresponds to the short relaxation time τ_s obtained from the broadband dielectric measurements [68].

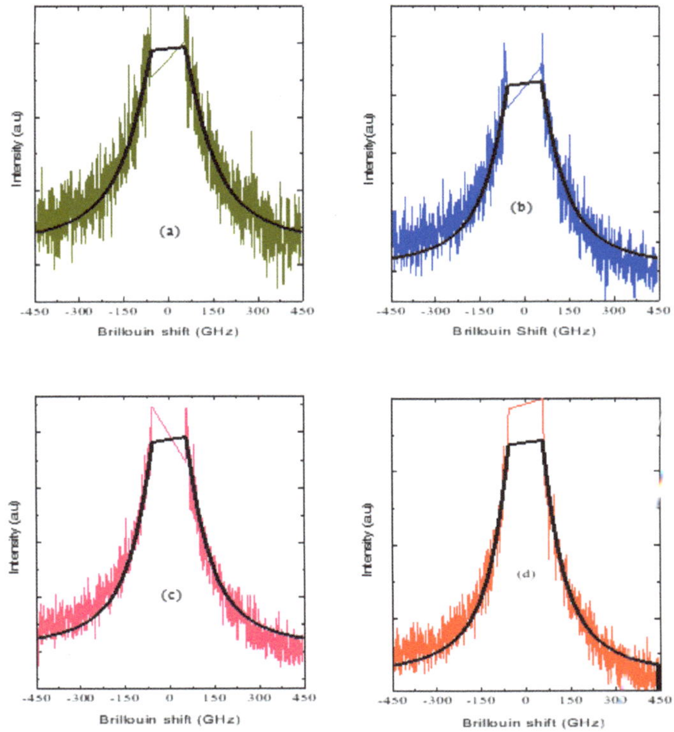

Figure 4. (**a**–**d**)Typical broadband Brillouin spectra of PMT at (**a**) 293 K, (**b**) 273 K, (**c**) 253 K and (**d**) 223 K.The solid black line is the Lorentzian fit of the experimental data.

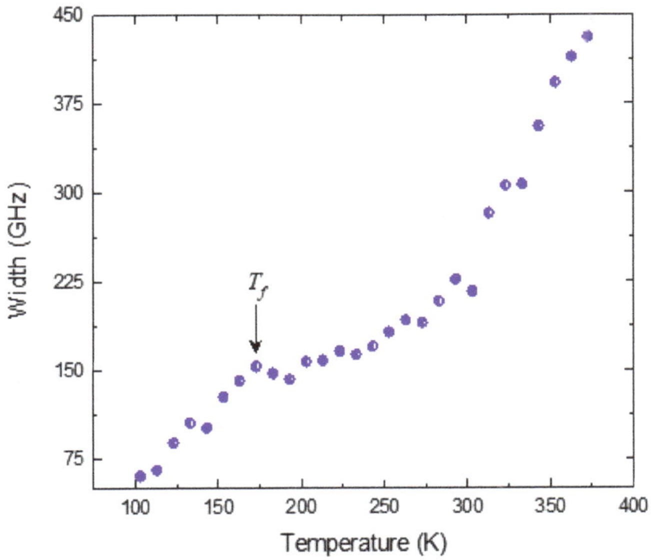

Figure 5. Temperature variation inthe width of the central peak of PMT (Reprinted with kind permission from Ref. [41]. 2015, the *European Physical Journal* (EPJ)).

To obtain a qualitative understanding of the relaxation behavior, the temperature dependence of relaxation time was analyzed using the simple Arrhenius law $\tau = \tau_0 \exp[E_a/kT]$, as shown in Figure 6. The parameters obtained from the fit are $\tau_0 = 2.0 \pm 0.14 \times 10^{-12}$S, $E_a = 220 \pm 9$ K for PMT. Though the fitting parameters are reasonable, the Arrhenius behavior does not agree well with the experimental data. The observed non-Arrhenius behavior suggests that the relaxation dynamics are not governed by dipolar fluctuations alone. Broadband dielectric studies in Pb-based relaxor systems suggest that the relaxation dynamics of PNRs are not governed by a single relaxation mechanism [68–70]. According to these studies, the dominant mechanism of relaxation between T^* and T_f is the dipole flipping, while that below T_f is the volume fluctuations (breathing) of PNRs [68–70]. The strong increase in the relaxation time below T_f suggests that the relaxation mechanism above T_f is different from that below T_f. Therefore, the observed critical slowing of the relaxation time of the central peak between T^* and T_f can be attributed to the dipole flipping and below T_f to the volume fluctuations of PNRs.

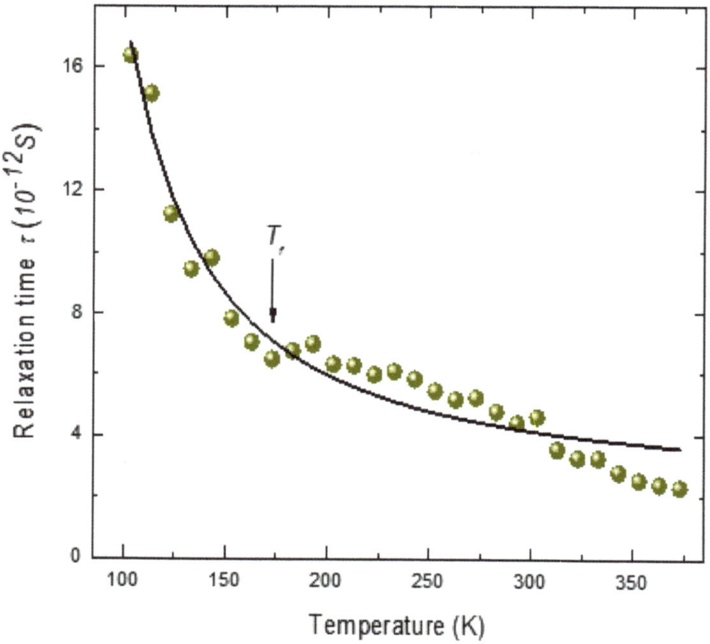

Figure 6. Fit of the Arrhenius law (solid line) with the relaxation time τ (solid sphere) (Reprinted with kind permission from Ref. [41]. 2015, the *European Physical Journal* (EPJ)).

This interpretation seems to be quite consistent with the existence of different stages of relaxation dynamics of PNRs as revealed by Raman scattering studies [26]. The increase in correlations between PNRs in the quasi-static regime between T^* and T_f increase the relaxation time. In the temperature range below T_f, due to the freezing, the dipolar flipping in PNRs is no longer possible. Therefore, the observed anomalous increase in the relaxation time below T_f could be due to the volume fluctuations of PNRs.

4.2. *Anomalous Behavior of Longitudinal Acoustic Phonons in PST*

Figure 7 shows the powder X-ray diffraction pattern of PST. The broad superlattice reflection due to the B-site ordering is observed at $2\theta = 19^{\circ}$.

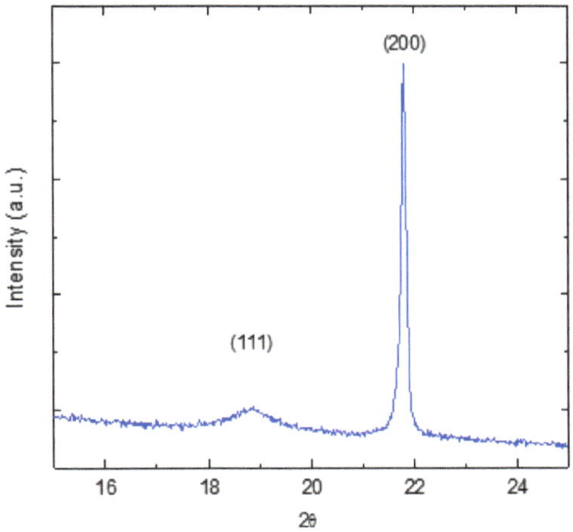

Figure 7. Powder X-ray diffraction of PST.

The degree of B-site ordering $S = \left[\left(\frac{I_{111}}{I_{200}}\right)^{1/2}\right]_{\exp} / \left[\left(\frac{I_{111}}{I_{200}}\right)^{1/2}\right]_{theo}$ was calculated from the ratio of the integrated intensities of the superstructure (111) and the adjacent fundamental (200) peaks [70]. Here, $\left(\frac{I_{111}}{I_{200}}\right)_{exp}$ is the experimental intensity ratio and $\left(\frac{I_{111}}{I_{200}}\right)_{theo}$ is the theoretical intensity ratio for the completely ordered PST. The theoretical intensity ratio I_{111}/I_{200} for the completely ordered PST is estimated to be 1.33 [71]. The S parameter for the as-grown PST single crystals is found to be $S = 0.55$. This indicates that the as-grown PST single crystal is partially disordered. Figure 8 shows the low-frequency dielectric dispersion of partially ordered PST. Since it is known that an increase in B-site ordering suppresses the relaxor behavior, the partial ordering in PST results in the weak relaxor-like behavior with T_m = 297 K.

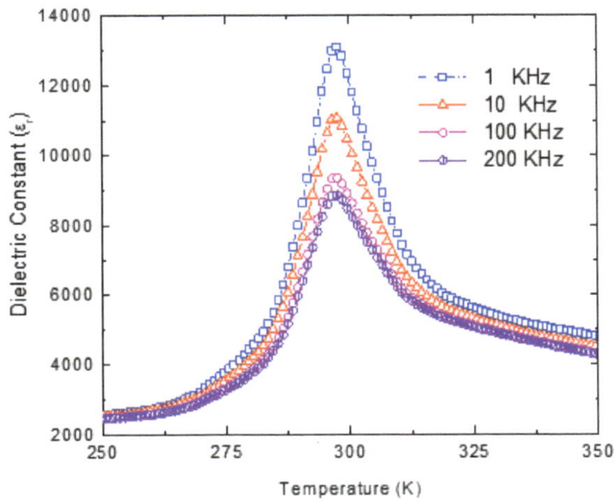

Figure 8. Low frequency dielectric dispersion of PST.

The typical Brillouin spectra of PST measured in the FSR range of 75 GHz at different temperatures are displayed in the Figure 9. In the cubic phase above 295 K, only the LA phonon mode could be observed at the backscattering geometry. Since PST exhibits a spontaneous phase transition from cubic to rhombohedral structure [72], a transverse acoustic (TA) phonon mode at 27 GHz, allowed by symmetry, appears below 295 K in the rhombohedral phase. However, the intensity of the TA phonon was too weak to obtain a reliable value for the frequency and width. Hence, the temperature dependence of TA mode is not discussed further.

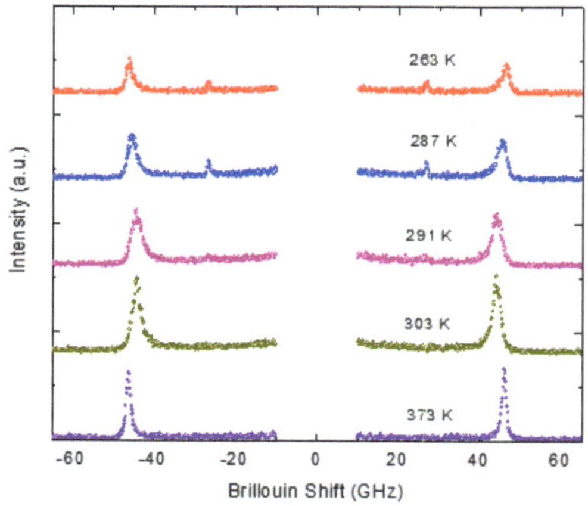

Figure 9. Brillouin spectra of PST measured in the FSR range of 75 GHz.

Figure 10 displays the temperature variation in frequency ω_{LA} and the width Γ_{LA} of the LA phonon. Above 600 K, the frequency and the width of the LA phonon are almost independent of temperature. The LA phonon mode begins to deviate from the normal high-temperature behavior below 600 K and exhibits marked softening with a sharp minimum at T_c = 295 K, the transition temperature. The width of the phonon mode begins to show an anomalous increase at 450 K with a sharp maximum at T_c.

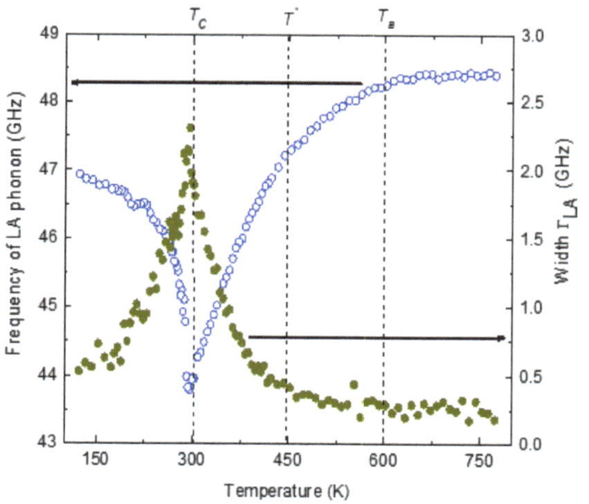

Figure 10. Temperature variation in frequency (open circles) and the width(closed circles) of the LA phonon of PST.

Interaction between the acoustic phonon and polarization affects the static and dynamic parts of the complex elastic constant C. Depending on the form of the interaction energy, the interaction can be classified as either linear or nonlinear. For bilinear coupling, the interaction energy $F_c = \beta SP$, where S is the strain, P is the polarization and β is the piezoelectric coefficient. $F_c = \gamma SP^2$ for electrostrictive coupling, where γ is the electrostrictive coefficient. The coupling between the strain and polarization affects both the static and dynamic parts of the complex elastic constant C, as given by [73–76]:

$$C^{*}(\omega) = C_{\infty} - \Delta C \tag{10}$$

In Equation (10), C_∞ is the elastic constant far from the transition temperature and ΔC is the change in elastic constant due to the coupling between strain and polarization. For bilinear coupling, the elastic constant softens as $/T - T_c/^{-1}$ on both sides of the transition. For electrostrictive coupling, the elastic constant abruptly decreases to a lower value at T_c [73,76]. Brillouin scattering measures the dynamic part of C according to Equation (11):

$$C(\omega) = \rho V^2(\omega) \tag{11}$$

where ρ is the bulk density of the crystal and V is the corresponding velocity of the acoustic phonon. Assuming a Debye-type relaxation, the dynamic behavior of C is given by the following form [74,75]:

$$C^{*}(\omega) = C_{\infty} - \frac{\Delta C}{1 - i\omega\tau} \tag{12}$$

where τ is the relaxation time. The frequency of the acoustic phonon is related to the real part of the elastic constant, while the width is related to the imaginary part of the above equation, as given by:

$$Re\, C(\omega) = C_{\infty} - \frac{\Delta C}{1 + \omega^2\tau^2} \tag{13}$$

$$\Gamma = \frac{q}{2\pi\rho V}(Im\, C) = \Gamma_{\infty} + \frac{\Delta C \tau q^2}{2\pi\rho(1 + \omega^2\tau^2)} \tag{14}$$

The relaxation mechanism is also known as Landau–Khalatnikov (LK) mechanism. The line shape of the width of the LA phonon, as shown in Figure 10, is markedly asymmetric about T_c, as expected from the LK mechanism.

Using Equations (5), (13) and (14), the relaxation time can be expressed as:

$$\frac{1}{2\pi\tau} = \frac{v_{\infty}{}^2 - v^2(T)}{\Gamma(T) - \Gamma_{\infty}} \tag{15}$$

where $v(\Gamma)$ is the frequency (width) of the acoustic phonon at a given temperature and $v_\infty(\Gamma_\infty)$ is the corresponding value of the acoustic phonon far above the transition temperature. In the relaxor-like system $Rb_{1-x}(NH_4)_xH_2PO_4$with x= 0.35, the temperature dependence of the width of the LA phonon in the quasi-static region was shown to be equivalent toan LK-type mechanism [62]. Similarly, in the relaxor phase of the PST, the anomalous increase in the width of the LA phonon below $T^*$is evidently connected with the development of the quasi-static part δP_s. Therefore, a linear coupling between the polarization and strain should exist in the quasi-static region below T^* [62]. Moreover, according to Young and Scott, Equation (12) applies to any form of coupling, such as bilinear or electrostrictive coupling and the coupling between the acoustic phonon mode and other excitations, which are altogether not connected with the order parameter [76]. This relaxation mechanism was used to describe the relaxation behavior of the acoustic phonon mode across commensurate–incommensurate phase transition in $Ba_2NaNb_5O_{15}$ where the exact form of the coupling was not known [76,77]. Based on the above studies in $Rb_{1-x}(NH_4)_xH_2PO_4$ and $Ba_2NaNb_5O_{15}$, we calculated the relaxation time both in the ferroelectric phase below T_c and in the quasi-static region below T^* (in the temperature range between T^* and T_c) using the LK relaxation mechanism. Figure 11 shows the temperature variation inthe relaxation time $\tau_{LA}$of the LA phonon.

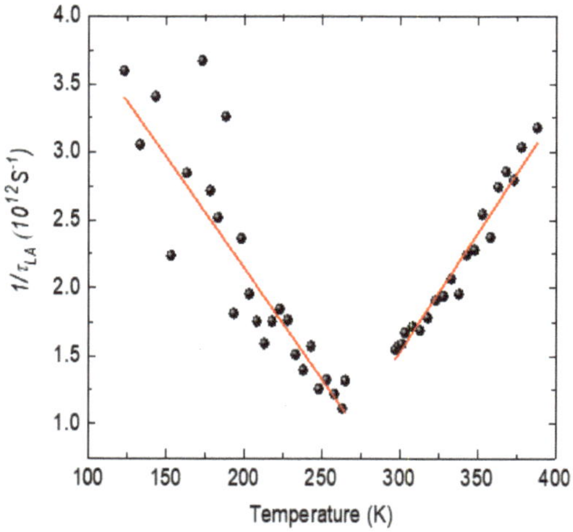

Figure 11. Plot showing $1/\tau_{LA}$ as a function temperature for the LA phonon of PST. Points are the experimental data and the solid line is the fit of Equation (16).

The relaxation time exhibits a critical slowing on approaching T_c from the ferroelectric and the relaxor phase. As shown in Figure 11, the temperature variation inthe relaxation rate exhibits approximately linear behavior where $1/\tau_{LA}$ satisfies the following relation:

$$\frac{1}{\tau_{LA}} = \frac{1}{\tau_o}\frac{|T - T_c|}{T_c} \quad (16)$$

The solid line is the linear fit to the experimental data. Best fit of the experimental data with Equation (16) yields τ_o = 0.22 ps in the ferroelectric phase and 0.19 ps in the relaxor phase. The relaxation time τ_o for PST is of a similar order of magnitude as observed for typical order–disorder ferroelectrics such as potassium dihydrogen phosphate KH_2PO_4 (KDP) and triglycine sulfate $(NH_2CH_2COOH)_3H_2SO_4$ (TGS); τ_o is 0.13 ps for KDP [78] and 0.10 ps for TGS [79]. Since the structural phase transition in PST does not involve the condensation of soft mode, the transition can be of the order–disorder-type. The local structure of PNRs in Pb-based relaxors is known to be rhombohedral [7] and sufficiently strong interaction between the quasi-static PNRs in the temperature range below T^* may eventually lead to the formation of long-range ferroelectric order with the macroscopic rhombohedral structure below T_c.

4.3. Brillouin Spectroscopy of Cubic–Tetragonal Structural Phase Transition in a 0.65PIN-0.35PT Single Crystal

PIN-PT solid solution, like PMN-PT and PZN-PT, shows the sequence of structural transition with temperature and PT content and has MPB for x = 0.37 [80]. PIN-PT composition near MPB shows a structural phase transition from cubic to tetragonal (T_{ct}) and tetragonal to rhombohedral (T_{tr}) phases [43,81]. While the dynamical behavior of PNRs has been widely studied in the PMN-PT and PZN-PT systems, no such studies have been carried out for PIN-PT solid solutions. In this review, the Brillouin scattering studies of the relaxation dynamics of PNRs in 0.65PIN-0.35PT single crystals are discussed.

The Brillouin spectra measured at various temperatures in the FSR range of 75 GHz are shown in Figure 12. The Brillouin doublet of the LA phonon can be seen clearly. A faint and broad TA phonon mode can also be observed in the cubic phase. This could be due to the presence of tetragonal-structure PNRs. The TA mode, however, could not be followed for all the temperatures and hence is not discussed further. Figure 13 shows the frequency and the width $\Gamma_{LA}$of the LA phonon as a function of temperature. Above 700 K, the minor temperature dependence in the frequency and the width of the LA phonon is attributed to the normal anharmonic process. Below 700 K, the frequency of the LA phonon shows substantial softening, followed by a strong increase in its width on approaching cubic to tetragonal phase transition temperature T_{ct} = 540 K. Also clear are the LA anomalies that can be seen at the tetragonal–rhombohedral transition temperature T_{tr} = 460 K.

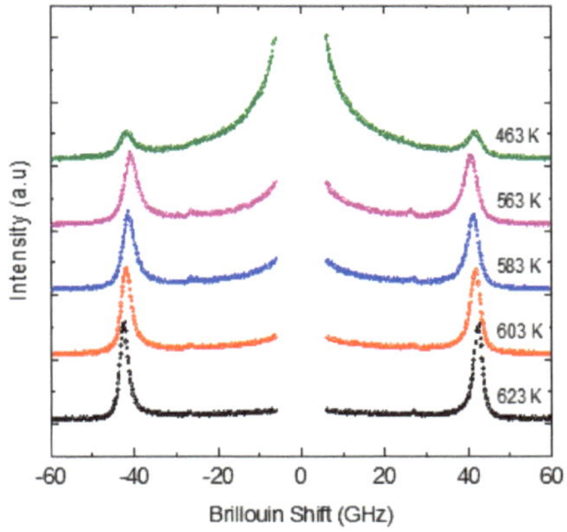

Figure 12. Brillouin spectra of 0.65PIN-0.35PT at various temperatures.

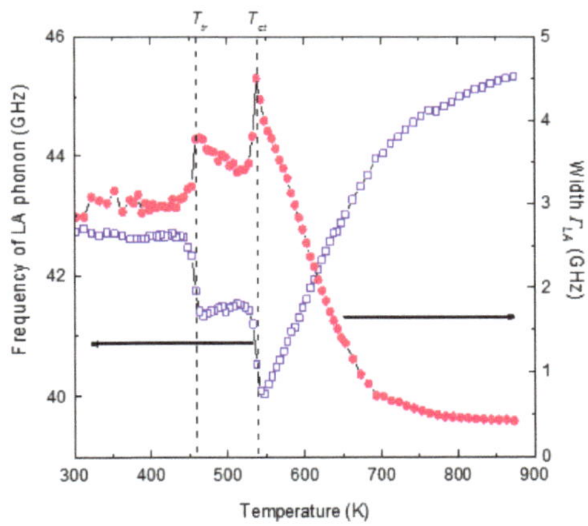

Figure 13. Temperature variation in the frequency and the width of the 0.65PIN-0.35PT LA phonon.

The observed anomalous behavior in the frequency and the width of the LA phonon below 700 K clearly suggests that T_B~700 K. As discussed in the Introduction, the strong electrostrictive coupling between the local polarization and strain in PNRs leads to the anomalous behavior of the LA phonon below T_B. As can be seen from Figure 12, a narrow CP (NCP) begins to appear at about 620 K and its intensity grows further substantially upon cooling. Figure 14a–d show the Brillouin spectra measured in the FSR range of 400 GHz. A broad CP begins (BCP) to appear below T_B and its intensity increases with a further decrease in temperature. The width of both NCP and BCP decreases upon cooling toward T_{ct}, reflecting the growth of the PNRs and the corresponding slowing of the relaxation time. The relaxation time of the LA phonon was calculated using Equation (15). We use the highest temperature value of the frequency and width of the LA mode for ν_∞ and Γ_∞. In Figure 15, the calculated value of the relaxation time from the above expression is shown along with the relaxation times obtained for narrow and broad central peaks above T_{ct}.

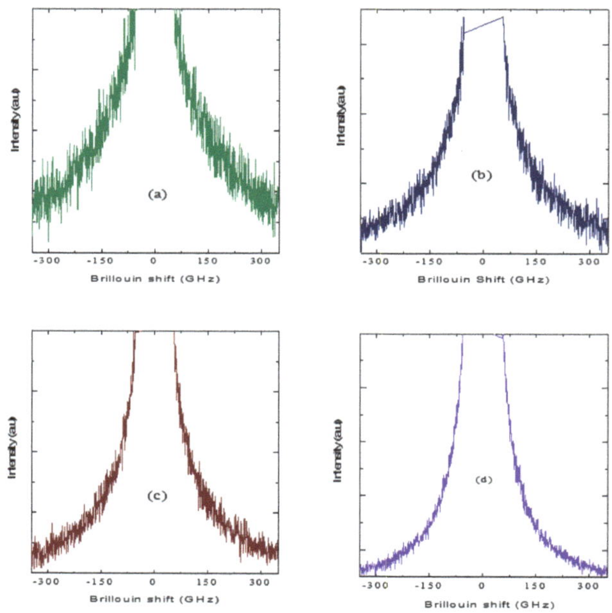

Figure 14. (**a**–**d**)Broad central peak of 0.65PIN-0.35PT at (**a**) 633 K, (**b**) 593 K, (**c**) 573 K and (**d**) 463 K.

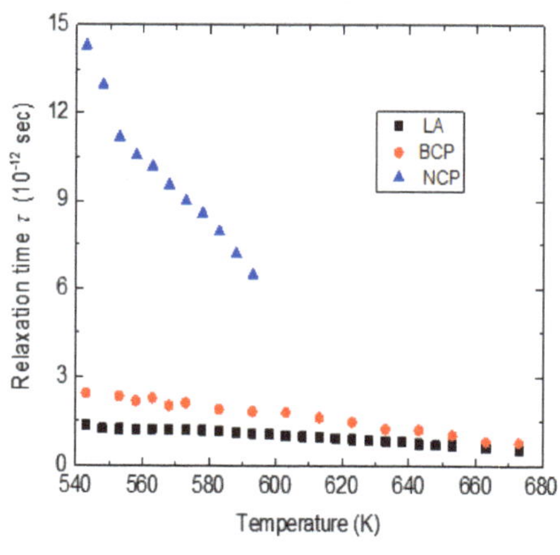

Figure 15. Temperature dependence of relaxation time of LA (solid squares), broad (solid circles) and narrow (solid triangles) central peaks of 0.65PIN-0.35PT.

The relaxation time for the LA phonon in the ferroelectric phase was not calculated due to the lack of sufficient data points. Below 600 K, the relaxation time of the local strain, τ_{LA} qualitatively agrees with the polarization fluctuations (as determined by BCP) of PNRs. Though the broad central peak appears below T_B, it effectively becomes coupled with the LA phonon only from 600 K. This result implies that the relaxation process of PNRs below 600 K is determined by the polarization fluctuations coupled with the local strain fluctuations. A similar result was obtained for 0.71PNN-0.29PT [82]. In this solid solution system, broad and narrow central peaks were observed around T^* = 500 K. The relaxation time of the LA phonon agrees well with that of the broad central peak. Considering the similarity of relaxation dynamics with 0.71PNN-0.29PT, we suggest that for 0.65PIN-0.35PT, the temperature 600 K at which the narrow central peak appears can be taken as T^*. At this point, we would like to remark on the two-component (narrow and broad) central peaks observed in 0.65PIN-0.35PT. Two-component central peaks appear to be the common feature in all the PT-substituted relaxor solid solutions [44,74,82]. We suggest the possible origin of two-component central peaks in the relaxor–PT solid solution. Owing to the high polarizability of the Ti^{4+} ion, the substitution of PT may lead to the formation of PNRs with a tetragonal structure in addition to the already existing PNRs with rhombohedral structure. With decrease in temperature, the size of tetragonal PNRs may grow faster and eventually become larger than the size of the rhombohedral one. In this case, the activation energy required for the flipping of dipoles along equivalent directions in tetragonal PNRs becomes higher. The higher activation energy results in a longer relaxation time for tetragonal PNRs. Therefore, we attribute the narrow component of the central peak to the relaxation of tetragonal PNRs and the broad component to the rhombohedral PNRs. It may be noted that in both PST and PSN, increased correlations between rhombohedral PNRs anddecrease in temperature results in long-range ferroelectric order with the macroscopic rhombohedral structure below T_c. In both PSN and PST, only a broad central peak corresponding to rhombohedral PNRs was observed below T^*[46,51]. Therefore, we suggest that in 0.65PIN-0.35PT, the phase transition from cubic relaxor phase to the tetragonal ferroelectric phase could be due to the nucleation and growth of tetragonal PNRs, whose relaxation dynamics are different from that of the rhombohedral PNRs.

4.4. Brillouin Light Scattering Studies of Acoustic Phonon Anomalies in SCT-0.03 Quantum Ferroelectrics

The Brillouin spectra of SCT-0.03 in some temperature ranges are displayed in Figure 16a,b. A longitudinal acoustic (LA) phonon peak around 75 GHz and a transverse acoustic (TA) mode at 45 GHz can be seen. As shown in Figure 16b, the TA mode splits into two components (TA1 and TA2) below 280 K (indicated by the arrow mark). The temperature variation in the frequency and the width of the LA phonon is shown in Figure 17. In the temperature range 500–323 K, a linear increase in the frequency of the LA phonon with the near constancy of its width is ascribed to the normal anharmonic process. Below 323 K, the LA phonon shows anomalous behavior with a sharp minimum in the frequency followed by a sharp maximum in width at the ferroelastic phase transition temperature T_{ct} = 280 K [83]. Similar elastic anomalies are observed in nondoped $SrTiO_3$ crystal by picosecond ultrasound spectroscopy [84]. The remarkable increase in T_{ct} by Ca doping can be caused by the smaller ionic radius of Ca ions, which induces the antiferro distortive rigid rotation of the oxygen octahedral [85]. A similar increase in T_{ct} was also reported in La-doped $SrTiO_3$ [86].

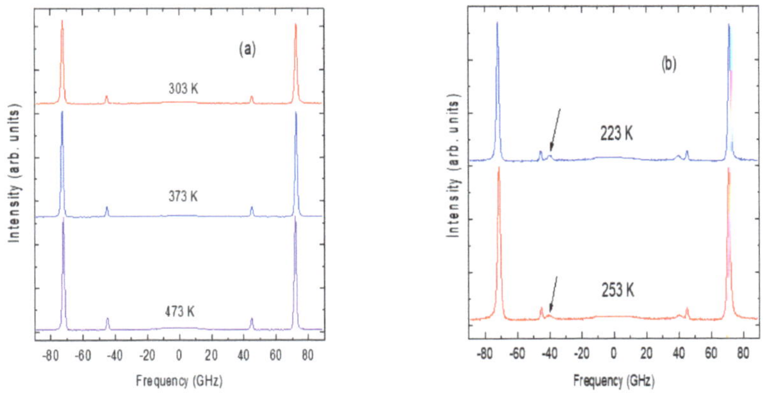

Figure 16. Brillouin spectra of SCT-0.03 measured with FSR of 100 GHz at (**a**) 473, 373 and 303 K and (**b**) 253 and 223 K. (Figure 16a. Reprinted with kind permission from Ref. [83]. Copyright (2020) The Japan Society of Applied Physics).

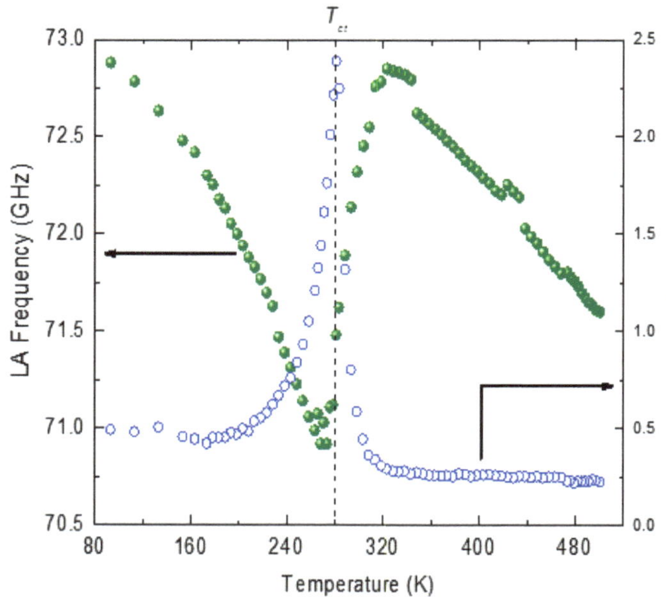

Figure 17. Temperature variation inthe frequency and the width of the SCT-0.03LA phonon. (Reprinted with kind permission from Ref. [83]. Copyright (2020) The Japan Society of Applied Physics).

The width of the LA phonon is markedly asymmetric in the vicinity of the transition temperature as expected from the LK mechanism. In the X-ray and the neutron diffraction studies of SCT-0.04, the cubic to tetragonal phase transition with the transition temperature of 210 K was reported [20]. The phase transition in SCT-0.04 was observed to be a ferroelastic-type as it was not associated with any dielectric anomalies. Therefore, the observed anomalous behavior of the LA phonon at 280 K in SCT-0.03 in the present study is attributed to the ferroelastic-type structural phase transition from cubic to tetragonal phase. The higher transition temperature in SCT-0.03 could be due to the lower Ca content. The observed anomalies of the LA phonon in the phase transition region are caused by the interaction between the phonon mode and the order parameter. It can be seen from Figure 17 that the order parameter interaction with the LA phonon does not agree with the bilinear coupling that requires $/T - T_{ct}/^{-1}$-type temperature dependence on either side of the phase transition. Slonczewski and Thomas reported that the ferroelastic phase transition at 110 K in $SrTiO_3$ can be described by coupling between the strain S and the order parameter Q in the form $F_c = \delta SQ^2$, where the order parameter Q is the soft mode coordinate and δ is the coupling coefficient [87]. The ferroelastic transition in $SrTiO_3$ could be well described by Landau theory, taking into account the interaction between the strain and the soft optical phonon mode.We suggest that in SCT also, the observed acoustic anomaly could be due to the coupling between the LA phonon and the soft mode. The temperature dependence of the relaxation time τ_{LA} of the LA phonon, calculated using Equation (15), is plotted in Figure 18. The relaxation time of the LA phonon exhibits critical slowing when approaching T_{ct} from both sides of the transition. The value of $\tau_0$obtained from Equation (16) is 0.077 ps below T_{ct} and 0.057 ps above T_{ct}. This is about half of the value of KDP and TGS and is very similar to that obtained in $PbHPO_4$ [75]. This suggests the weaker coupling of the LA phonon with the soft mode of the paraelectric host.

The temperature dependence of the frequency and width of the TA1 and TA2 phonons are shown in Figures 19 and 20. The frequency and the width of the TA1 phonon mode exhibit similar behavior as that of the LA phonon mode. However, a large scattering of the data observed in the width of the TA1 phonon mode limits further detailed discussion on its temperature dependence. The TA2 mode appears just near T_{ct}. The experimental data, however, could be unambiguously fitted only below 265 K. The TA2 mode shows a significant softening with the decrease in temperature in the ferroelastic phase. In the Brillouin scattering study of phase transition in Cs_2CdBr_4, the splitting of TA mode into two components was observed on the phase transition from orthorhombic to monoclinic phase [88]. One of the components was purely TA mode (C_{66}) corresponding to the parent orthorhombic phase, while the other was ascribed to the quasi-transverse mode γ_4(b), propagating along the b axis of the crystal. The frequency of the γ_4(b) mode was observed to decrease with decrease in temperature in the monoclinic phase. In SCT-0.03, the splitting of the TA phonon (TA1 and TA2 in the tetragonal phase) at T_{ct} and the temperature dependence of TA2 mode in the ferroelastic phase appears to be quite similar to that observed in Cs_2CdBr_4. We therefore tentatively ascribe the TA2 phonon as the quasi-transverse mode propagating along the *c* axis of SCT-0.03 in the tetragonal phase. However, the correct description of TA2 mode requires detailed Brillouin light scattering studies in various polarization geometries.

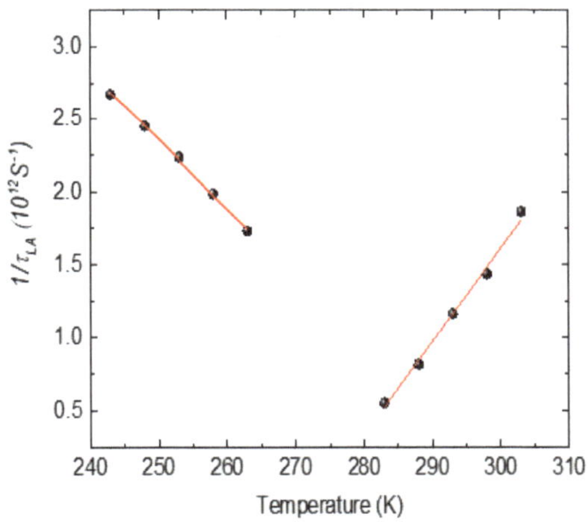

Figure 18. Plot showing $1/\tau_{LA}$ as a function temperature for the LA phonon of SCT-0.03. The solid line shows the fit of the experimental data points.

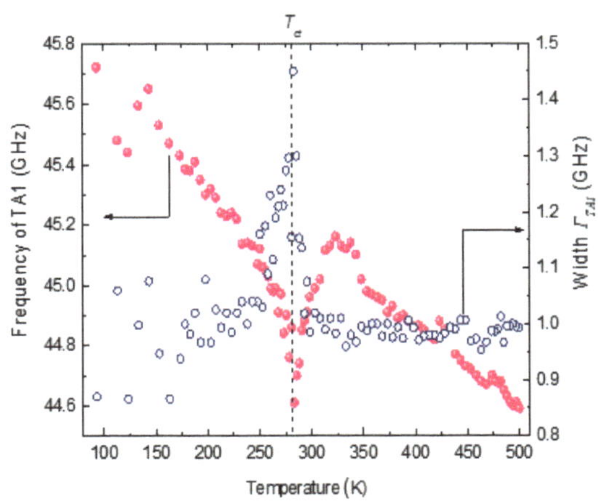

Figure 19. Temperature variation in thefrequency and the width of TA1 phonon mode. (Reprinted with kind permission from Ref. [83]. Copyright (2020) The Japan Society of Applied Physics).

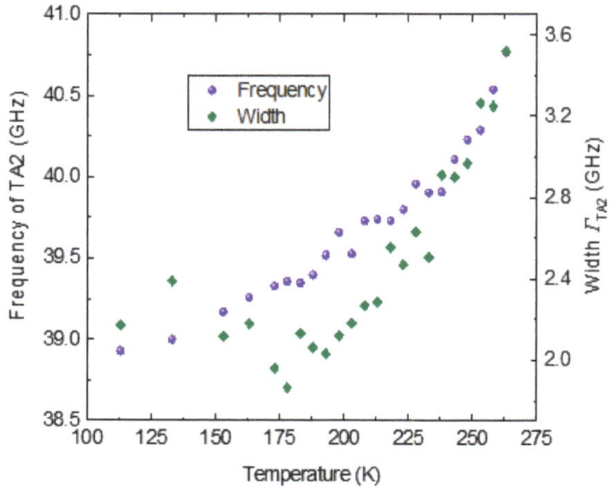

Figure 20. The frequency and the width of the TA2 phonon as a function of temperature.

5. Conclusions

Brillouin light scattering studies on the relaxation dynamics of disordered perovskite compounds have been reviewed. In PMT, the elastic constant C_{11} clearly reveals the existence of two distinct dynamical regimes of polarization characterized by the Burns temperature T_B and the intermediate temperature T^*. In the quasi-dynamic phase between T_B and T^*, the elastic constant obeys the mean field model and softens as $(T-T_o)^{0.5}$. The reduced softening of the elastic constant C_{11} and the increased attenuation in the quasi-static region below $T^*$are attributed to the strong coupling between the local polarization and strain in PNRs. The temperature dependence of the relaxation time of the central peak was found to be non-Arrhenius. The observed critical slowing of the relaxation time between T^* and T_f is attributed to dipole flipping and the volume fluctuations of PNRs below T_f. In the partially disordered PST, a sharp anomaly was observed in the frequency and the width of the LA phonon upon the transition from cubic relaxor to rhombohedral ferroelectric phase at 295 K. The calculated relaxation time of the LA phonon suggested the order–disorder nature of the structural phase transition. In 0.65PIN-0.35PT single crystal, anomalous behavior in the frequency and the width of the LA phonon was observed at cubic–tetragonal and tetragonal–rhombohedral phase transition temperatures. Two-component quasi-elastic (narrow and broad) central peaks were suggested to originate from two types of PNRs with different local structures. While the rhombohedral-type PNRs were pertinent to the parent relaxor PIN, the tetragonal-type PNRs areinduced by PT substitution. The broad central peak was suggested to associate with the relaxation of rhombohedral PNRs, while that of the narrow central peak to the relaxation of tetragonal PNRs. The relaxation time of the LA phonon was found to be qualitatively similar to that of the broad quasi-elastic central peak. The phase transition from cubic to tetragonal structure was suggested to originate from nucleation and growth of tetragonal PNRs. In SCT-0.03, the LA phonon showed critical anomalies at the structural phase transition temperature T_{ct} = 280 K. The Landau–Khalatnikov-type asymmetry was observed in the width of the LA phonon. The relaxation time of the LA phonon showed a critical slowing in the vicinity of the transition.The observed acoustic anomaly was suggested to be due to the coupling between the LA phonon and the soft mode of $SrTiO_3$. In the ferroelastic phase, the quasi-transverse TA (TA2) phonon mode was found to exhibit softening behavior with decrease in temperature.

Author Contributions: V.S.: conceptualization, methodology and writing—original draft; S.K.: supervision; S.K. and S.G.: writing—review and editing. All authors have read and agreed to the published version of the manuscript.

Funding: This research received no external funding.

Informed Consent Statement: Not applicable.

Data Availability Statement: The data sets generated during and/or analyzed during the current study are available from the corresponding author on request.

Acknowledgments: V.S. is grateful to the Japanese Society for Promotion of Science Invitation Fellowship Program for Research in Japan and to the Division of Materials Science, University of Tsukuba, where part of the work was carried out. The corresponding author acknowledges S.G. for help in growing a PMT single crystal and S.K. for providing a SCT single crystal.

Conflicts of Interest: The authors declare no conflict of interest.

References

1. Lines, M.E.; Glass, A.M. *Principles and Applications of Ferroelectrics and Related Materials*; Clarendon Press: Oxford, UK, 1977; pp. 24–58.
2. Samara, G.A. Ferroelectricity Revisited-Advances in Materials and Physics. *Solid State Phys.* **2001**, *56*, 252–343.
3. Hochli, U.T.; Maglione, M. Dielectric relaxation spectroscopy and the ground state of $K_{1-x}Li_xTaO_3$. *J. Phys. Condens. Matter* **1989**, *1*, 2241–2256. [CrossRef]
4. Toulouse, J.; Pattnaik, R. Collective behaviours in disordered ferroelectrics KLT and KTN. *J. Korean Phys. Soc.* **1998**, *32*, S942–S946.
5. Samara, G.A. Nature of the phase transition in $KTaO_3$ with random site impurities. *Jpn. J. Appl. Phys.* **1985**, *24*, 80–84. [CrossRef]
6. Viehland, D.; Jang, S.J.; Cross, L.E.; Wuttig, M. Freezing of the polarization fluctuations in lead magnesium niobate relaxors. *J. Appl. Phys.* **1990**, *68*, 2916–2921. [CrossRef]
7. de Mathan, N.; Husson, E.; Calvarin, G.; Gavarri, J.R.; Hewat, A.W.; Morell, A. A structural model for the relaxorPb($Mg_{1/3}Nb_{2/3}$)O_3 at 5 K. *J. Phys. Condens. Matter* **1991**, *3*, 8159–8171. [CrossRef]
8. Cross, L.E. Relaxor Ferroelectrics. *Ferroelectrics* **1987**, *76*, 241. [CrossRef]
9. Knauss, L.A.; Wang, X.M.; Toulouse, J. Polarization-strain coupling in the mixed ferroelectric $KTa_{1-x}Nb_xO_3$. *Phys. Rev. B* **1995**, *52*, 13261–13268. [CrossRef] [PubMed]
10. Sviteskiy, O.; Toulouse, J.; Yong, G.; Ye, Z.-G. Polarized Raman study of the phonon dynamics in Pb($Mg_{1/3}Nb_{2/3}$)O_3crystal. *Phys. Rev. B* **2003**, *68*, 104107. [CrossRef]
11. Lushnikov, S.G.; Fedoseev, A.I.; Gvasalia, S.N.; Kojima, S. Anomalous dispersion of the elastic constants at the phase transformation of the $PbMg_{1/3}Nb_{2/3}O_3$relaxor ferroelectric. *Phys. Rev. B* **2008**, *77*, 104122. [CrossRef]
12. Doussineau, P.; Farssi, Y.; Frenois, C.; Levelut, A.; McEnaney, K.; Toulouse, J.; Ziolkiewicz, S. Dielectric study of the phase transition in $K_{1-x}Li_xTaO_3$. *Europhys. Phys. Lett.* **1993**, *24*, 415–420. [CrossRef]
13. DiAntonio, P.; Vugmeister, B.E.; Toulouse, J.; Boatner, L.A. Polar fluctuations and first-order Raman scattering in highly polarisable $KTaO_3$ crystals with off-center Li and Nb ions. *Phys. Rev. B* **1993**, *47*, 5629–5637. [CrossRef]
14. Toulouse, J.; DiAntonio, P.; Vugmeister, B.E.; Wang, X.M.; Knauss, L.A. Precursor effects and ferroelectric macroregions in $KTa_{1-x}Nb_xO_3$ and $K_{1-y}Li_yTaO_3$. *Phys. Rev. Lett.* **1992**, *68*, 232–235. [CrossRef] [PubMed]
15. Nuzhnyy, D.; Bovtun, V.; Buixaderas, E.; Petzelt, J.; Savinov, M.; Kempa, M.; Pasciak, M.; Rafalovskyi, I.; ChelodPaingad, V.; Kuzel, P.; et al. Unusual dynamics of ferroelectric phase transition in $K_{1-x}Li_xTaO_3$ crystals. *Phys. Rev. B* **2022**, *105*, 184103. [CrossRef]
16. Bianchi, U.; Dec, J.; Kleemann, W.; Bednorz, J.G. Cluster and domain-state dynamics of ferroelectric $Sr_{1-x}Ca_xTiO_3$ (x=0.007). *Phys. Rev. B* **1995**, *51*, 8737–8746. [CrossRef] [PubMed]
17. Kleemann, W.; Albertini, A.; Kuss, M.; Lindner, R. Optical detection of symmetry breaking on a nanoscale in $SrTiO_3$:Ca. *Ferroelectrics* **1997**, *203*, 57–74. [CrossRef]
18. Bednorz, J.G.; Muller, K.A. $Sr_{1-x}Ca_xTiO_3$: An XY quantum ferroelectric with transition to randomness. *Phys. Rev. Lett.* **1984**, *52*, 2289–2292. [CrossRef]
19. Ranjan, R.; Pandey, D.; Lalla, N.P. Novel features of $Sr_{1-x}Ca_xTiO_3$ phase diagram: Evidence for Competing Antiferroelectric and Ferroelectric Interactions. *Phys. Rev. Lett.* **2000**, *84*, 3726. [CrossRef] [PubMed]
20. Geneste, G.; Kiat, J.-M. Ground state Ca-doped strontium titanate: Ferroelectricity versus polar nanoregions. *Phys. Rev. B* **2008**, *77*, 174101. [CrossRef]
21. Rischau, C.W.; Lin, X.; Grams, C.P.; Finck, D.; Harms, S.; Engelmayer, J.; Lorentz, T.; Gallais, Y.; Fauque, B.; Hemberger, J.; et al. A ferroelectric quantum phase transition inside the superconducting dome of $Sr_{1-x}Ca_xTiO_{3-\delta}$. *Nat. Phys. Lett.* **2017**, *13*, 643–648.
22. Viehland, D.; Jang, S.J.; Cross, L.E.; Wuttig, M. Deviation from Curie-Weiss behaviour in relaxor ferroelectrics. *Phys. Rev. B* **1992**, *46*, 8003–8006. [CrossRef] [PubMed]
23. Viehland, D.; Li, J.F.; Jang, S.J.; Cross, L.E.; Wuttig, M. Glassy polarization behaviour of relaxor ferroelectrics. *Phys. Rev. B* **1992**, *46*, 8013–8017. [CrossRef] [PubMed]
24. Moriya, Y.; Kawaji, H.; Tojo, T.; Atake, T. Specific heat anomaly caused by ferroelectric nanoregions in Pb($Mg_{1/3}Nb_{2/3}$)O_3 and Pb($Mg_{1/3}Ta_{2/3}$)O_3 relaxors. *Phys. Rev. Lett.* **2003**, *90*, 205901. [CrossRef] [PubMed]
25. Cowley, R.A.; Gvasaliya, S.N.; Lushnikov, S.G.; Roessli, B.; Rotaru, G.M. Relaxing with relaxors: A review of relaxor ferroelectrics. *Adv. Phys.* **2011**, *60*, 229–327. [CrossRef]

26. Toulouse, J.; Jiang, F.; Sviteskiy, O.; Chen, W.; Ye, Z.-G. Temperature evolution of the relaxor dynamics in $Pb(Zn_{1/3}Nb_{2/3})O_3$: A critical Raman analysis. *Phys. Rev. B* **2005**, *72*, 184106. [CrossRef]
27. Sviteskiy, O.; La-Orauttapong, D.; Toulouse, J.; Chen, W.; Ye, Z.-G. $PbTiO_3$ addition and internal dynamics in $Pb(Zn_{1/3}Nb_{2/3})O_3$ crystal studied by Raman spectroscopy. *Phys. Rev. B* **2005**, *72*, 172106. [CrossRef]
28. Dul'kin, E.; Roth, M.; Janolin, P.E.; Dkhil, B. Acoustic emission study of phase transitions and polar nanoregions in relaxor-based systems: Application to the family of $Pb(Zn_{1/3}Nb_{2/3})O_3$ family of single crystals. *Phys. Rev. B* **2006**, *73*, 012102. [CrossRef]
29. Roth, M.; Mojaev, E.; Dul'kin, E.; Gemeiner, P.; Dkhil, B. Phase transition at a nanometer scale detected by acoustic emission within the cubic phase $Pb(Zn_{1/3}Nb_{2/3})O_3$-$x$$PbTiO_3$relaxor ferroelectrics. *Phys. Rev. Lett.* **2007**, *98*, 265701. [CrossRef]
30. Dkhil, B.; Gemeiner, P.; Al-Barakaty, A.; Bellaiche, L.; Dul'kin, E.; Mojaev, E.; Roth, M. Intermediate temperature scale T^* in lead-based relaxor systems. *Phys. Rev. B* **2009**, *80*, 064103. [CrossRef]
31. Park, S.E.; Shrout, T.R. Ultra high strain and piezoelectric behaviour in relaxor based ferroelectric single crystals. *J. Appl. Phys.* **1997**, *82*, 1804–1811. [CrossRef]
32. Cox, D.E.; Noheda, B.; Shirane, G.; Useu, Y.; Fujishiro, K.; Yamada, Y. Universal phase diagram for high-piezoelectric perovskite systems. *Appl. Phys. Lett.* **2001**, *79*, 400–402. [CrossRef]
33. Yasuda, N.; Ohwa, H.; Kume, M.; Yamashita, Y. Piezoelectric properties of a high Curie temperature $Pb(In_{1/2}Nb_{1/2})O_3$-$PbTiO_3$ binary system single crystal near a morphotropic phase boundary. *Jpn. J. Appl. Phys.* **2000**, *39 Pt 2*, L66–L68. [CrossRef]
34. Setter, N.; Cross, L.E. The role of B-site cation disorder in diffuse phase transition behaviour of perovskite ferroelectrics. *J. Appl. Phys.* **1980**, *51*, 4356–4360. [CrossRef]
35. Chu, F.; Reaney, I.M.; Setter, N. Spontaneous (zero-field) relaxor-to-ferroelectric phase transition in disordered $Pb(Sc_{1/2}Nb_{1/2})O_3$. *J. Appl. Phys.* **1995**, *77*, 1671–1676. [CrossRef]
36. Chu, F.; Setter, N.; Taganstev, A.K. The spontaneous relaxor-ferroelectric transition of $Pb(Sc_{0.5}Ta_{0.5})O_3$. *J. Appl. Phys.* **1993**, *74*, 5129–5134. [CrossRef]
37. Malibert, C.; Dkhil, B.; Kiat, J.M.; Durand, D.; Berar, J.F.; Spasojevic-de Bire, A. Order and disorder in the relaxor ferroelectric perovskite $Pb(Sc_{1/2}Nb_{1/2})O_3$ (PSN): Comparison with simple perovskites $BaTiO_3$ and $PbTiO_3$. *J. Phys. Condens. Matter* **1997**, *9*, 7485–7500. [CrossRef]
38. Yasuda, N.; Ohwa, H.; Oohashi, J.; Nomura, K.; Terauchi, H.; Iwata, M.; Ishibashi, Y. The temperature and pressure dependence of the dielectric properties of disordered and ordered $Pb(In_{1/2}Nb_{1/2})O_3$ single crystals. *J. Phys. Soc. Jpn.* **1998**, *67*, 3952–3957. [CrossRef]
39. Blinc, R.; Gregorovic, A.; Zalar, B.; Pirc, R.; Laguta, V.V.; Glinchuk, M.D. Nuclear magnetic resonance study of the relaxor ferroelectric $Pb(Sc_{1/2}Nb_{1/2})O_3$. *J. Appl. Phys.* **2001**, *89*, 1349–1354. [CrossRef]
40. Laguta, V.V.; Glinchuk, M.D.; Bykov, I.P.; Blinc, R.; Zalar, B. NMR study of ionic shifts and polar ordering in the relaxor ferroelectric $Pb(Sc_{1/2}Nb_{1/2})O_3$. *Phys. Rev. B* **2004**, *69*, 054103. [CrossRef]
41. Sivasubramanian, V.; Ganesamoorthy, S. Elastic behaviour and dynamical aspects of polarization in $Pb(Mg_{1/3}Nb_{2/3})O_3$ and $Pb(Mg_{1/3}Ta_{2/3})O_3$ studied by Brillouin spectroscopy. *Eur. Phys. J. Appl. Phys.* **2015**, *69*, 10701. [CrossRef]
42. Ko, J.-H.; Kojima, S.; Lushnikov, S.G. Dynamic behaviours of $Pb(Mg_{1/3}Nb_{2/3})O_3$and $Ba(Mg_{1/3}Nb_{2/3})O_3$single crystals studied by micro-Brillouin scattering and dielectric spectroscopy. *App. Phys. Lett.* **2003**, *82*, 4128. [CrossRef]
43. Shabir, G.; Kojima, S. Central peak and acoustic phonon anomalies of the relaxor ferroelectric lead magnesium niobate-lead titanate single crystal studied by the micro-Brillouin scattering. *App. Phys. Lett.* **2007**, *91*, 062911. [CrossRef]
44. Ko, J.-H.; Kim, D.H.; Kojima, S. Central peak, acoustic modes, and the dynamics of polar nanoregions in $Pb[(Zn_{1/3}Nb_{2/3})_{1-x}Ti_x]O_3$ single crystals studied by Brillouin spectroscopy. *Phys. Rev. B* **2008**, *77*, 104110. [CrossRef]
45. Tsukada, S.; Hikada, Y.; Kojima, S.; Bokov, A.A.; Ye, Z.-G. Development of nanoscale polarization fluctuations in relaxor based $(1-x)Pb(Sc_{1/2}Nb_{1/2})O_3$-$x$$PbTiO_3$ ferroelectrics studied by Brillouin scattering. *Phys. Rev. B* **2013**, *87*, 014101. [CrossRef]
46. Ahart, M.; Hushur, A.; Ping, Y.; Ye, Z.-G.; Hemley, R.J.; Kojima, S. Critical slowing down of relaxation dynamics near the Curie temperature in the relaxor $Pb(Sc_{0.5}Nb_{0.5})O_3$. *Appl. Phys. Lett.* **2009**, *94*, 142906. [CrossRef]
47. Ko, J.-H.; Kim, D.H.; Tsukada, S.; Kojima, S.; Bokov, A.A.; Ye, Z.-G. Crossover in the mechanism of ferroelectric phase transition of $Pb[(Mg_{1/3}Nb_{2/3})_{1-x}Ti_x]O_3$ single crystals studied by Brillouin light scattering. *Phys. Rev. B* **2010**, *82*, 104110. [CrossRef]
48. Tsukada, S.; Ike, Y.; Kano, J.; Sekiya, T.; Shimojo, Y.; Wang, R.; Kojima, S. Broadband inelastic light scattering of a relaxor ferroelectric $0.71Pb(Ni_{1/3}Nb_{2/3})O_3$-$0.29PbTiO_3$. *Appl. Phys. Lett.* **2006**, *89*, 212903. [CrossRef]
49. Ko, J.-H.; Kojima, S.; Bokov, A.A.; Ye, Z.-G. Light scattering study of acoustic phonon modes and central peaks in $Pb[(Mg_{1/3}Nb_{2/3})_{0.45}Ti_{0.55}]O_3$ single crystals. *Appl. Phys. Lett.* **2007**, *91*, 252909. [CrossRef]
50. Sivasubramanian, V.; Tsukada, S.; Kojima, S. Anomalous behaviour of acoustic phonon and central peak in $Pb[(In_{1/2}Nb_{1/2})_{0.65}Ti_{0.35}]O_3$ single crystal. *J. Appl. Phys.* **2009**, *105*, 014108. [CrossRef]
51. Sivasubramanian, V.; Kojima, S. Brillouin scattering studies of acoustic phonon modes and central peak in single crystal $Pb(Sc_{1/2}Ta_{1/2})O_3$. *Phys. Rev. B* **2012**, *85*, 054104. [CrossRef]
52. Sivasubramanian, V.; Subramanian, V.; Kojima, S. Classical ferroelectric like behaviour of highly ordered $Pb(Sc_{1/2}Nb_{1/2})O_3$ studied by dielectric and Brillouin scattering spectroscopy. *Phys. Rev. B* **2016**, *93*, 054115. [CrossRef]
53. Kojima, S.; Ko, J.-H. Broadband micro-Brillouin scattering spectroscopy of $Pb(B_{1/3}B'_{2/3})O_3$ based relaxor ferroelectrics. *Curr. Appl. Phys.* **2011**, *11*, S22–S32. [CrossRef]

54. Kojima, S.; Muhtar, A.; Sivasubramanian, V.; Volkov, A.A.; Ye, J.C. Precursor dynamics of $Pb(B_{1/3}B'_{2/3})O_3$-type relaxor ferroelectrics studied by micro-Brillouin scattering. *J. Adv. Dielec.* **2012**, *2*, 1241004. [CrossRef]
55. Hayes, W.; Loudon, R. *Scattering of Light by Crystals*; Dover Publications: New York, NY, USA, 2004.
56. Sandercock, J.R. The design and use of a stabilized multipass interferometer of high contrast ratio. In *Light Scattering Solids*; Flammarion Sciences: Paris, France, 1972; Volume 9.
57. Kania, A. Flux growth of $Pb(Mg_{1/3}Ta_{2/3})O_3$ single crystals. *J. Cryst. Growth* **2007**, *300*, 343–346. [CrossRef]
58. Setter, N.; Cross, L.E. Flux growth of Lead Scandium tantalate $Pb(Sc_{0.5}Ta_{0.5})O_3$ and Lead Magnesium Niobate $Pb(Mg_{1/3}Nb_{2/3})O_3$ single crystals. *J. Cryst. Growth* **1980**, *50*, 555–556. [CrossRef]
59. Scheel, H.J.; Bednorz, J.G.; Dill, P. Crystal growth of strontium titanate $SrTiO_3$. *Ferroel. Lett.* **1976**, *13*, 507–509. [CrossRef]
60. Lushnikov, S.G.; Fedoseev, A.I.; Gvasalia, S.N.; Ko, J.-H.; Kojima, S. Anomalous behaviour of acoustic phonon velocity and elastic constants of relaxor ferroelectric $Pb(Mg_{1/3}Ta_{2/3})O_3$. *J. Phys. Condens. Matter* **2007**, *19*, 496206. [CrossRef]
61. Courdille, J.M.; Dumas, J.; Ziolkiewicz, S.; Joffrin, J. Soft mode characteristics in the KTN ferroelectric transition. *J. Phys.* **1977**, *12*, 1519–1525. [CrossRef]
62. Courtens, E.; Vacher, R.; Dagorn, Y. Brillouin spectroscopy of polarization fluctuations in a $Rb_{1-x}(NH_4)_xH_2PO_4$glass *Phys. Rev. B* **1986**, *33*, 7625–7636. [CrossRef]
63. Courtens, E. Hypersound observations of glass fluctuations. *Phys. Rev. Lett.* **1985**, *55*, 722–725. [CrossRef]
64. Blinc, R.; Dolinsek, J.; Gregorovic, A.; Zalar, B.; Filipic, C.; Kutnjak, Z.; Levstik, A.; Pirc, R. Local polarization distribution and Edwards-Anderson order parameter of relaxor ferroelectrics. *Phys. Rev. Lett.* **1999**, *83*, 424–427. [CrossRef]
65. Burns, G.; Dacol, F.H. Glassy polarization behaviour in ferroelectric compounds $Pb(Mg_{1/3}Nb_{2/3})O_3$ and $Pb(Zn_{1/3}Nb_{2/3})O_3$. *Solid State Commun.* **1983**, *48*, 853–856. [CrossRef]
66. Nagata, Y.; Tsujimi, Y.; Katsuraya, K.; Iwata, M.; Yagi, T. Dynamical slowing down of polar nanoregion in relaxor based ferroelectric $0.89Pb(Zn_{1/3}Nb_{2/3})O_3$-$0.11PbTiO_3$. *Appl. Phys. Lett.* **2006**, *89*, 022903. [CrossRef]
67. Siny, I.G.; Lushnikov, S.G.; Katiyar, R.S.; Rogacheva, E.A. Central peak in light scattering from the relaxor ferroelectric $Pb(Mg_{1/3}Nb_{2/3})O_3$. *Phys. Rev. B* **1997**, *56*, 7962–7966. [CrossRef]
68. Kamba, S.; Nuzhnyy, D.; Veljkov, S.; Bovtun, V.; Petzelt, J.; Wang, Y.L.; Setter, N.; Levoska, J.; Tyunina, M.; Macutkevic, J.; et al. Dielectric relaxation and polar phonon softening in relaxor ferroelectric $PbMg_{1/3}Ta_{2/3}O_3$. *J. Appl. Phys.* **2007**, *102*, 074106. [CrossRef]
69. Bovtun, V.; Kamba, S.; Pashkin, A.; Savinov, M.; Samoukhina, P.; Petzelt, J.; Bykov, I.P.; Glinchuk, M.D. Central-peak components and polar soft mode in relaxor$PbMg_{1/3}Nb_{2/3}O_3$ crystals. *Ferroelectrics* **2004**, *298*, 23–30. [CrossRef]
70. Macutkevic, J.; Kamba, S.; Banys, J.; Brilingas, A.; Pashkin, A.; Petzelt, J.; Bormanis, K.; Sternberg, A. Polar and chemical states at a nanometer scale in a $Pb(Sc_{1/2}Nb_{1/2})O_3$-$PbTiO_3$ system investigated by Raman spectroscopy. *Phys. Rev. B* **2006**, *74*, 104106. [CrossRef]
71. Stenger, C.G.F.; Burggraaf, A.J. Order-disorder reaction in the ferroelectric perovskites $Pb(Sc_{1/2}Nb_{1/2})O_3$ and $Pb(Sc_{1/2}Ta_{1/2})O_3$. *Phys. Stat. Sol. (A)* **1980**, *61*, 275–285. [CrossRef]
72. Woodward, P.M.; Baba-Kishi, K.Z. Crystal structures of the relaxor oxide $Pb_2(ScTa)O_6$ in the paraelectric and ferroelectric states. *J. Appl. Cryst.* **2002**, *35*, 233–242. [CrossRef]
73. Rehwald, W. Study of structural phase transitions by means of ultrasonic experiments. *Adv. Phys.* **1973**, *22*, 721–755. [CrossRef]
74. Tsukada, S.; Kojima, S. Broadband light scattering of two relaxation processes in relaxor ferroelectric 0.93 $Pb(Zn_{1/3}Nb_{2/3})O_3$-$0.07PbTiO_3$ single crystals. *Phys. Rev. B* **2008**, *78*, 144106. [CrossRef]
75. Maczka, M.; Hanuza, J.; Majchrowski, A.; Kojima, S. Heat capacity and high-resolution Brillouin scattering studies of phase transitions in $K_2MgWO_2(PO_4)_2$: Observation of the coupled soft optic and acoustic mode. *Phys. Rev. B* **2007**, *75*, 214105. [CrossRef]
76. Young, P.W.; Scott, J.F. Brillouin spectroscopy of the incommensurate-commensurate phase transition in Barium Sodium Niobate. *Phase Transit.* **1986**, *6*, 175–234. [CrossRef]
77. Zhang, M.S.; Yagi, T.; Oliver, W.F.; Scott, J.F. Brillouin spectroscopy of incommensurate $Ba_2NaNb_5O_{15}$at its 105-K transition. *Phys. Rev. B* **1986**, *33*, 1381–1385. [CrossRef]
78. Taksuzaki, I.; Kasahara, M.; Tokunaga, M.; Tanaka, H. Polarization relaxation of $KH_{2(1-x)}D_{2x}PO_4$ studied by light scattering. *Ferroelectrics* **1981**, *39*, 1049–1052. [CrossRef]
79. Gammon, R.W.; Cummins, H.Z. Brillouin scattering dispersion in ferroelectric triglycine sulfate. *Phys. Rev. Lett.* **1966**, *17*, 193–195. [CrossRef]
80. Alberta, E.F.; Bhalla, A.S. Piezoelectric properties of $Pb(InNb)_{1/2}O_3$-$PbTiO_3$ solid solution ceramics. *J. Korean Phys. Soc.* **1998**, *32*, S1265–S1267.
81. Ko, J.-H.; Kim, D.H.; Kojima, S. Low frequency central peaks in relaxor ferroelectric PZN-9%PT single crystals. *Ferroelectrics* **2007**, *347*, 25–29. [CrossRef]
82. Tsukada, S.; Ike, Y.; Kano, J.; Sekiya, T.; Shimojo, Y.; Wang, R.; Kojima, S. Dynamical properties of polar nanoregions of relaxor ferroelectric $Pb(Ni_{1/3}Nb_{2/3})O_3$-$0.29PbTiO_3$. *J. Phys. Soc. Jpn.* **2008**, *77*, 033707. [CrossRef]
83. Sivasubramanian, V.; Kojima, S. Acoustic phonon anomalies in Ca doped $SrTiO_3$ quantum ferroelectrics as studied by Brillouin light scattering. *Jp. J. Appl. Phys.* **2020**, *59*, SKKA10. [CrossRef]

84. Nagakubo, A.; Yamamoto, A.; Tanigaki, K.; Ogi, H.; Nakamura, N.; Hirao, M. Monitoring of longitudinal-wave velocity and attenuation of $SrTiO_3$ at low temperatures using picoseconds ultrasound spectroscopy. *Jpn. J. Appl. Phys.* **2012**, *51*, 07GA09. [CrossRef]
85. Mattheiss, L.F. Effect of the 110 K phase transition on the $SrTiO_3$ conduction band. *Phys. Rev. B* **1972**, *6*, 4740–4753. [CrossRef]
86. Ozakia, T.; Kusunose, K.; Yamaguchi, H.; Kajiwara, K.; Chikaura, Y. Spontaneous strain of conductive strontium titanate $Sr_{1-x}La_xTiO_3$ measured by using X-ray topographic domain contrast. *Phase Transit.* **2011**, *84*, 837–842. [CrossRef]
87. Slonczewski, J.C.; Thomas, H. Interaction of elastic strain with the structural transition of Strontium Titanate. *Phys. Rev. B* **1970**, *1*, 3599–3608. [CrossRef]
88. Kuzel, P.; Moch, P.; Gomez-Cuevas, A.; Dvorak, V. Brillouin and ultrasonic studies of phase transitions in Cs_2CdBr_4. I. Experimental. *Phys. Rev. B* **1994**, *49*, 6553–6562. [CrossRef]

materials

MDPI

Article

Brillouin Scattering Study of Ferroelectric Instability of Calcium–Strontium–Barium Niobate Single Crystals

Seiji Kojima [1,*], Md Aftabuzzaman [1,2], Jan Dec [3] and Wolfgang Kleemann [4]

1 Division of Materials Science, University of Tsukuba, Tsukuba 305-8573, Japan
2 Department of Physics, Pabna University of Science and Technology, Pabna 6600, Bangladesh
3 Institute of Materials Science, University of Silesia, PL-40-007 Katowice, Poland
4 Angewandte Physik, University of Duisburg-Essen, D-47048 Duisburg, Germany
* Correspondence: kojima@ims.tsukuba.ac.jp

Abstract: Uniaxial ferroelectrics with tetragonal tungsten bronze structure are important functional materials with photorefractive, electrooptic, piezoelectric, and pyroelectric properties. $Sr_xBa_{1-x}Nb_2O_6$ (SBN100*x*) with $x > 50$ is known as a typical uniaxial relaxor ferroelectric, while $Ca_xBa_{1-x}Nb_2O_6$ (CBN100*x*) undergoes nearly normal ferroelectric phase transitions. Single crystals of CSBN100*x* = [*x*(CBN28) + (1 − *x*) (SBN61)] = $xCa_{0.28}Ba_{0.72}Nb_2O_6 + (1-x)\,Sr_{0.61}Ba_{0.39}Nb_2O_6$ with nominal *x* = 0.00, 0.25, 0.50, 0.75, and 1.00 were studied to clarify the dynamical properties at the crossover from relaxor (*x* = 0) to normal (*x* = 1) ferroelectric behavior. The longitudinal acoustic (LA) and transverse acoustic (TA) modes and a central peak (CP) related to the relaxation process of polarization fluctuations along the polar *c*-axis were studied in uniaxial ferroelectric CSBN single crystals as a function of temperature via Brillouin scattering spectroscopy. A CBN28 (*x* = 1.00) crystal shows the sharp elastic anomaly of the LA mode in the gigahertz range toward Curie temperature, T_c. However, those of CSBN25 (*x* = 0.25) and SBN61 (*x* = 0.00) crystals show diffusive anomalies due to stronger random fields. The relaxation time determined from the width of a CP shows a critical slowing down in the vicinity of T_c. The elastic anomaly and slowing down of relaxation time of CSBN100*x* crystals become diffusive in the vicinity of T_c as the CBN28 content decreases. The origin of the crossover from relaxor to normal ferroelectric phase transitions is discussed in terms of the difference in the A1 and A2 sites' occupancies.

Keywords: Brillouin scattering; ferroelectric; relaxor; acoustic modes; elastic properties; slowing down

Citation: Kojima, S.; Aftabuzzaman, M.; Dec, J.; Kleemann, W. Brillouin Scattering Study of Ferroelectric Instability of Calcium–Strontium–Barium Niobate Single Crystals. *Materials* **2023**, *16*, 2502. https://doi.org/10.3390/ma16062502

Academic Editor: Georgios C. Psarras

Received: 16 February 2023
Revised: 16 March 2023
Accepted: 19 March 2023
Published: 21 March 2023

1. Introduction

Ferroelectricity is defined by the existence of a spontaneous polarization, the direction of which is switchable by an external electric field [1]. Uniaxial ferroelectric materials with a tetragonal tungsten bronze (TTB) structure are technologically important for optical applications involving electro-optic, nonlinear optic, photorefractive, pyroelectric, and piezoelectric properties [2]. In light of recent environmental problems, Pb-free ferroelectrics and their functional properties have become important [3]. In TTB ferroelectrics, the direction of spontaneous polarization is restricted to the polar *c*-axis, which is why they are also called uniaxial ferroelectrics [4]. The structural formula of TTB ferroelectrics is expressed by $(A1)_2(A2)_4(C)_4$-$(B1)_2(B2)_8O_{30}$, with corner-sharing distorted BO_6 octahedra as shown in Figure 1. The smallest C site is occupied only by Li, such as in $K_3Li_2Nb_5O_{15}$(KLN). In ferroelectric $Ba_2NaNb_5O_{15}$ (BNN), which is well known by its excellent second harmonic generation, all the A1 and A2 sites are occupied by Ba^{2+} and Na^{1+} ions, respectively. BNN belongs to the filled TTB ferroelectrics due to their complete occupation of A1 and A2 sites and undergoes a normal ferroelectric phase transition with a sharp dielectric anomaly in the vicinity of a ferroelectric Curie temperature, T_c, which is the highest among TTB ferroelectrics [2,5]. In contrast, the A1 sites of $Sr_xBa_{1-x}Nb_2O_6$ (SBN100*x*) are occupied in part by Sr^{2+} ions, and the A2 sites are occupied in part by both Ba^{2+} and Sr^{2+} ions. Since 1/6 (A1 + A2) sites remain unoccupied, it belongs to the so-called unfilled (open) TTB ferroelectrics due to the incomplete occupancy of A1 and A2 sites. The empty A1 and/or A2 sites cause the charge disorder and are the main sources of quenched random fields (RFs), which enhance the relaxor features characterized by the frequency dispersion of dielectric susceptibility and diffusive phase transition [6,7]. In SBN, the strength of RFs increases as the Sr content of smaller (1.12 Å) Sr^{2+} ions increases, and SBN has attracted much attention as a typical uniaxial relaxor ferroelectric [8].

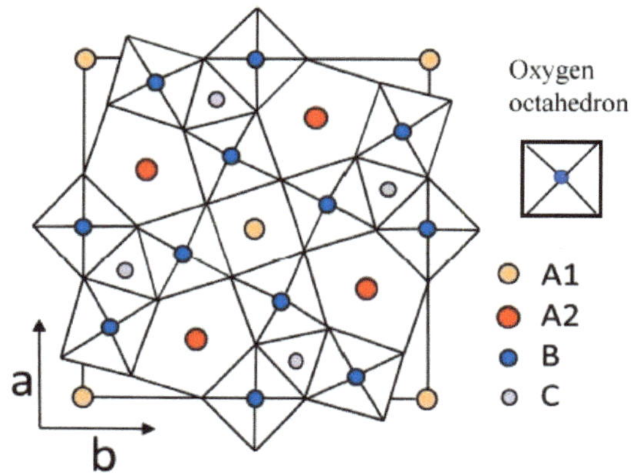

Figure 1. Crystal structure of *ab* plane, which is perpendicular to the polar *c*-axis, in tetragonal tungsten bronze ferroelectrics. There are three different interstices (two square A1, four pentagonal A2, and four trigonal C sites in a unit cell).

The disadvantage of SBN in application is the relatively low Curie temperatures. The $Ca_xBa_{1-x}Nb_2O_6$ (CBN100x) compounds also belong to the unfilled TTB structure and show quite similar physical properties to SBN, while their Curie temperatures are much higher than those of SBN [9]. Therefore, the excellent optical and ferroelectric properties of CBN make them potential candidates for applications at relatively high temperatures [10]. Recently, CBN nanopowders were synthesized. Their analysis of the various optical properties, especially the photorefractive effect, suggests that CBN nanopowders can be potentially applied for ultrahigh-density optical data storage [11]. In CBN, most of the smaller (0.99 Å) Ca^{2+} ions occupy A1 sites, whereas the relatively larger (1.34 Å) Ba^{2+} ions predominantly occupy the A2 sites [12,13]. Thus, in contrast to SBN with strong RFs, the lower degree of disorder of the Ca^{2+} and Ba^{2+} ions of CBN causes the weak RFs, and the diffusive nature of CBN is weaker than that of SBN [14]. In the Brillouin scattering study of CBN28, the intense central peak (CP) caused by polarization fluctuations along the *c*-axis was clearly observed in the vicinity of T_c. The relaxation time determined by the CP width clearly shows critical slowing down towards T_c, reflecting a weakly first-order phase transition under weak RFs [15].

Within the quasi-ternary $CaNb_2O_6$-$SrNb_2O_6$-$BaNb_2O_6$ system, selected $Ca_xSr_yBa_{1-x-y}Nb_2O_6$ (CSBN) compounds were grown via the Czochralski method in a tungsten–bronze-type structure [16]. The solid solution of CSBN is technologically important due to the coexistence of the high T_c of CBN and excellent functionality of SBN. In fundamental science, the crossover of CSBN from weak RFs of CBN with a nearly normal ferroelectric nature to strong RFs of SBN with a relaxor nature is very interesting regarding control of the strength of RFs only by Ca content. Since there are three kinds of cations, Ca^{2+}, Sr^{2+}, and Ba^{2+}, with increasing ionic radii occupying A1 and A2 sites, the degree of freedom to control physical properties increases. The Curie temperatures of xCBN28-(1 − x) SBN61 (CSBN100x) are shown in Figure 2. As the Ca content increases in CSBN, T_c monotonically increases, and the ferroelectric phase transition becomes less diffusive due to the decrease of disorder, which indicates the suppression of the RFs [17,18]. In addition, it is worth mentioning here that the above composition formula of the CSBN system may be expressed in a more compact fashion: (i) x = 0, $Sr_{0.61}Ba_{0.39}Nb_2O_6$ for SBN61, (ii) x = 0.25, $Ca_{0.07}Sr_{0.458}Ba_{0.472}Nb_2O_6$ for CSBN25, (iii) x = 0.50, $Ca_{0.14}Sr_{0.305}Ba_{0.555}Nb_2O_6$ for CSBN50, (iv) x = 0.75, $Ca_{0.21}Sr_{0.153}Ba_{0.637}Nb_2O_6$ for CSBN75, and (v) x = 1.00, $Ca_{0.28}Ba_{0.72}Nb_2O_6$ for CBN28.

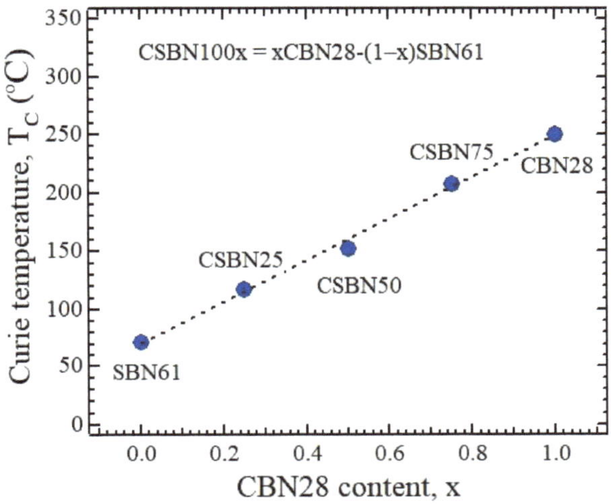

Figure 2. Ferroelectric Curie temperatures of calcium strontium barium niobate crystals with TTB structure. The dotted line is a linear fitting.

Brillouin scattering is the inelastic light scattering by thermally excited sound waves and has been used as a non-contact and non-destructive method to obtain elastic constants in the gigahertz range [19]. In the present study, the elastic properties and dynamical instability of CSBN100x crystals were investigated by using Brillouin scattering spectroscopy.

2. Experimental Methods

CSBN100x single crystals were grown via the Czochralski method [16] for the nominal compositions, x = 0.00, 0.25, 0.5, 0.75, 1.00. The CSBN single crystals were grown at the Institute of Electronic Materials Technology (Poland) under the guidance of Prof. T. Lukasiewicz. Their structural measurements were published [20,21]. The real composition of the crystals obtained was checked with the use of ICP-OES (inductively coupled plasma–optical emission spectroscopy) method and the result was published [22]. It has appeared that the real composition was, within uncertainties, satisfactorily close to the established one. Single crystalline plates were cut along [100] (a-plate) and [001] (c-plate) with optically polished 5 × 5 mm^2 surfaces and 1 mm thickness. Brillouin scattering spectra were measured at the back scattering geometry using a high-contrast 3 + 3 passes tandem Fabry–Perot interferometer, as shown in Figure 3 [19]. The exciting source was a diode-pumped solid state (DPSS) laser with a wavelength of 532 nm and a power of 100 mW. Scattered light was detected using a photon counting system. The specimen's temperature was controlled using a cooling/heating stage (Linkham, THMS600, Salfords, UK) with a stability of ±0.1 °C. Using an a-plate, longitudinal acoustic (LA) and transverse acoustic (TA) modes–which propagate along the a-axis—and the CP of polarization fluctuations along the ferroelectric c-axis were measured. Using a c-plate, LA and TA modes—which propagate along the c-axis—were measured.

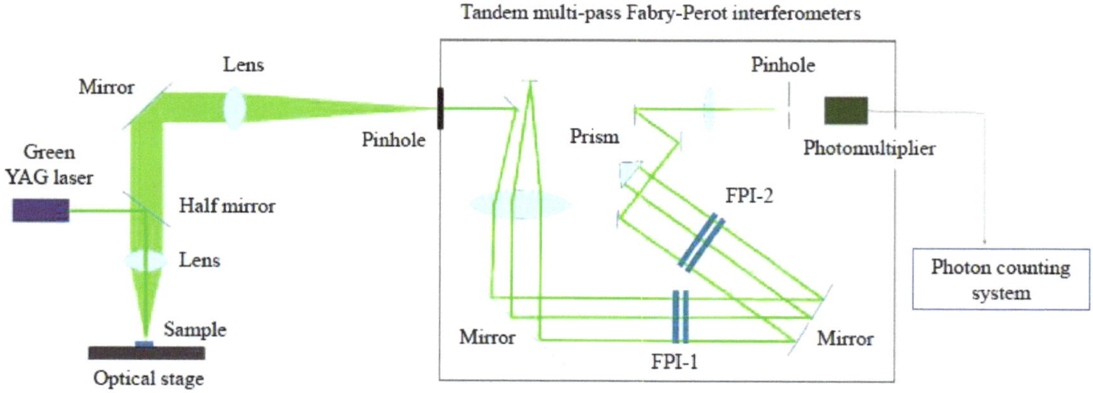

Figure 3. Schematic illustration of experimental setup of Brillouin scattering measurement with tandem multipass Fabry–Perot interferometers.

3. Results and Discussion

3.1. Elastic Anomaly of LA Modes

The temperature dependences of Brillouin scattering spectra of a CSBN50 crystal measured at the backward scattering geometry using *a*- and *c*-plates are shown in Figures 4 and 5, respectively. The direction of the wave vector of a scattered phonon is parallel to the *a*-axis for an *a*-plate and to the *c*-axis for a *c*-plate. The spectrum at 160 °C in Figure 4 shows doublets of TA and LA modes which propagate along the *a*-axis at about 32 and 59 GHz, respectively. The broad Rayleigh wings observed at 0 GHz are a CP with $A_1(z)$ symmetry, which is related to the polarization fluctuations along the ferroelectric *c*-axis. The maximum of intensity of a CP was observed at 152 °C. The spectrum at 160 °C in Figure 5 shows TA and LA modes which propagate along the *c*-axis at about 32 and 48 GHz, respectively. However, a CP related to the polarization fluctuations in the *ab*-plane, which is perpendicular to the ferroelectric *c*-axis, was not observed at all the temperatures.

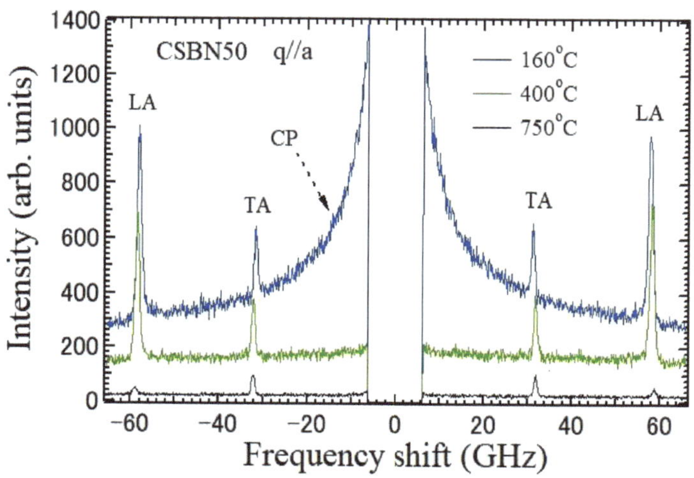

Figure 4. Temperature dependences of Brillouin scattering spectra measured by backward scattering geometry using the *a*-plate of a CSBN50 crystal. The wave vector *q* of the scattered phonon is parallel to the *a*-axis. Doublets of TA and LA modes which propagate along the *a*-axis were also observed. An intense CP was observed at zero frequency shift.

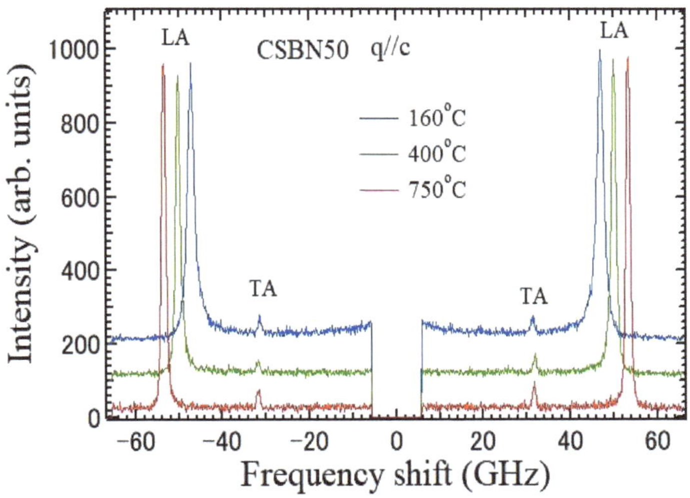

Figure 5. Temperature dependence of Brillouin scattering spectra measured by backward scattering geometry using the *c*-plate of a CSBN50 crystal. The wave vector q of the scattered phonon is parallel to the *c*-axis. Doublets of TA and LA modes which propagate along the *c*-axis were observed. No CP was observed.

The measured Brillouin spectra were fitted using Voigt functions, a convolution of Lorentzian and Gaussian functions, for which the width of the Gaussian function was fixed as an instrumental function. The temperature dependence of frequency shift and width of the LA mode, which propagates along the ferroelectric *c*-axis, were determined by the fitting, as shown in Figure 6. Upon cooling from the high temperature above the Burns temperature, T_B = 520 °C, the LA frequency shows the remarkable softening toward T_c = 152 °C. Upon cooling from the high temperature, the LA mode width shows a remarkable increase toward T_c. Such an elastic anomaly is related to the temperature evolution of polar nanoregions (PNRs) triggered by the RFs [15,23]. In the ferroelectric phase, the width gradually decreases due to the freezing of PNRs into stable nanodomains.

In lead-based relaxor ferroelectrics with perovskite structure, it is known that T_B = 427 °C and the intermediate temperature, T^* = 227 °C, are unaffected by compositions [24]. In SBN, T_B = 350 °C and T^* = 190 °C are unaffected by compositions of Sr ions, and in CBN, T_B = 517 °C and T^* = 367 °C are unaffected by compositions of Ca ions [23]. In CSBN—if we assume linear change in T_B and T^* for the composition dependence—T_B = 434 °C and T^* = 279 °C are expected for CSBN50, as shown in Figure 6. Upon cooling, the dynamic–static transition and rapid growth PNRs occur at T^*, and the remarkable decrease of the LA frequency and an increase in LA width occur towards T_C due to the scattering of LA phonons by PNRs.

The sound velocity, V, is determined by the frequency shift ν_B in the Brillouin scattering spectrum using the equation:

$$V = \frac{\lambda_i v_B}{2n \sin\frac{\theta}{2}} \quad (1)$$

where λ_i, θ, and n are the wavelength of an incident beam, the scattering angle, and the refractive index of the sample, respectively. The velocity is determined from the frequency shift. The attenuation, α, is determined using

$$\alpha = \frac{\pi\Gamma}{V} \quad (2)$$

where Γ is the FWHM of the Brillouin peak [25]. The dispersions of refractive indices and the Curie temperatures were determined in CSBN100x crystals grown using the Czochralski method [26]. The temperature dependences of LA velocity and LA attenuation calculated from the LA shift and width for five CSBN crystals using the values of refractive indices [26] are shown in Figures 7 and 8, respectively. The LA velocity and attenuation of CBN28 with very weak RFs show remarkable changes in the vicinity of T_C = 254 °C and a sharp minimum in the LA velocity. In contrast, in SBN61, with strong RFs, the temperature dependence of the LA velocity and the attenuation in the vicinity of T_C = 72 °C are diffusive in the vicinity of T_C. These differences can be caused by the variation in the strength of RFs, which suppress the sharp changes in the vicinity of T_C. The temperature dependences of LA velocity and attenuation gradually change as the CBN28 content decreases from normal ferroelectric, such as CBN28, to relaxor SBN61. This crossover was also reported on the dielectric properties of CSBN ceramics [27]. Such a crossover from normal to relaxor ferroelectrics was also studied in SBN from Ba- to Sr- rich regions [7,28,29]. Recently, the crossover from normal to relaxor nature in SBN and CSBN ceramics was studied by the Rietveld refinement of X-ray diffraction and Raman spectroscopy [18]. The observed site occupancies at A1 and A2 sites were analyzed using residual entropy calculations. It was concluded that the origin of the crossover is directly related to the majority of A2 sites being occupied by Ba^{2+}.

Among elastic constants, the coupling between c_{33} and the polarization fluctuations along the ferroelectric c-axis is very strong, while the coupling between other elastic constants and the order parameters of a ferroelectric phase transition is very week. Consequently, the elastic anomaly of other elastic constants is very small, and it is difficult to discuss the dynamical properties of a phase transition using the other two elastic constants. Therefore, the elastic anomaly of c_{33} was analyzed. The difference in temperature dependences of the elastic stiffness constant $c_{33} = \rho V_{LA}^2$ between CBN28 and CSBN75 is shown in Figure 9a,b, respectively, where ρ and V_{LA} are the density and LA velocity of Figure 7, respectively. In comparison with the remarkable softening of elastic constant of CBN28 towards T_c, that of CSBN25 is suppressed, and the anomaly becomes diffusive. The elastic anomaly of perovskite ferroelectrics was analyzed by the following equation in the paraelectric phase.

$$c_{ij}(t) = c_{ij}^0 + c_{ij}^1 T - c_{ij}^2 \left(\frac{T - T_0}{T_0} \right)^{-n}, \text{for } T_0 \le T_c \le T. \quad (3)$$

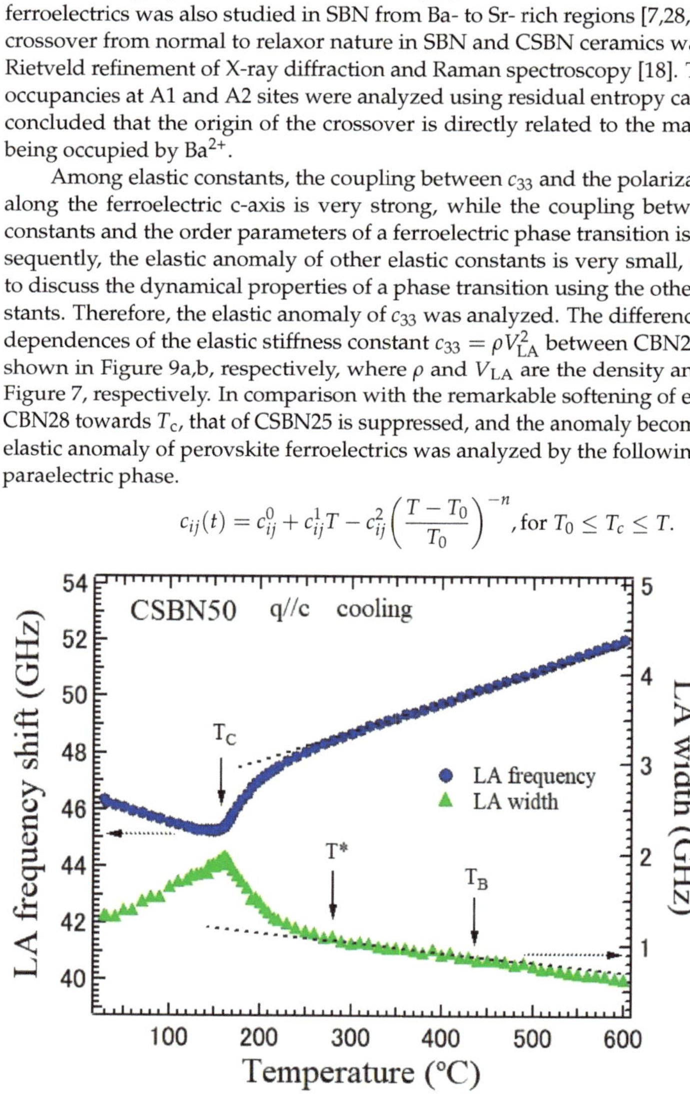

Figure 6. Temperature dependence of LA frequency shift and width along the ferroelectric *c*-axis of a CSBN50 crystal on cooling. The dotted lines show the nearly linear temperature dependences at high temperatures. T*and T_B are the intermediate and Burns temperatures, respectively. The dotted line arrows indicate the corresponding *y*-axis.

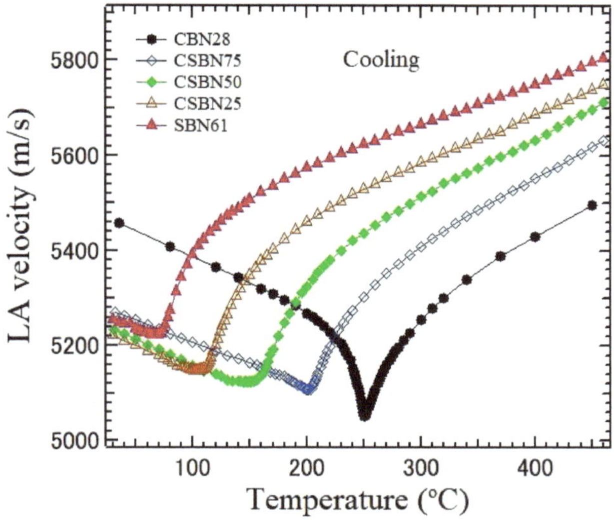

Figure 7. Temperature dependences of LA velocity which propagates along the ferroelectric *c*-axis of five CSBN100*x* crystals on cooling. As the CBN28 content decreases, the change in LA velocity becomes diffusive in the vicinity of T_c.

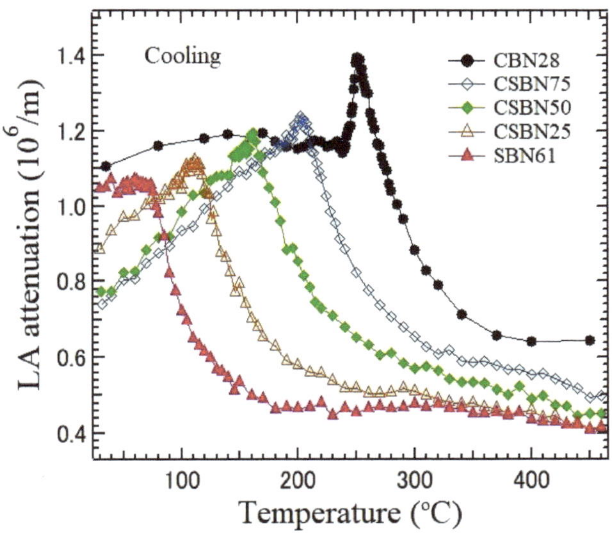

Figure 8. Temperature dependences of LA attenuation, which propagates along the ferroelectric *c*-axis of five CSBN100*x* crystals on cooling. As the CBN28 content decreases, the change in LA attenuation becomes diffusive in the vicinity of T_c.

Here, c_{ij}^0, c_{ij}^1, and c_{ij}^2 are constants. On the right-hand side, the second term is the anharmonic effect and the third term is the elastic anomaly caused by the fluctuations in the order parameter. The critical exponent, *n*, was predicted to be 0.5 for three-dimensional fluctuations, 1.0 for two-dimensional fluctuations, and 1.5 for one-dimensional fluctuation of the order parameters [30]. For example, for undoped and Li-doped $K(Ta_{0.6}Nb_{0.4})TiO_3$ crystals with perovskite structure, the observed exponent of $n = 0.5$ indicates the three-dimensional fluctuations of polarization, which are related to the 8-site model of the off-center of B-site ions along the eight equivalent [111] directions [31]. In CBN28, the fitted value of $n = 1.52$ indicates the one-dimensional fluctuations in polarization along the [001] axis, which is suitable for a uniaxial ferroelectric phase transition. In CSBN75, the fitted value $n = 2.17$ shows the deviation from the value of theoretical model, in which structural disorder was not considered. The origin of the deviation from $n = 1.5$ in CSBN75 is probably attributed to the diffusive nature induced by the RFs.

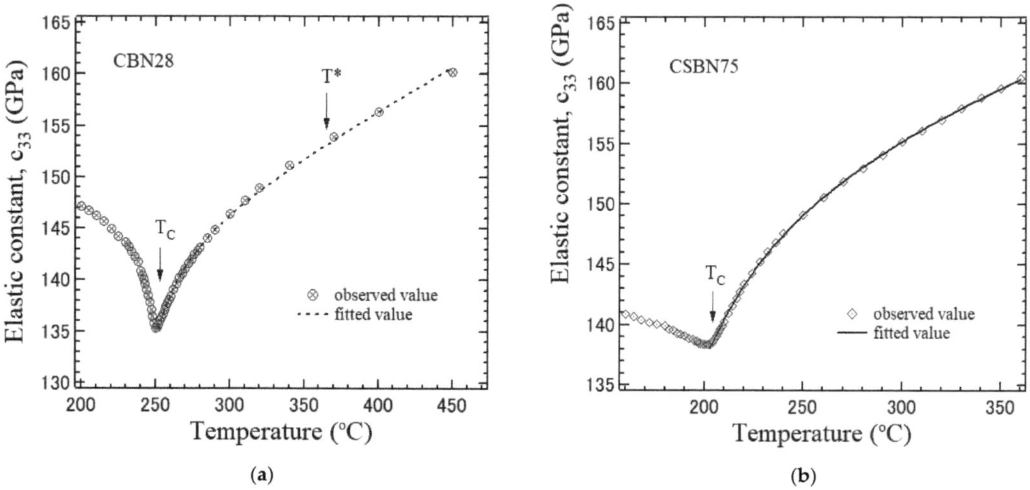

Figure 9. Difference in the temperature dependences of the elastic stiffness constant c_{33} between (**a**) CBN28 and (**b**) CSBN75 crystals. The elastic constant of CBN28 shows a sharp anomaly in the vicinity of T_C, while that of CSBN75 shows the diffusive nature and that the elastic anomaly is smaller. The solid lines in a paraelectric phase are curves fitted using Equation (3).

3.2. Critical Slowing down Observed by a Central Peak

The relaxation process of the polarization fluctuations along a ferroelectric *c*-axis has been observed from the width of the narrow CP in CBN28 [15]. Temperature dependences of relaxation time determined from the width of a narrow central peak observed using *a*-plates of five CSBN crystals were analyzed. In CBN28, the temperature dependences of relaxation time of polarization fluctuations determined by a central peak along the *c*-axis show the critical slowing down toward T_C. However, as the CBN28 content decreases, the slowing down is stretched by the strengthened RFs. In relaxor ferroelectrics, the dielectric constant ε obeys the extended Curie–Weiss law based on the compositional heterogeneity model as below [32].

$$\frac{1}{\varepsilon} = \frac{1}{\varepsilon_0} + \frac{1}{\varepsilon_1}\left(\frac{T - T_C}{T_C}\right)^{\gamma}, \quad \text{for } T > T_c. \tag{4}$$

Here, ε_0 and ε_1 are constants and γ is the diffuseness exponent. The case of $\gamma = 1$ refers to the Curie–Weiss law for a normal ferroelectric phase transition, and that of $\gamma = 2$ is a typical relaxor phase transition. For a partially disordered ferroelectric phase transition, it holds $1 < \gamma \leq 2$. In Nb-doped $Pb(Zr_{0.75}Ti_{0.25})O_3$ ceramics, $\gamma = 1.5$ was reported [33]. The dielectric properties of CSBN crystals showed that a relaxor nature with noticeable dielectric dispersion was observed in CSBN25 [22]. The origin of its relaxor properties was attributed to lower excess oxygen in CSBN25.

In an order–disorder ferroelectric phase transition, the relaxation time τ shows critical slowing down towards T_c. However, in the ferroelectric phase transitions of relaxor ferroelectrics, the diffusive nature was observed in the critical slowing down [34]. In $0.93Pb(Zn_{1/3}Nb_{2/3})O_3$-$0.07PbTiO_3$ (PZN-7PT), the slowing down was suppressed below the intermediate temperature T^* and typical critical slowing down was not observed near T_c [35]. The local transition from dynamic to static PNRs at T^* suppresses further slowing down. To describe such a suppressed slowing down by RFs, an empirical equation of the stretched slowing down was used in the vicinity of T_c, as given by the following equation [35],

$$\frac{1}{\tau_{CP}} = \frac{1}{\tau_0} + \frac{1}{\tau_1}\left(\frac{T - T_C}{T_C}\right)^{\beta}, \quad \text{for } T > T_c. \tag{5}$$

where β is the stretched exponent. In the case of β = 1.0, Equation (5) gives the critical slowing down of normal ferroelectrics. Regarding TTB ferroelectrics, β = 1.0 was reported for BNN [36] and CBN28 [15]. In the case of β > 1.0, the slowing down of relaxation time is suppressed and/or stretched by the increase of the strength of RFs. In PMN-17PT and PMN-56PT, the observed values are β = 2.12 and 1.43, respectively [37]. RFs of PMN-17PT are stronger than those of PMN-56PT.

The difference in the temperature dependences of the stretched slowing down is shown in Figure 10 between CSBN75 and CSBN50 crystals. The dotted lines are values fitted using Equation (5) with β = 1.53 for CSBN75 and β = 1.88 for CSBN50 in a paraelectric phase. Therefore, as the CBN28 content decreases, the stretched exponent increases monotonically from β = 1 of the normal ferroelectric case.

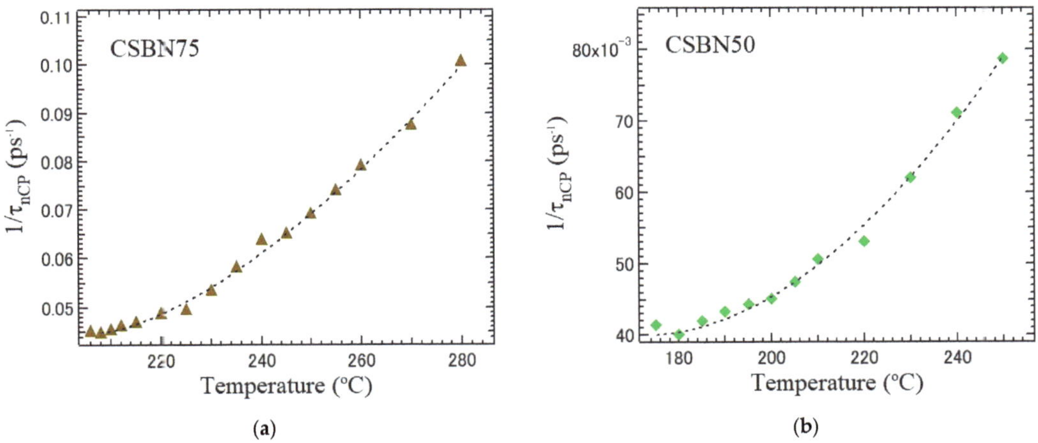

Figure 10. Difference in the temperature dependences of the stretched slowing down between (**a**) CSBN75 and (**b**) CSBN50 crystals. The brown triangles and green box are the observed values of CSBN75 and CSBN50, respectively. The dotted lines are fitted values to Equation (5) with β = 1.53 for CSBN75 and β = 1.88 for CSBN50.

The CBN28 content dependence of stretched index of five CSBN crystals is shown in Figure 11. The values of the diffuseness exponent determined by dielectric measurement using Equation (4) are also plotted for the comparison [27]. These exponents increase from 1.0 as the CBN28 content decreases. These similar dependences in the stretched index and diffuseness exponent indicate a crossover from a normal ferroelectric-like transition to a relaxor ferroelectric transition in CSBN.

Recently, the origin of the crossover from normal ferroelectric to relaxor in SBN and CSBN was discussed on the basis of site occupancy at A1 and A2 sites and residual entropy calculation [18]. The origin of the relaxor nature is attributed to the larger occupancy ratio of Sr:Ba at A2 sites. Differently from Sr ions, smaller Ca ions occupy only the A1 site. Therefore, the larger occupancy ratio decreases as the Ca content increases and the relaxor nature of SBN is suppressed. As another reason, lower excess oxygen in CSBN25 was also suggested [22]. To clarify the microscopic origin of the relaxor nature in TTB ferroelectrics, further studies are necessary.

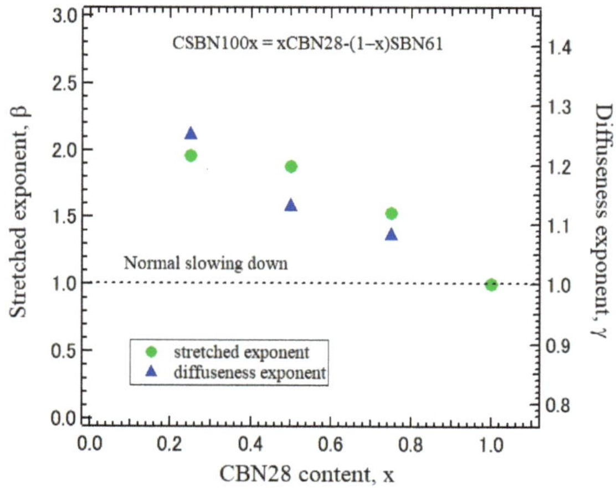

Figure 11. CBN28 content dependence of the exponent of stretched slowing down of Equation (5) in CSBN crystals. The values of the diffuseness exponent of dielectric constant of Equation (4) [27] are also plotted by way of the comparison.

4. Conclusions

Three dimensional relaxor ferroelectrics have been extensively studied with respect to their physical properties and to their origin of relaxor nature. Within this context, the results on uniaxial relaxor ferroelectrics have not completely been satisfactory. In this paper, the ferroelectric phase transitions of uniaxial ferroelectrics with tetragonal tungsten bronze structure were studied. Using Brillouin scattering spectroscopy, the elastic anomaly and a central peak (CP) with $A_1(z)$ symmetry were investigated. $Sr_xBa_{1-x}Nb_2O_6$ (SBN100x) with strong random fields (RFs) undergoes a relaxor ferroelectric phase transition, while $Ca_xBa_{1-x}Nb_2O_6$ (CBN100x) with weak RFs undergoes a nearly normal ferroelectric phase transition. The crossover from normal to relaxor ferroelectric behaviors was investigated in $x$$Ca_{0.28}Ba_{0.72}Nb_2O_6$-$(1-x)$$Sr_{0.61}Ba_{0.39}Nb_2O_6$ (CSBN100x) crystals. In a CBN28, crystal, the sharp elastic anomaly of the LA mode was observed in the vicinity of T_C. As the CBN28 content decreases, the anomaly of CSBN100x crystals becomes diffusive. A CSBN0 (SBN61) crystal shows a typical relaxor nature. In CBN28, the relaxation time determined from the width of a CP shows critical slowing down in the vicinity of T_C. While, as the CBN28 content decreases, slowing down of relaxation time in the vicinity of T_C becomes diffusive. In the gigahertz range, it was observed that two dynamical processes show a crossover from normal to relaxor nature in uniaxial ferroelectric CSBN100x crystals. The origins of the crossover from relaxor to normal ferroelectric phase transitions are discussed with regards to the difference in the A1 and A2 sites occupancies.

Author Contributions: Investigation, S.K., M.A., J.D. and W.K. All authors have read and agreed to the published version of the manuscript.

Funding: This research received no external funding.

Institutional Review Board Statement: Not applicable.

Informed Consent Statement: Not applicable.

Data Availability Statement: Data are contained within the articles.

Acknowledgments: The investigated crystals were grown at the Department of Oxide Single Crystals Technology in the Institute of Electronic Materials Technology, Warsaw under the leadership of T. Lukasiewicz. The authors are thankful to K. Matsumoto and K. Suzuki for the Brillouin scattering measurements.

Conflicts of Interest: The authors declare no conflict of interest.

References

1. Valasek, J. Piezo-Electric and Allied Phenomena in Rochelle Salt. *Phys. Rev.* **1921**, *17*, 475–481. [CrossRef]
2. Geusic, J.E.; Levinstein, H.J.; Rubin, J.J.; Singh, S.; Van Uitert, L.G. The Nonlinear Optical Properties of Ba2NaNb5O15. *Appl. Phys. Lett.* **1967**, *11*, 269–271. [CrossRef]
3. Bell, A.J.; Deubzer, O. Lead-free piezoelectrics—The environmental and regulatory issues. *MRS Bull.* **2018**, *43*, 581–587. [CrossRef]
4. Abrahams, S.C.; Jamieson, P.B.; Bernstein, J.L. Ferroelectric Tungsten Bronze-Type Crystal Structures. III. Potassium Lithium Niobate $K_{(6-x-y)}Li_{(4+x)}Nb_{(10+y)}O_{30}$. *J. Chem. Phys.* **1971**, *54*, 2355–2363. [CrossRef]
5. Ota, S.; Matsumoto, K.; Suzuki, K.; Kojima, S. Elastic anomaly and order-disorder nature of multiferroic barium sodium niobate studied by broadband Brillouin scattering. *IOP Conf. Ser. Mater. Sci. Eng.* **2014**, *54*, 012018. [CrossRef]
6. Lines, M.E.; Glass, A.M. *Principles and Applications of Ferroelectrics and Related Materials*; Clarendon Press: Oxford, UK, 1979.
7. Shvartsman, V.V.; Dec, J.; Miga, S.; Łukasiewicz, T.; Kleemann, W. Ferroelectric Domains in $Sr_xBa_{1-x}Nb_2O_6$ Single Crystals ($0.4 \leq x \leq 0.75$). *Ferroelectrics* **2008**, *376*, 1–8. [CrossRef]
8. Aftabuzzaman, M.; Helal, M.A.; Paszkowski, R.; Dec, J.; Kleemann, W.; Kojima, S. Electric field and aging effects of uniaxial ferroelectrics $Sr_xBa_{1-x}Nb_2O_6$ probed by Brillouin scattering. *Sci. Rep.* **2017**, *7*, 11615. [CrossRef]
9. Eßer, M.; Burianek, M.; Klimm, D.; Mühlberg, M. Single crystal growth of the tetragonal tungsten bronze $Ca_xBa_{1-x}Nb_2O_6$ (x = 0.28; CBN-28). *J. Cryst. Growth* **2002**, *240*, 1–5. [CrossRef]
10. Song, H.; Zhang, H.; Jiang, Q.; Xu, X.; Lu, C.; Hua, X.; Wang, J.; Jiang, M. Growth, dielectric, ferroelectric and optical properties of $Ca_{0.28}Ba_{0.72}Nb_2O_6$ single crystals. *J. Cryst. Growth* **2006**, *290*, 431–435. [CrossRef]
11. Rubel, M.H.K.; Islam, M.S.; Mahmuda, U.S.M.; Rahaman, M.M.; Hossain, M.E.; Parvez, M.S.; Hossain, K.M.; Hossain, M.I.; Hossain, J.; Yamanaka, J.; et al. $Ca_xBa_{1-x}Nb_2O_6$ Ferroelectric Nanopowders for Ultrahigh-Density Optical Data Storage. *ACS Appl. Nano Mater.* **2018**, *1*, 6289–6300. [CrossRef]
12. Niemer, A.; Pankrath, R.; Betzler, K.; Burianek, M.; Müehlberg, M. Dielectric Properties and the Phase Transition of Pure and Cerium Doped Calcium-Barium-Niobate. *World J. Condens. Matter Phys.* **2012**, *2*, 80–84. [CrossRef]
13. Liu, H.; Xu, Y. The influence of sintering temperature on structural and electrical properties of x $(Ca_{0.28}Ba_{0.72})Nb_2O_6$-(1−x)$(Sr_{0.61}Ba_{0.39})Nb_2O_6$ ceramics. *J. Aust. Ceram. Soc.* **2020**, *56*, 1405–1411. [CrossRef]
14. Qi, Y.J.; Lu, C.J.; Zhu, J.; Chen, X.B.; Song, H.L.; Zhang, H.J.; Xu, X.G. Ferroelectric and dielectric properties of $Ca_{0.28}Ba_{0.72}Nb_2O_6$ single crystals of tungsten bronzes structure. *Appl. Phys. Lett.* **2005**, *87*, 082904. [CrossRef]
15. Suzuki, K.; Matsumoto, K.; Dec, J.; Łukasiewicz, T.; Kleemann, W.; Kojima, S. Critical slowing down and elastic anomaly of uniaxial ferroelectric $Ca_{0.28}Ba_{0.72}Nb_2O_6$ crystals with tungsten bronze structure. *Phys. Rev. B* **2014**, *90*, 064110. [CrossRef]
16. Lukasiewicz, T.; Swirkowicz, M.A.; Dec, J.; Hofman, W.; Szyrski, W. Strontium–barium niobate single crystals, growth and ferroelectric properties. *Crystal Growth* **2008**, *310*, 1464–1469. [CrossRef]
17. Kojima, S.; Aftabuzzaman, M.; Dec, J.; Kleemann, W. Brillouin Scattering Study of Ferroelectric Instability of Calcium Strontium Barium Niobate Crystals. *Proc. Ultrason. Electron. Symp.* **2022**, *3*, 89–90.
18. Liu, H.; Dkhil, B. Origin of the crossover from ferroelectric to relaxor in tetragonal tungsten bronzes. *J. Alloys Comp.* **2022**, *929*, 167314. [CrossRef]
19. Kojima, S. Gigahertz Acoustic Spectroscopy by Micro-Brillouin Scattering. *Jpn. J. Appl. Phys.* **2010**, *49*, 07HA01. [CrossRef]
20. Wieteska, K.; Wierzchowski, W.; Malinowska, A.; Lefeld-Sosnowska, M.; Świrkowicz, M.; Łukasiewicz, T.; Paulmann, C. Synchrotron diffraction topography of SrxBa1−xNb$_2$O$_6$ (SBN), $Ca_xBa_{1-x}Nb_2O_6$ (CBN) and mixed $(Ca_{0.28}Ba_{0.72})_y(Sr_{0.61}Ba_{0.39})_{1-y}Nb_2O_6$ (CSBN) crystals. *Radiat. Phys. Chem.* **2013**, *93*, 87–91. [CrossRef]
21. Wierzchowski, W.; Wieteska, K.; Malinowska, A.; Wierzbicka, E.; Romaniec, M.; Lefeld-Sosnowska, M.; Świrkowicz, M.; Łukasiewicz, T.; Sakowska, H.; Paulmann, C. "Ghost" segregation pattern and ferroelectric domains in mixed calcium-strontium-barium niobates. *X-ray Spectrom.* **2015**, *44*, 356–362. [CrossRef]
22. Malyshkina, O.; Ivanova, A.; Malyshkin, Y.; Folomeeva, A.; Shashkov, M.; Dec, J. Effect of Ca, Sr and Ba distribution on the relaxor properties of CSBN single crystals. *Ferroelectrics* **2017**, *511*, 76–81. [CrossRef]
23. Aftabuzzaman, M.; Dec, J.; Kleemann, W.; Kojima, S. Field dependent elastic anomaly in uniaxial tungsten bronze relaxors. *Jpn. J. Appl. Phys.* **2016**, *55*, 10TC01. [CrossRef]
24. Roth, M.; Mojaev, E.; Dul'kin, E.; Gemeiner, P.; Dkhil, B. Phase Transition at a Nanometer Scale Detected by Acoustic Emission within the Cubic Phase $Pb(Zn_{1/3}Nb_{2/3})O_3-xPbTiO_3$ Relaxor Ferroelectrics. *Phys. Rev. Lett.* **2007**, *98*, 265701. [CrossRef]
25. Hays, W.; Loudon, R. *Scattering of Light by Crystals*; Dover Publishing: New York, NY, USA, 1978.
26. Mühlberg, M.; Burianek, M.; Joschko, B.; Klimm, D.; Danilewsky, A.; Gelissen, M.; Bayarjargal, L.; Gorler, G.P.; Hildmann, B.O. Phase equilibria, crystal growth and characterization of the novel ferroelectric tungsten bronzes $Ca_xBa_{1-x}Nb_2O_6$ (CBN) and $Ca_xSr_yBa_{1-x-y}Nb_2O_6$ (CSBN). *J. Crys. Growth* **2008**, *310*, 2288–2294. [CrossRef]
27. Xu, Y.; Liu, H. Crystal structure effect on the phase transition of $(Ca_{0.28}Ba_{0.72})x$ $(Sr_{0.61}Ba_{0.39})_{1-x}Nb_2O_6$ ceramics. *J. Mater. Sci. Mater. Electron.* **2020**, *31*, 5221–5226. [CrossRef]
28. Kim, M.S.; Wang, P.; Lee, J.H.; Kim, J.J.; Lee, Y.L.; Cho, S.H. Site Occupancy and Dielectric Characteristics of Strontium Barium Niobate Ceramics: Sr/Ba Ratio Dependence. *Jpn. J. Appl. Phys.* **2002**, *41*, 7042–7047. [CrossRef]
29. Kojima, S.; Aftabuzzaman, M.; Dec, J.; Kleemann, W. Ferroelectric phase transitions of uniaxial $Sr_{1-x}Ba_xNb_2O_6$ and their composition variation. *Jpn. J. Appl. Phys.* **2019**, *58*, SLLA02. [CrossRef]

30. Knauss, L.A.; Wang, X.M.; Toulouse, J. Polarization-strain coupling in the mixed ferroelectric $KTa_{1-x}Nb_xO_3$. *Phys. Rev. B* **1995**, *52*, 13261–13268. [CrossRef]
31. Rahaman, M.M.; Imai, T.; Kobayashi, J.; Kojima, S. Effect of Li-doping on polar-nanoregions in $K(Ta_{1-x}Nb_x)O_3$ single crystals. *Jpn. J. Appl. Phys.* **2015**, *54*, 10NB01. [CrossRef]
32. Uchino, K.; Nomura, S. Critical exponents of the dielectric constants in diffused-phase-transition crystals. *Ferroelectrics* **1982**, *44*, 55–61. [CrossRef]
33. Dong, X.L.; Kojima, S. Dielectric and resonance frequency investigations of phase transitions in Nb-doped PZT95/5 and 75/25 ceramics. *J. Phys. Conden. Matter.* **1997**, *9*, L171. [CrossRef]
34. Tsukada, S.; Kojima, S. Broadband light scattering of two relaxation processes in relaxor ferroelectric $0.93Pb(Zn_{1/3}Nb_{2/3})O_3-0.07PbTiO_3$ single crystals. *Phys. Rev. B* **2008**, *78*, 144106. [CrossRef]
35. Kojima, S.; Tsukada, S. Micro-Brillouin Scattering of Relaxor Ferroelectrics with Perovskite Structure. *Ferroelectrics* **2010**, *405*, 32. [CrossRef]
36. Kojima, S. Broadband Brillouin scattering study of ferroelectric instability of barium sodium niobate. *Condens. Matter Phys.* **2022**, *25*, 43702. [CrossRef]
37. Helal, M.A.; Aftabuzzaman, M.; Svirskas, S.; Banys, J.; Kojima, S. Temperature evolution of central peaks and effect of electric field in relaxor ferroelectric $0.83Pb(Mg_{1/3}Nb_{2/3})O_3$–$0.17PbTiO_3$ single crystals. *Jpn. J. Appl. Phys.* **2017**, *56*, 10PB03. [CrossRef]

materials

Article

Acoustic Anomalies and the Critical Slowing-Down Behavior of $MAPbCl_3$ Single Crystals Studied by Brillouin Light Scattering

Jeong Woo Lee [1], Furqanul Hassan Naqvi [1], Jae-Hyeon Ko [1,*], Tae Heon Kim [2] and Chang Won Ahn [2]

[1] School of Nano Convergence Technology, Nano Convergence Technology Center, Hallym University, Chuncheon 24252, Korea; dlwjddn1992@naver.com (J.W.L.); furqanhassan05@gmail.com (F.H.N.)
[2] Department of Physics and Energy Harvest-Storage Research Center (EHSRC), University of Ulsan, Ulsan 44610, Korea; thkim79@ulsan.ac.kr (T.H.K.); cwahn@ulsan.ac.kr (C.W.A.)
* Correspondence: hwangko@hallym.ac.kr

Abstract: Inelastic light scattering spectra of organic–inorganic halide perovskite $MAPbCl_3$ single crystals were investigated by using Brillouin spectroscopy. Sound velocities and acoustic absorption coefficients of longitudinal and transverse acoustic modes propagating along the cubic [100] direction were determined in a wide temperature range. The sound velocities exhibited softening upon cooling in the cubic phase, which was accompanied by the increasing acoustic damping. The obtained relaxation time showed a critical slowing-down behavior, revealing the order–disorder nature of the phase transition, which is consistent with the growth of strong central peaks upon cooling toward the phase transition point. The temperature dependences of the two elastic constants C_{11} and C_{44} were obtained in the cubic phase for the first time. The comparison of C_{11} and C_{44} with those of other halide perovskites showed that C_{11} of $MAPbCl_3$ is larger and C_{44} is slightly smaller compared to the values of $MAPbBr_3$ and $MAPbI_3$. It suggests that $MAPbCl_3$ has a more compact structure (smaller lattice constant) along with stronger binding forces, causing larger C_{11} and bulk modulus in this compound, and that the shear rigidity is exceedingly small similar to other halide perovskites. The reported elastic constants in this study may serve as a testbed for theoretical and calculational approaches for $MAPbCl_3$.

Keywords: halide perovskite; Brillouin scattering; elastic constant; sound velocity

Citation: Lee, J.W.; Naqvi, F.H.; Ko, J.-H.; Kim, T.H.; Ahn, C.W. Acoustic Anomalies and the Critical Slowing-Down Behavior of $MAPbCl_3$ Single Crystals Studied by Brillouin Light Scattering. *Materials* **2022**, *15*, 3692. https://doi.org/10.3390/ma15103692

Academic Editor: Efrat Lifshitz

Received: 28 April 2022
Accepted: 19 May 2022
Published: 21 May 2022

1. Introduction

Perovskites with a generic ABX_3 structure are being employed in several research fields [1,2]. Hybrid organic–inorganic halide perovskites (HOPs) have arisen as key materials for application in optoelectronic and photovoltaic devices [3,4], such as light-emitting diodes [5], photodetectors [6] and, especially, solar cells [7], which have shown a drastic increase in the solar power conversion efficiency from 3.9% to 25.7% in a few years [8]. Various experimental and theoretical approaches have been adopted to understand the structural properties, phase transition behaviors, and device performances of HOPs [9]. However, a deeper understanding of the fundamental intrinsic properties and phase transition behaviors of these materials are prerequisites to further improvements of the devices based on HOPs. In this context, single crystals may be ideal for deeper investigation from a fundamental point of view due to the lack of any effects from grain boundaries, morphologies, and defects. Single crystals can be made using a variety of techniques, including solvent evaporation, temperature lowering, inverse crystallization, and antisolvent vapor-assisted methods [10]. We chose the solvent evaporation method among these to synthesize single crystals of high optical quality.

One of the fundamental properties of solids is their elastic properties, including various elastic moduli and acoustic absorption coefficients [11,12]. Elastic constants are directly related to the interatomic forces in the crystal and provide an experimental platform for testing theoretical/calculational models. There are several experimental methods for probing these properties, such as low-frequency indentation and ultrasonic methods [13], etc. Among them, Brillouin light scattering is a nondestructive, noncontact method that is a powerful tool in the determination of the elastic constants, sound velocities, and acoustic attenuation coefficients of acoustic waves propagating in condensed matters [14–20]. Despite extensive studies on HOPs in terms of various experimental techniques, there are only a few acoustic studies on these materials [13,21–27]. Various acoustic properties of $CH_3NH_3PbX_3$ ($MAPbX_3$ with X = Cl, Br, I) and $CH(NH_2)_2PbX_3$ ($FAPbX_3$ with X = Cl, Br, I) have been reported at room temperature or as a function of temperature. In addition, there are a few calculation studies on the mechanical properties of these compounds [12,28].

Ferreira et al. used coherent inelastic neutron scattering and Brillouin scattering techniques to probe the elastic properties of several lead-based halide perovskites [24]. They found a very low shear modulus and softness in these compounds. Significant elastic anisotropy was confirmed in $MAPbBr_3$ measured by the laser ultrasonic technique [13]. Harwell et al. used resonant ultrasound spectroscopy to investigate the temperature dependence of acoustic resonant frequencies and the acoustic losses of methylammonium lead halide perovskites [25]. They found that acoustic losses are substantially large in the tetragonal phase of $MAPbBr_3$ and $MAPbI_3$, which was attributed to the mobility of the ferroelastic twin walls. The overall changes in the elastic properties were explained in terms of the MA order and disorder in each phase [25]. Zhevstovskikh et al. used the ultrasonic technique to study the changes in the sound velocity and the attenuation coefficient of both the longitudinal and transverse acoustic waves of $MAPbI_3$ single crystals in the MHz range [27]. This study revealed a step-like change in the sound velocity and a sharp attenuation peak near the tetragonal–orthorhombic phase transition point. These anomalies were interpreted based on the phenomenological theory of the order parameter–strain coupling and the fluctuations of the order parameter.

Another interesting study on the high-frequency dynamics of $MAPbX_3$ was reported by Anusca et al. [23]. They showed that a fast relaxation process appears in the GHz range probed by broadband dielectric spectroscopy, the relaxation time of which exhibited a critical slowing-down behavior near the phase transition point, indicating the order–disorder nature of the phase transition. They also reported significant changes in the sound velocity and the attenuation coefficient at both phase transitions of the cubic-to-tetragonal and tetragonal-to-orthorhombic phases. Since the GHz relaxation process is exactly overlapped with the frequency window of the Brillouin spectroscopy, it would be interesting to find out the correlation between the dielectric relaxation process and the Brillouin scattering results.

The information on the elastic properties of halide perovskites is important for the investigation of their stability in device applications or material preparation. The purpose of this study is to investigate the temperature dependence of the acoustic properties of $MAPbCl_3$ single crystals in the hypersonic range by using Brillouin spectroscopy. Especially, the temperature dependence of some elastic constants will be reported for the first time and their correlation with the phase transition will be discussed in detail. $MAPbCl_3$ is the most stable halide perovskite, with a bandgap energy of ~2.9 eV [29]. It has been studied in terms of various experimental methods [30–32], but the acoustic study is very rare [23,25]. The obtained results will be compared to previous studies, especially the high-frequency dielectric properties reported by Anusca et al. [23].

2. Materials and Methods

Lead chloride ($PbCl_2$, 99.999%), methylamine (CH_3NH_2, 40% in water), hydrochloric acid (HCl, 37%, ACS reagent), dimethyl sulfoxide (DMSO, anhydrous $\geq$ 99.9%), diethyl ether (HPLC grade, $\geq$99.9%), and ethanol (anhydrous 99.5%) were purchased from Sigma Aldrich (St. Louis, MO, USA). Methylammonium chloride (MACl) was synthesized by dropwise addition of HCl into ice-bath-cooled methylamine in the 1.2:1 molar ratio followed by stirring until complete dissolution. The obtained solution was then dried into a rotary evaporator at 60 °C under a vacuum. After 3 h, the solution had completely dried, and a white shiny MACl powder was obtained. MACl powder was then purified by dissolving it in ethanol (200 mL) at 40 °C. Next, diethyl ether (200 mL) was added to achieve precipitation in this solution. The purification step was repeated twice. Finally, the obtained powder was dried overnight in a vacuum oven at 60 °C.

Equimolar solutions of the obtained white MACl powder (0.01 M) and $PbCl_2$ (0.01 M) were dissolved in DMSO (10 mL) by stirring at 60 °C. After complete dissolution, the solution was filtered through a 0.22 µm syringe filter into a crystallization dish. The dish was covered with aluminum foil and left undisturbed for 1–2 days at a constant temperature of 100 °C. After 1–2 days, transparent $MAPbCl_3$ crystals with a typical size of $5 \times 5 \times 2$ mm^3 were obtained. Afterward, the crystals were washed with acetone and dried overnight in a vacuum oven at 60 °C. Figure 1a shows the photo of the grown single crystal which was used for the Brillouin measurement.

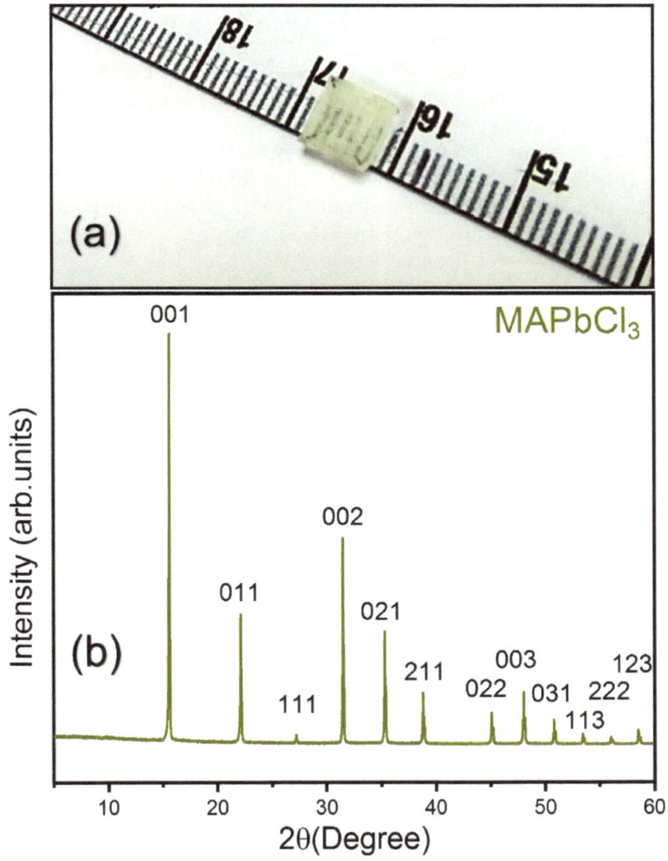

Figure 1. (Color online) (**a**) A picture of the grown single crystal. (**b**) The X-ray diffraction pattern of the powder $MAPbCl_3$.

A high-resolution powder X-ray diffractometer (PANalytical X'pert Pro MPD, Malvern, UK) was used to record X-ray diffraction (XRD) patterns at room temperature, with Cu Kβ radiations with λ = 1.5406 Å as an excitation source to probe the crystal at a scan rate of 3° per minute in a scan range of 10° < 2θ < 60°.

Brillouin spectra were measured by using a typical tandem multi-pass Fabry–Perot interferometer (TFP-2, JRS Co., Zürich, Switzerland). The free spectral range was 50 GHz, and the scan range was ±35.7 GHz. The typical finesse of the interferometer was around 100. A modified microscope (BH-2, Olympus, Tokyo, Japan) including a small prism for redirecting the laser beam, was used for the backscattering experiment. A diode-pumped solid-state single-mode laser (Excelsior 532−300, Spectra Physics, Santa Clara, CA, USA) at a wavelength of 532 nm was used as an excitation light source. A conventional photon-counting system combined with a multichannel analyzer (1024 channels) was used to detect and average the signal. The sample was put in a cryostat stage (THMS 600, Linkam, Tadworth, UK) for temperature control. All the measurements were taken in the temperature range from −196 °C to 40 °C.

3. Results and Discussion

The powder X-ray diffraction for the obtained single crystal, which is shown in Figure 1b, was consistent with previous reports [33]. Figure 2a shows a temperature dependence of the Brillouin spectrum of the $MAPbCl_3$ single crystal. In the high-temperature cubic phase, the spectrum consists of two Brillouin doublets appearing at ~25 GHz and ~8 GHz. The high- and the low-frequency peaks correspond to the longitudinal acoustic (LA) and the transverse acoustic (TA) modes, respectively. As temperature decreases, a quasi-elastic central peak begins to appear at approximately −80 °C and grows upon further cooling. This is evident from the intensity color plot, as shown in Figure 2b. The LA mode is split at about −116 °C, which corresponds to the tetragonal-to-orthorhombic phase transition point. The TA mode is split as well at the same temperature, which is, however, not clearly seen in Figure 2a.

For quantitative analysis, the measured spectrum was curve-fitted by using a Voigt function. It is a convolution function of the Lorentzian function (an approximate response function of the damped harmonic oscillator) and the Gaussian instrumental function. The acoustic-mode frequency (ν_B) and the full width at half maximum (FWHM, Γ_B) was derived as a function of temperature. Figure 3a,b show the temperature dependences of the ν_B and the Γ_B of the LA mode, respectively, for the cooling process. The temperature dependence of the ν_B, as well as its splitting at −116 °C, was reproducible by subsequent measurements. $MAPbCl_3$ has been known to undergo two structural phase transitions, from the cubic-to-tetragonal phase at approximately −95 °C (=$T_{C\text{-}T}$), and then from the tetragonal-to-orthorhombic phase at about −116 °C (=$T_{T\text{-}O}$). The LA mode shows gradual softening and then a slight upward hardening upon cooling in the cubic phase. The ν_B exhibits a small cusp, while the Γ_B displays a clear maximum near $T_{C\text{-}T}$. Figure S1a,b in the Supplementary Materials show the comparison of the Brillouin data and the real part of the dielectric permittivity reported in [34]. the $T_{C\text{-}T}$ is characterized by the maxima of both the permittivity and the Γ_B, while the permittivity and the ν_B drop precipitously at $T_{T\text{-}O}$. These comparisons indicate that the acoustic anomalies are strongly correlated with the changes in the dielectric properties.

The Brillouin spectra were recorded upon heating as well. The comparison between cooling and heating processes is shown in Figure S2 in the Supplementary Materials. The splitting of the LA mode at low temperatures and the softening of the LA mode in addition to a significant damping factor in the cubic phase were confirmed in the heating process. However, the data are a little bit scattered, which may be probably due to the sample degradation caused by the long exposure to the ambient condition.

Both LA- and TA-mode frequencies are split at $T_{T\text{-}O}$, as described above. Especially, the Γ_B of the LA mode exhibits a sharp damping peak at $T_{T\text{-}O}$. The half-widths are small in the orthorhombic phase, while the mode frequencies show slight hardening upon further cooling below $T_{T\text{-}O}$. There are small but noticeable anomalies near −140 °C, the origin of which is not clear. However, previous acoustic studies by resonant ultrasound spectroscopy revealed a clear acoustic damping peak in the same temperature range, which was tentatively attributed to some freezing process of the MA rotation and/or the hopping of Cl ions between vacancies [25]. However, this small anomaly does not appear during the heating process (Figure S2), indicating that the relevant structural change may be metastable.

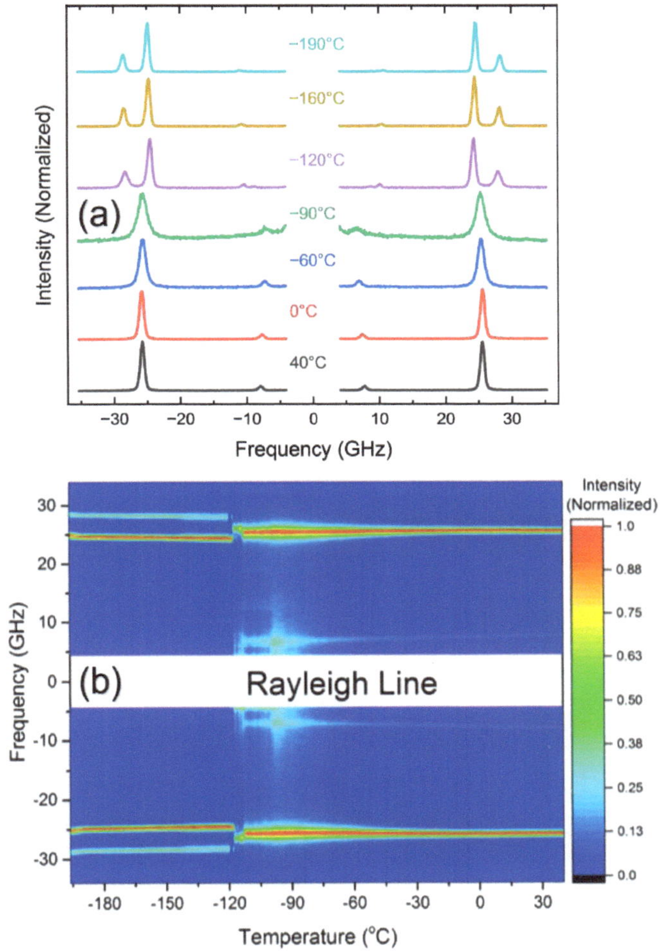

Figure 2. (Color online) (**a**) Temperature dependence of the Brillouin spectrum and (**b**) the intensity color plot of the measured spectra.

Figure 4 shows the temperature dependence of the ν_B of the TA mode for the cooling process. It displays a gradual softening upon cooling in the cubic phase, a broad minimum at $T_{C\text{-}T}$, and a sudden increase along with splitting at $T_{T\text{-}O}$. The Γ_B of the TA mode, which is shown in Figure S3b in the Supplementary Materials, is rather scattered but shows a small peak near $T_{T\text{-}O}$. Splitting of the TA mode at and below $T_{T\text{-}O}$ is seen and is very similar to the case of the LA mode. The splitting of the acoustic-mode frequencies was confirmed in $MAPbBr_3$ as well and ascribed to the presence of multiply-strained domains [22]. In this case, there may exist several multiply-strained domains in the focal point of the probe laser beam, where differently strained areas with different refractive indices would give rise to split Brillouin doublets. Therefore, a quantitative analysis for the orthorhombic phase was difficult to carry out. This result, combined with [22], indicates that the formation of multiply-strained domains in the low-temperature orthorhombic phase seems to be a common phenomenon in MA-based halide perovskite single crystals.

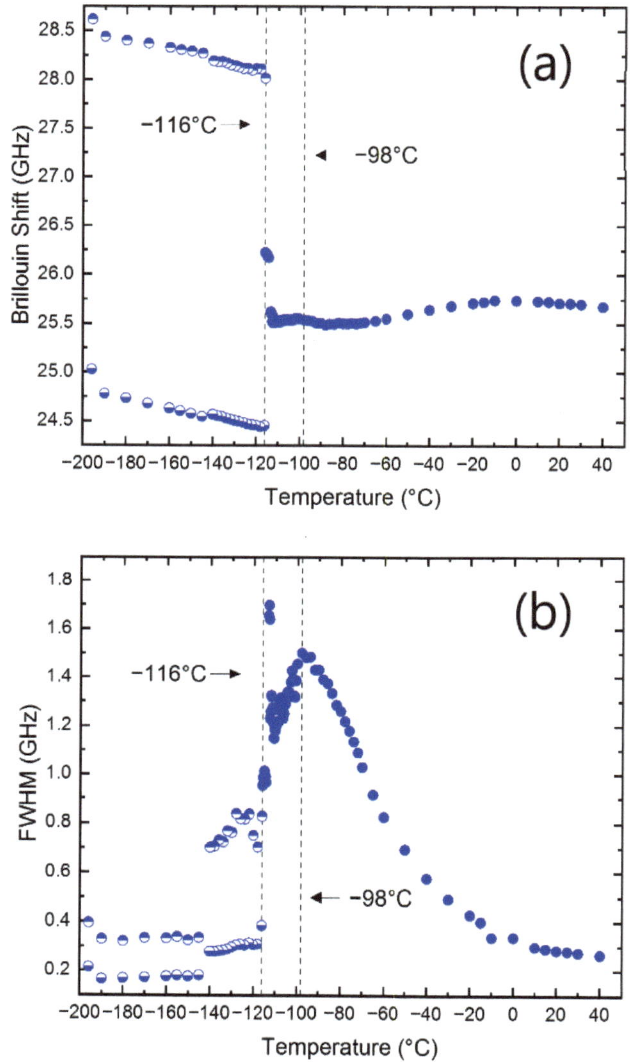

Figure 3. (Color online). Temperature dependences of (**a**) the Brillouin frequency shift and (**b**) the FWHM of the LA mode propagating along the [100] direction, measured upon cooling. The half-filled circles represent the split peaks. The two symbols at low temperatures indicate the data from the split LA mode.

The data of ν_B and Γ_B were transformed into the sound velocity V and the acoustic absorption coefficient α via the following equations valid for the backscattering geometry;

$$V = \frac{\nu_B \lambda}{2n}, \quad (1)$$

$$\alpha = \frac{\pi \Gamma_B}{V}. \quad (2)$$

In these equations, λ is the laser wavelength (=532 nm) and n is the refractive index. Figure 5a,b show the temperature dependence of the sound velocity and the absorption coefficient of the LA mode, respectively. The refractive index of $n = 1.9$ at 532 nm, reported in the previous study [35], was used for the calculation. The temperature dependence of n was not considered. The longitudinal sound velocity is approximately 3599 m/s at room temperature and shows a mild softening upon cooling. This elastic softening is accompanied by a drastic increase in the absorption coefficient, a more than five times increase upon the change of T from room temperature to $T_{C\text{-}T}$. It indicates that the longitudinal sound wave is strongly coupled to other degrees of freedom associated with the cubic-to-tetragonal phase transition and that the dissipative energy exchange between them becomes very active near $T_{C\text{-}T}$. There are two reports on the longitudinal sound velocities of $MAPbCl_3$ [23,26]. The longitudinal sound velocity of ceramic $MAPbCl_3$ is approximately 3450 m/s at 0 °C [23], which is comparable to the result of this study. However, the other result on the single-crystal $MAPbCl_3$ shows that the sound velocity is 4000 m/s at room temperature [26], which is much larger than the other two results, including the present study. The sound velocity of the TA mode is shown in Figure 6. It decreases from ~1098 m/s to ~967 m/s at $T_{C\text{-}T}$ and then increases in the tetragonal phase drastically. The sound velocity exhibits two branches in the orthorhombic phase corresponding to the two acoustic modes (split TA mode) shown in Figure 4. The transverse sound velocity of this study is much smaller compared to the previous report of 1770 m/s [26]. Since the phonon propagation direction was not completely determined and the laser wavelength was different in the previous work [26], it is not clear whether the difference in the sound velocities is due to the dispersion effect or the difference in the phonon direction.

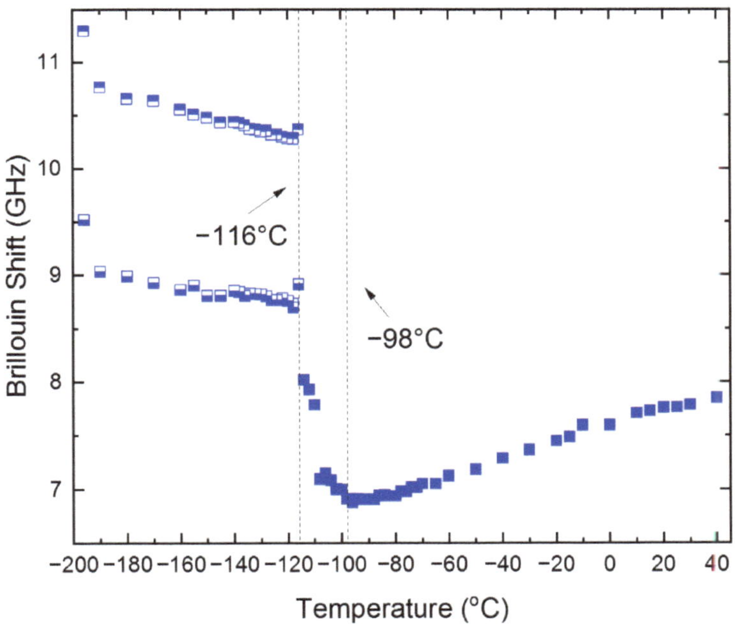

Figure 4. (Color online). Temperature dependences of the Brillouin frequency shift of the TA mode propagating along the [100] direction, measured upon cooling. The two symbols at low temperatures indicate the data from the split TA mode.

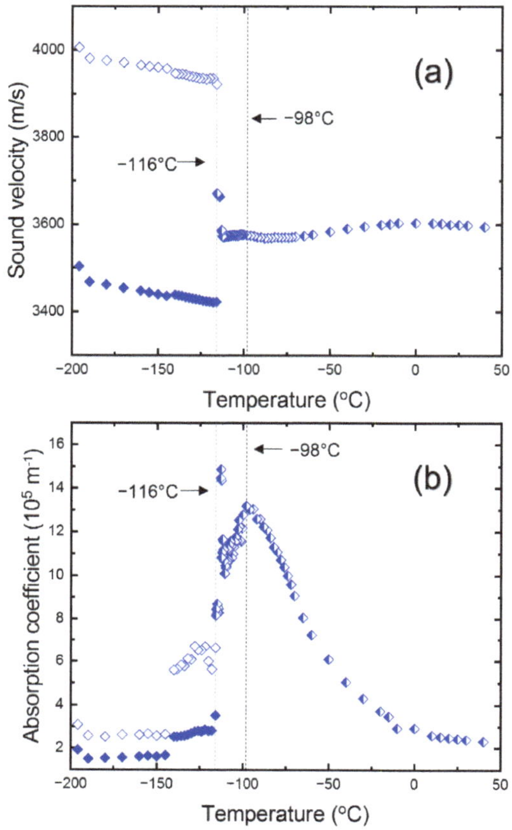

Figure 5. (Color online). Temperature dependences of (**a**) the sound velocity and (**b**) the absorption coefficient of the LA mode. Phase boundaries are shown as vertical dotted lines along with the phase transition temperatures. The two symbols at low temperatures indicate the data from the split LA mode.

The coupling between the strain caused by the LA mode and the other degrees of freedom relevant to the phase transition is more clearly seen from the elastic constant data. $MAPbCl_3$ maintains a cubic symmetry at high temperatures down to ~ $T_{C\text{-}T}$, at which the cubic phase changes into a tetragonal one. There are three independent elastic constants in the cubic phase, namely C_{11}, C_{12}, and C_{44} [14]. Since the phonon propagation direction of the present experiment is the [100] direction, the LA and TA modes shown in Figure 2 correspond to the C_{11} and C_{44}, respectively. The reported density of ρ = 3171 kg/m^3 [36] was used to obtain the two elastic constants as ρV^2 in the cubic phase. Table 1 shows the comparison of the sound velocities and the two elastic constants of the MA-based halide perovskites at room temperature. There is no experimental result for the elastic constant of $MAPbCl_3$. Only theoretical calculations provided the estimations for the elastic constants [12,28]. The reported values for C_{11} from these theoretical studies are 39.5 and 42.1 GPa, which are very similar to the present result of 41.0 GPa. C_{44} has been theoretically estimated to be 2.9 or 6.5 GPa, and our result of 3.74 GPa is located in this range. These consistent results suggest both theoretical studies are reliable, although the difference in predicted C_{44} values is rather large, and that our experimental results provide a reliable testbed for further theoretical works.

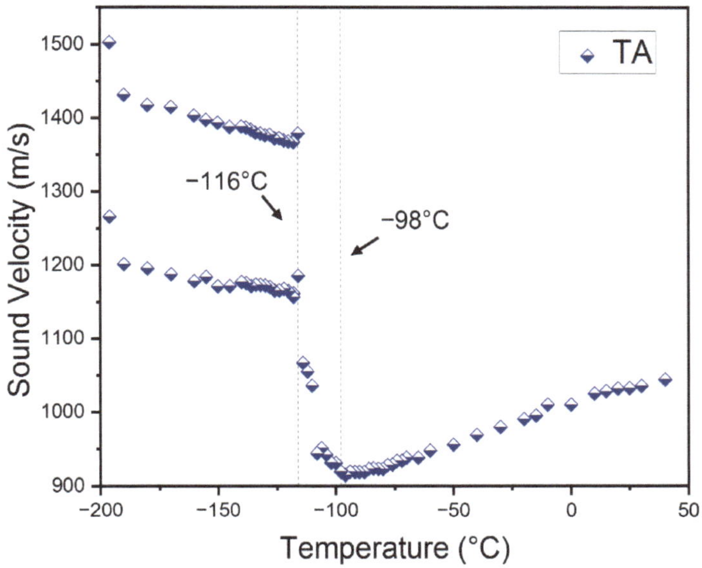

Figure 6. (Color online). Temperature dependences of the sound velocity of the TA mode. Phase boundaries are shown as vertical dotted lines along with the phase transition temperatures.

Table 1. Comparison of sound velocities and elastic constants of several halide perovskites.

Parameters	$MAPbCl_3$ (This Work)	$MAPbCl_3$ (Other Works)	$MAPbBr_3$ (Other Works)	$MAPbI_3$ (Other Works)
V_{LA} (m/s)	3599	4000 [26]	3075 [22]	2300 [24]
V_{TA}(m/s)	1086	1770 [26]	1010 [22]	1330 [24]
C_{11} (GPa)	41.0	39.5 [12]	35.9 [22]	32.3 [28]
C_{44} (GPa)	3.74	2.9 [12]	3.9 [22]	4.8 [28]

Figure 7 shows the change in the elastic constants as a function of the lattice constant (and, thus, X in $MAPbX_3$) in lead-based halide perovskites. Since the elastic bulk modulus cannot be obtained from the present result due to the lack of C_{12}, theoretical values of [28] were included for comparison. The C_{11} and the bulk modulus increase with the decreasing lattice constant, while C_{44} values are relatively similar and very small. A more compact structure at a smaller lattice constant is responsible for the larger strength of the binding interactions, which was attributed to the more symmetric MA cations and the resulting steric effect [24]. On the other hand, the exceedingly small C_{44} values of the three compounds indicate their very unstable structure against the shear forces, resulting in large elastic anisotropy and an exceptionally low shear modulus.

The temperature dependences of the obtained elastic constants in the cubic phase are shown in Figure 8. Both elastic constants display softening upon cooling toward $T_{C\text{-}T}$. C_{11} shows a maximum near ~0 °C and begins to be softened below this temperature, which indicates that the coupling of the LA mode and other degrees of freedom begins near this temperature. C_{11} changes from ~41.2 GPa at about 0 °C to ~40.4 GPa at −90 °C. In the case of C_{44}, it continuously decreases from ~3.82 GPa at 40 °C to ~2.96 GPa at $T_{C\text{-}T}$. The surprisingly small C_{44} was also observed from other halide perovskites, such as $MAPbBr_3$ [13,22]. The C_{44} of $MAPbBr_3$ is around 3.5 GPa at room temperature and shows the smallest value along the [100] direction in the slowness curve [13]. The C_{44} of $MAPbI_3$ is reported to be approximately 7.3 GPa [24] or 4.8 GPa [28]. On the other hand, the FA-based halide perovskites exhibit even smaller C_{44} values than MA-based materials [24].

It indicates that the extremely low shear rigidity may be a common characteristic in MA- and FA-based halide perovskites, with the latter being softer to the shear stress.

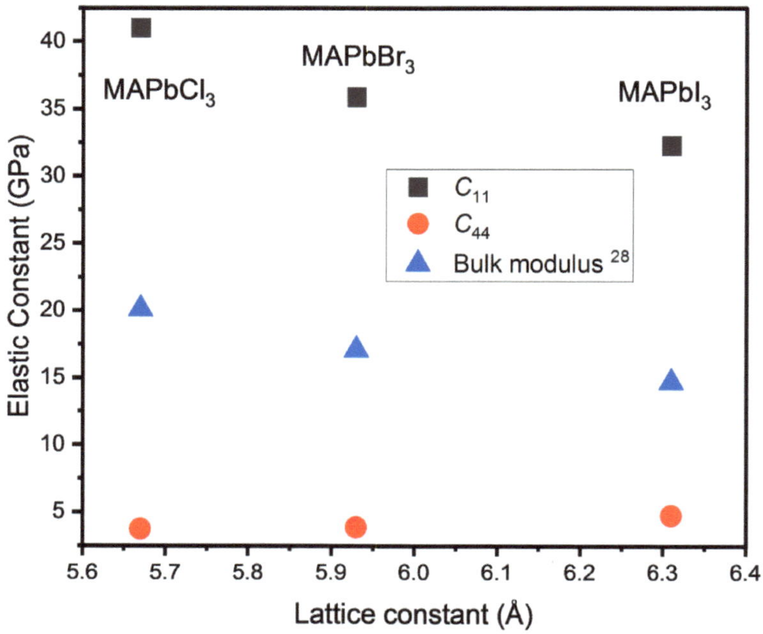

Figure 7. (Color online). Comparison of two elastic constants, C_{11} and C_{44}, and the bulk modulus of three MA-based halide perovskites at room temperature. The C_{11} and C_{44} of $MAPbCl_3$ are from the present work and other values are from several references, as shown in Table 1.

The above results clearly showed that the acoustic mode is coupled to other degrees of freedom, which are associated with the cubic-to-tetragonal phase transition, in the cubic phase. Létoublon et al. suggested that the disordered reorientational motions of the MA cations are responsible for the translation–rotation coupling, which induces the elastic softening [22]. They revealed that the low-frequency Brillouin spectra of $MAPbBr_3$ are characterized by the central peak coupled to the LA mode, which is another indication of the order–disorder characteristic of the cubic-to-tetragonal phase transition. In another work, the rotations and tilts of the corner-sharing oxygen octahedra were suggested to be the origin of the acoustic softening [13]. Especially, the rotational motions may be responsible for the extremely low shear modulus. We assume that the translation–rotation coupling between the strains caused by the acoustic modes and the relevant degrees of freedom can be explained in terms of the phenomenological approach based on the Ginzburg–Landau free energy expansion [37]. In this model, which can be applied to both cases of linear and quadratic couplings, the relaxation process coupled to the strain is dynamic and has frequency-dependent properties. The relaxation time of the relevant process that couples to the acoustic waves can quantitatively be analyzed based on an assumption of a single relaxation process. This assumption may be justified in the high-temperature range, where the distribution of relaxation times is expected to be narrow. This narrow distribution is supported by the previous results from broadband dielectric spectroscopy [23]. The relaxation time τ_{LA} is expressed by the following equation [38], which was derived from a simple acoustic dispersion relationship;

$$\tau_{LA} = \frac{\Gamma_B - \Gamma_\infty}{2\pi\left(\nu_\infty^2 - \nu_B^2\right)} \tag{3}$$

Here, ν_∞ and Γ_∞ are an unrelaxed Brillouin shift at the high-frequency limit and the high-frequency background damping, respectively. Both represent the quantities that are not related to the softening of the LA mode and the local or macroscopic phase transition.

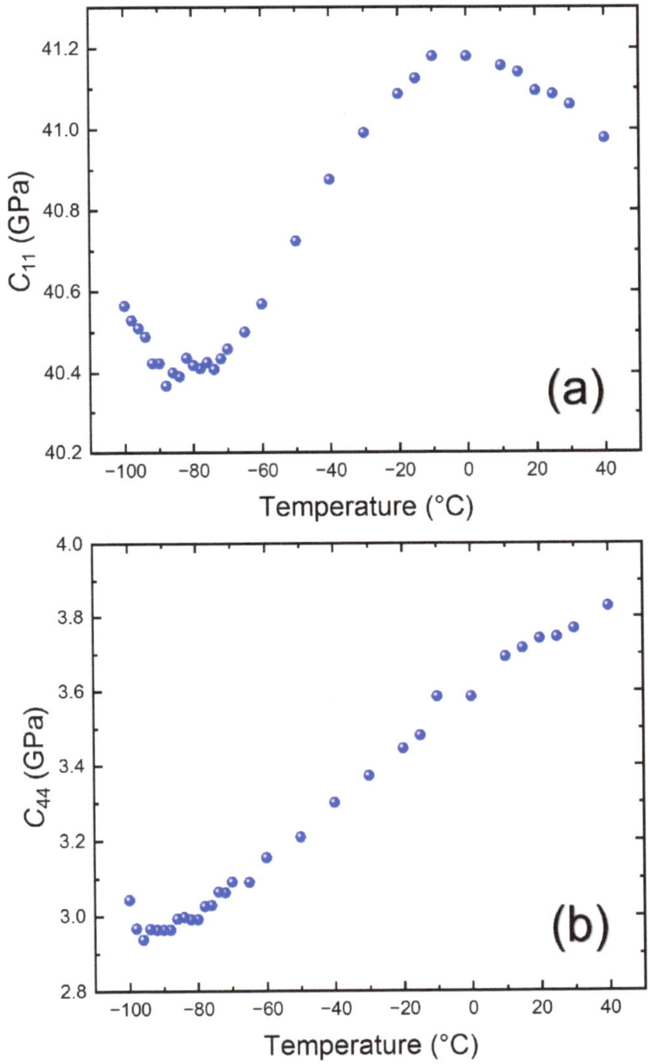

Figure 8. (Color online). Temperature dependences of (**a**) C_{11} and (**b**) C_{44} in the cubic phase.

The widest measurement temperature range covered by this study makes it possible for us to carry out a quantitative analysis for the LA modes. The ν_∞ and Γ_∞ were derived from the high-temperature values of the ν_B and the Γ_B. The calculated relaxation times are shown in Figure 9. The τ_{LA} increases from ~16 ps to ~24 ps when the temperature decreases from −60 °C to −100 °C. The orders of magnitude of τ_{LA} are consistent with the value of 28 ps derived from the central peak reported in the previous work [19]. The breakpoint of approximately −100 °C is nearly the same as, but slightly lower than, $T_{C\text{-}T}$. We could observe the central peak as well, but the quantitative analysis was difficult to carry out due to the strong Rayleigh line, as described below.

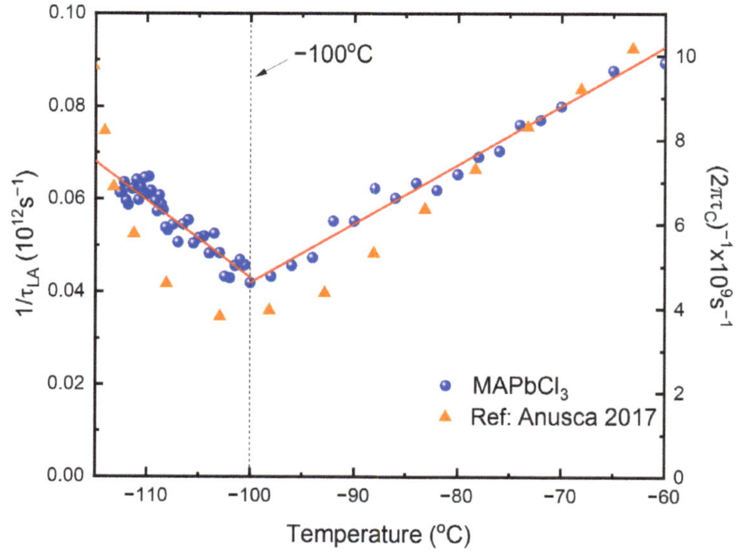

Figure 9. (Color online). Temperature dependence of the inverse relaxation time obtained by using the acoustic anomalies of the LA mode and Equation (3) [20]. The solid line is the best-fitted result by using Equation (4). The inverse dielectric relaxation time τ_c taken from [23] is shown for comparison.

If the relevant degrees of freedom coupled to the acoustic modes are relaxation processes, they can be probed in terms of broadband dielectric spectroscopy as well. Anusca et al. reported broadband dielectric data where they could observe strong dielectric relaxation in the GHz range [23]. The obtained relaxation time exhibited a critical slowing-down behavior, which was attributed to the rotational motions of the MA cations [23]. Figure 9 includes the dielectric relaxation time reported in this reference. Both acoustic and dielectric relaxation times exhibit quite similar temperature dependences near $T_{C\text{-}T}$, i.e., both values increase (slow-down) toward this transition temperature upon cooling in the cubic phase. This result clearly shows that the microscopic origin of both relaxation processes, acoustic and dielectric, is the same, i.e., the correlated MA motions [23], which accompany both dipole changes under an oscillating electric field and polarization fluctuations.

The linear temperature dependence of the relaxation time is a typical characteristic of the critical slowing-down behavior. The two linear parts in Figure 9 were fitted in terms of the following equation [39]:

$$\frac{1}{\tau_{LA}} = \frac{1}{\tau_0}\left(\frac{T - T_0}{T_0}\right) \tag{4}$$

The obtained fitting parameters are summarized in Table 2. This kind of critical slowing-down behavior indicates that the nature of the phase transition is of an order–disorder type; thus, the relaxation time increases upon approaching the phase transition point due to the growing volume (or correlation length) of the correlated MA cations.

Table 2. Fitting parameters for the relaxation time estimated by using Equation (4).

Fitting Parameter	$T < -100$ °C	$T > -100$ °C
τ_0 (ps)	8.07	5.92
T_0 (°C)	−74.2 (199 K)	−133.0 (140 K)

Yamamuro et al. discussed the nature of the cubic-to-tetragonal phase transition from their calorimetric, infrared, and dielectric measurements [40,41], according to which the

MA ions are suggested to be located in equivalent states in the high-temperature prototype phase. The phase transition is accompanied by a partial ordering of the disordered MA configurations in the cubic phase. The estimated entropy indicates that the phase transition is of the order–disorder type, which is consistent with the critical slowing-down behavior observed in this study and the dielectric study [23]. Similar behaviors were observed from other systems, such as ferroelectric $BaTiO_3$ [36] and relaxors [42], etc., where nanoscale precursor polar clusters or polar nanoregions grow upon cooling toward the transition temperature or freezing temperature.

The fast relaxation process is in many cases responsible for the formation of quasi-elastic central peaks in the inelastic light scattering spectrum. A low-frequency Raman study on $MAPbBr_3$ revealed the existence of the central peak centered at a zero frequency [22]. Brillouin spectra also exhibit a low-frequency quasi-elastic central peak, as shown in Figure 10a, which was measured in a wide-frequency range of ±546.86 GHz (±18.26 cm^{-1}). The central peak grows significantly as temperature decreases but becomes very small below $T_{T\text{-}O}$, as is revealed in Figure 10b. The central peak was curve-fitted by using the convolution of the Lorentzian function centered at a zero frequency (which represents a relaxation process of a Debye type) and the Gaussian instrumental function. The temperature dependences of the intensity and the FWHM of the central peak are plotted in Figure 10b. The intensity is large in the cubic phase, decreases significantly in the tetragonal phase, and then becomes very small in the orthorhombic phase. The drastic change in the intensity indicates that the relaxation process is very active in the cubic phase. The quantitative analysis of the relaxation time was difficult due to the strong Rayleigh peak, which distorts the low-frequency central peak. However, an approximate relaxation time τ_{CP} can be estimated from the FWHM via $\tau_{CP} = 1/\pi\Gamma_{CP}$, where Γ_{CP} is the FWHM of the central peak. Figure 10b shows that τ_{CP} decreases from ~1 ps to 3.2 ps upon cooling in the cubic phase. The general behavior mimicking the slowing-down behavior is similar to that calculated from the LA mode, but the absolute values are one order of magnitude smaller, which may be due to the insufficient spectral range, especially the low-frequency spectral features, which were distorted by the strong Rayleigh peak.

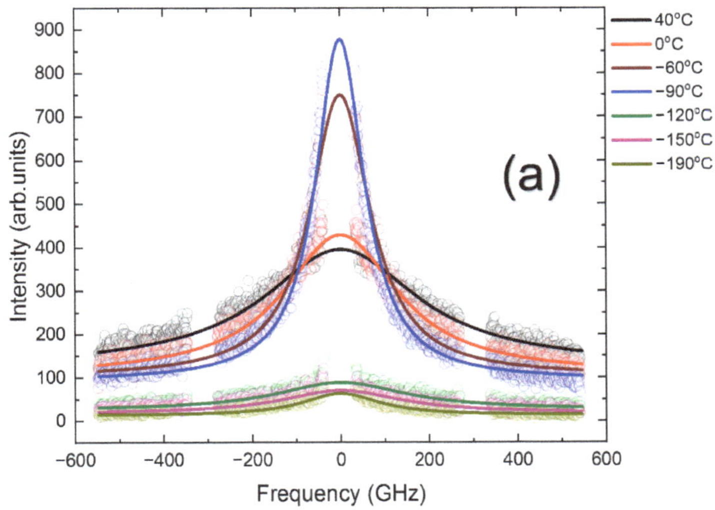

Figure 10. *Cont.*

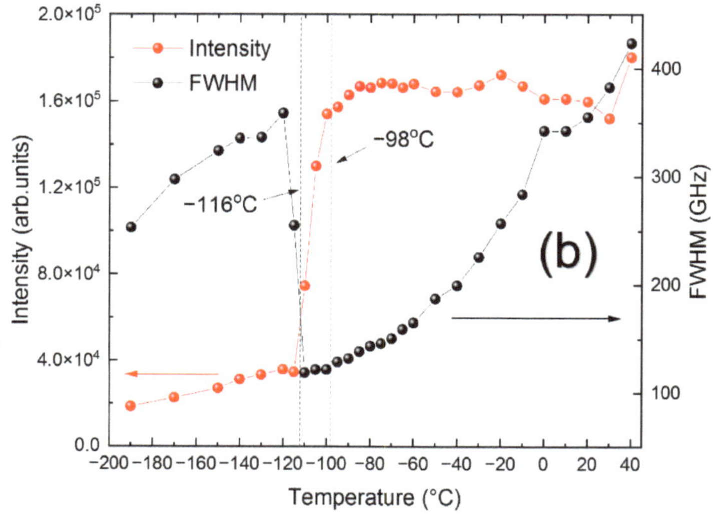

Figure 10. (Color online). (**a**) The central-peak spectra at several temperatures and (**b**) the temperature dependence of the central-peak intensity and its half-width.

The present results show clear evidence for the existence of precursor dynamics of $MAPbCl_3$ in the prototype cubic phase. The correlated MA regions couple to the acoustic waves, interact with them, and induce acoustic softening as well as the formation of a strong central peak near the phase transition point. The estimated relaxation time from the LA-mode anomalies follows the critical slowing-down behavior, which demonstrates the order–disorder nature of the corresponding phase transition. The first and detailed elastic constant data can be served as an experimental testbed for checking the reliability of theoretical approaches to this important class of materials.

4. Conclusions

The acoustic properties of $MAPbCl_3$ single crystals were investigated in a wide temperature range from 40 °C to −196 °C by using Brillouin spectroscopy. The sound velocities and the absorption coefficients of both the longitudinal and transverse acoustic modes propagating along the cubic [100] direction were obtained as a function of temperature, and the temperature dependences of the two elastic constants, C_{11} and C_{44}, could be obtained in the cubic phase for the first time. The two-phase transition points were clearly identified from the anomalous changes in the sound velocities and absorption coefficients. The low-temperature orthorhombic phase was characterized by split acoustic modes, which were attributed to the presence of multiply-strained domains. The C_{11} (=41.0 GPa) of $MAPbCl_3$ was larger than those of $MAPbBr_3$ and $MAPbI_3$, indicating a more compact crystal structure and stronger binding interactions, while the C_{44} (=3.74 GPa) value was exceedingly small, similar to other MA- and FA-based halide compounds. The anomalous changes in the acoustic properties indicated the coupling of the acoustic waves to the dynamic MA cations and their collective rotational motions in the prototype cubic phase. The obtained relaxation time followed the critical slowing-down behavior consistent with previous dielectric studies demonstrating that the cubic-to-tetragonal phase transition is of an order–disorder type. This is consistent with the observation of strong central peaks near the phase transition point, reflecting the existence of an active relaxation process, which can be probed by dielectric spectroscopy and coupled to the acoustic waves.

Supplementary Materials: The following supporting information can be downloaded at: https://www.mdpi.com/article/10.3390/ma15103692/s1, Figure S1: Comparison of (a) the Brillouin frequency shift and (b) the FWHM with the real part of the dielectric permittivity reported in [31]; Figure S2: Temperature dependences of (a) the mode frequency and (b) the FWHM of the TA mode propagating along the [100] direction measured upon cooling and heating; Figure S3: Temperature dependences of (a) the mode frequency and (b) the FWHM of the TA mode propagating along the [100] direction measured upon cooling and heating.

Author Contributions: Conceptualization, J.-H.K. and T.H.K.; Data curation, J.W.L. and F.H.N.; Formal analysis, J.W.L., F.H.N. and C.W.A.; Funding acquisition, J.-H.K.; Investigation, J.W.L., F.H.N., J.-H.K., T.H.K. and C.W.A.; Methodology, J.-H.K., T.H.K. and C.W.A.; Project administration, J.-H.K.; Resources, J.-H.K., T.H.K. and C.W.A.; Supervision, J.-H.K.; Validation, J.-H.K.; Visualization, J.W.L. and F.H.N.; Writing—original draft, J.-H.K.; Writing—review & editing, J.W.L., F.H.N., T.H.K. and C.W.A. All authors have read and agreed to the published version of the manuscript.

Funding: This work was supported by the National Research Foundation of Korea (NRF) grant funded by the Korean Government (MSIT) (No. 2020R1A2C101083111).

Institutional Review Board Statement: Not applicable.

Informed Consent Statement: Not applicable.

Data Availability Statement: Data presented in this article is available on request from the corresponding author.

Conflicts of Interest: The authors declare no conflict of interest.

References

1. Tishkevich, D.I.; Vorobjova, A.I.; Vinnik, D.A. Formation and corrosion behavior of nickel/alumina nanocomposites. *Solid State Phenom.* **2020**, *299*, 100–106. [CrossRef]
2. Zhivulin, V.E.; Trofimov, E.A.; Gudkova, S.A.; Punda, A.Y.; Valiulina, A.N.; Gavrilyak, A.M.; Zaitseva, O.V.; Tishkevich, D.I.; Zubar, T.I.; Sun, Z.; et al. Impact of the a-site rare-earth ions (Ln^{3+}-Sm^{3+}, Eu^{3+}, Gd^{3+}) on structure and electrical properties of the high entropy $LnCr_{0.2}Mn_{0.2}Fe_{0.2}Co_{0.2}Ni_{0.2}O_3$ perovskites. *Ceram. Int.* **2022**, *48*, 9239–9247. [CrossRef]
3. Zhao, Y.; Zhu, K. Organic-inorganic hybrid lead halide perovskites for optoelectronic and electronic applications. *Chem. Soc. Rev.* **2016**, *45*, 655–689 [CrossRef] [PubMed]
4. Stranks, S.D.; Snaith, H.J. Metal-halide perovskites for photovoltaic and light-emitting devices. *Nat. Nanotechnol.* **2015**, *10*, 391–402. [CrossRef] [PubMed]
5. Liu, X.K.; Xu, W.; Bai, S.; Jin, Y.; Wang, J.; Friend, R.H.; Gao, F. Metal halide perovskites for light-emitting diodes. *Nat. Mater.* **2021**, *20*, 10–21. [CrossRef] [PubMed]
6. Bao, C.; Yang, J.; Bai, S.; Xu, W.; Yan, Z.; Xu, Q.; Liu, J.; Zhang, W.; Gao, F. High performance and stable all-inorganic metal halide perovskite-based photodetectors for optical communication applications. *Adv. Mater.* **2018**, *30*, 1803422. [CrossRef] [PubMed]
7. Xiang, W.; Tress, W. Review on recent progress of all-inorganic metal halide perovskites and solar cells. *Adv. Mater.* **2019**, *31*, 1902851. [CrossRef] [PubMed]
8. NREL Efficiency Chart. Available online: https://www.nrel.gov/pv/cell-efficiency.html (accessed on 18 May 2022).
9. Leguy, A.M.A.; Azarhoosh, P.; Alonso, M.I.; Campoy-Quiles, M.; Weber, O.J.; Yao, J.; Bryant, D.; Weller, M.T.; Nelson, J.; Walsh, A.; et al. Experimental and theoretical optical properties of methylammonium lead halide perovskites. *Nanoscale* **2016**, *8*, 6317–6327. [CrossRef]
10. Murali, B.; Kolli, H.K.; Yin, J.; Ketavath, R.; Bakr, O.M.; Mohammed, O.F. Single crystals: The next big wave of perovskite optoelectronics. *ACS Mater. Lett.* **2020**, *2*, 184–214. [CrossRef]
11. Songvilay, M.; Giles-Donovan, N.; Bari, M.; Ye, Z.G.; Minns, J.L.; Green, M.A.; Xu, G.; Gehring, P.M.; Schmalzl, K.; Ratcliff, W.D.; et al. Common acoustic phonon lifetimes in inorganic and hybrid lead halide perovskites. *Phys. Rev. Mater.* **2019**, *3*, 093602. [CrossRef]
12. Faghihnasiri, M.; Izadifard, M.; Ghazi, M.E. DFT study of mechanical properties and stability of cubic methylammonium lead halide perovskites ($CH_3NH_3PbX_3$, X = I, Br, Cl). *J. Phys. Chem. C* **2017**, *121*, 27059–27070. [CrossRef]
13. Lomonosov, A.M.; Yan, X.; Sheng, C.; Gusev, V.E.; Ni, C.; Shen, Z. Exceptional elastic anisotropy of hybrid organic–inorganic perovskite $CH_3NH_3PBBR_3$ measured by laser ultrasonic technique. *Phys. Status Solidi-Rapid Res. Lett.* **2016**, *10*, 606–612. [CrossRef]
14. Vacher, R.; Boyer, L. Brillouin scattering: A tool for the measurement of elastic and photoelastic constants. *Phys. Rev. B* **1972**, *6*, 639–673. [CrossRef]
15. Cummins, H.Z. Brillouin scattering spectroscopy of ferroelectric and ferroelastic phase transitions. *Phil. Trans. R. Soc. Lond. A* **1979**, *293*, 393–405.

16. Polian, A. Brillouin scattering at high pressure: An overview. *J. Raman Spectrosc.* **2003**, *34*, 633–637. [CrossRef]
17. Kojima, S. Gigahertz acoustic spectroscopy by micro-brillouin scattering. *Jpn. J. Appl. Phys.* **2010**, *49*, 07HA01. [CrossRef]
18. Comez, L.; Masciovecchio, C.; Monaco, G.; Fioretto, D. *Progress in Liquid and Glass Physics by Brillouin Scattering Spectroscopy*, 1st ed.; Elsevier: Amsterdam, The Netherlands, 2012; Volume 63, ISBN 9780123970282.
19. Ko, J.-H.; Jeong, M.-S.; Lee, B.W.; Kim, J.H.; Ko, Y.H.; Kim, K.J.; Kim, T.H.; Kojima, S.; Ahart, M. Pressure dependence of acoustic properties of liquid ethanol by using high-pressure brillouin spectroscopy. *Korean J. Opt. Photonics* **2013**, *24*, 279–286. [CrossRef]
20. Speziale, S.; Marquardt, H.; Duffy, T.S. Brillouin scattering and its application in geosciences. *Spectrosc. Methods Mineral. Mater. Sci.* **2014**, *78*, 543–603. [CrossRef]
21. Rakita, Y.; Cohen, S.R.; Kedem, N.K.; Hodes, G.; Cahen, D. Mechanical properties of $APbX_3$ (A = Cs or CH3NH3; X = i or Br) perovskite single crystals. *MRS Commun.* **2015**, *5*, 623–629. [CrossRef]
22. Létoublon, A.; Paofai, S.; Rufflé, B.; Bourges, P.; Hehlen, B.; Michel, T.; Ecolivet, C.; Durand, O.; Cordier, S.; Katan, C.; et al. Elastic constants, optical phonons, and molecular relaxations in the high temperature plastic phase of the $CH_3NH_3PbBr_3$ Hybrid perovskite. *J. Phys. Chem. Lett.* **2016**, *7*, 3776–3784. [CrossRef]
23. Anusca, I.; Balčiūnas, S.; Gemeiner, P.; Svirskas, Š.; Sanlialp, M.; Lackner, G.; Fettkenhauer, C.; Belovickis, J.; Samulionis, V.; Ivanov, M.; et al. Dielectric response: Answer to many questions in the methylammonium lead halide solar cell absorbers. *Adv. Energy Mater.* **2017**, *7*, 1700600. [CrossRef]
24. Ferreira, A.C.; Létoublon, A.; Paofai, S.; Raymond, S.; Ecolivet, C.; Rufflé, B.; Cordier, S.; Katan, C.; Saidaminov, M.I.; Zhumekenov, A.A.; et al. Elastic softness of hybrid lead halide perovskites. *Phys. Rev. Lett.* **2018**, *121*, 085502. [CrossRef] [PubMed]
25. Harwell, J.R.; Payne, J.L.; Sajjad, M.T.; Heutz, F.J.L.; Dawson, D.M.; Whitfield, P.S.; Irvine, J.T.S.; Samuel, I.D.W.; Carpenter, M.A. Role of lattice distortion and a site cation in the phase transitions of methylammonium lead halide perovskites. *Phys. Rev. Mater.* **2018**, *2*, 065404. [CrossRef]
26. Kabakova, I.V.; Azuri, I.; Chen, Z.; Nayak, P.K.; Snaith, H.J.; Kronik, L.; Paterson, C.; Bakulin, A.A.; Egger, D.A. The effect of ionic composition on acoustic phonon speeds in hybrid perovskites from brillouin spectroscopy and density functional theory. *J. Mater. Chem. C* **2018**, *6*, 3861–3868. [CrossRef]
27. Zhevstovskikh, I.V.; Averkiev, N.S.; Sarychev, M.N.; Semenova, O.I.; Golyashov, V.A.; Tereshchenko, O.E. Correlation of elastic and optoelectronic properties near structural phase transition in organic-inorganic lead iodide perovskite single crystals. *J. Phys. Condens. Matter* **2020**, *33*, 045403. [CrossRef] [PubMed]
28. Roknuzzaman, M.; Ostrikov, K.; Chandula Wasalathilake, K.; Yan, C.; Wang, H.; Tesfamichael, T. Insight into lead-free organic-inorganic hybrid perovskites for photovoltaics and optoelectronics: A first-principles study. *Org. Electron.* **2018**, *59*, 99–106. [CrossRef]
29. Chen, Y.; Hou, X.; Tao, S.; Fu, X.; Zhou, H.; Yin, J.; Wu, M.; Zhang, X. Synthesis, Crystal structure and photoresponse of tetragonal phase single crystal $CH_3NH_3PbCl_3$. *Chem. Commun.* **2020**, *56*, 6404–6407. [CrossRef]
30. Wang, L.; Wang, K.; Xiao, G.; Zeng, Q.; Zou, B. Pressure-induced structural evolution and band gap shifts of organometal halide perovskite-based methylammonium lead chloride. *J. Phys. Chem. Lett.* **2016**, *7*, 5273–5279. [CrossRef]
31. Thu Nguyen, T.T.; Kim, Y.; Bae, S.; Bari, M.; Jung, H.R.; Jo, W.; Kim, Y.H.; Ye, Z.G.; Yoon, S. Raman Scattering studies of the structural phase transitions in single-crystalline $CH_3NH_3PbCl_3$. *J. Phys. Chem. Lett.* **2020**, *11*, 3773–3781. [CrossRef]
32. Maeda, M.; Hattori, M.; Hotta, A.; Suzuki, I. Dielectric studies on $CH_3NH_3PbX_3$. (X = Cl and Br) single cystals. *J. Phys. Soc. Jpn.* **1997**, *66*, 1508–1511. [CrossRef]
33. Maculan, G.; Sheikh, A.D.; Abdelhady, A.L.; Saidaminov, M.I.; Haque, M.A.; Murali, B.; Alarousu, E.; Mohammed, O.F.; Wu, T.; Bakr, O.M. $CH_3NH_3PbCl_3$. Single crystals: Inverse temperature crystallization and visible-blind UV-photodetector. *J. Phys. Chem. Lett.* **2015**, *6*, 3781–3786. [CrossRef] [PubMed]
34. Govinda, S.; Kore, B.P.; Swain, D.; Hossain, A.; De, C.; Row, T.N.G.; Sarma, D. Critical comparison of $FAPbX_3$ and $MAPbX_3$ (X = Br and Cl): How do they differ? *J. Phys. Chem. C* **2018**, *122*, 13758–13766. [CrossRef]
35. He, C.; Zha, G.; Deng, C.; An, Y.; Mao, R.; Liu, Y.; Lu, Y.; Chen, Z. Refractive index dispersion of organic–inorganic hybrid halide perovskite $CH_3NH_3PbX_3$ (X = Cl, Br, I) single crystals. *Cryst. Res. Technol.* **2019**, *54*, 1900011. [CrossRef]
36. Knop, O.; Wasylishen, R.E.; White, M.A.; Cameron, T.S.; Van Oort, M.J.M. Alkylammonium lead halides. Part 2. $CH_3NH_3PbX_3$ (X = C1, Br, I) perovskites: Cuboctahedral halide cages with isotropic cation reorientation. *Can. J. Chem.* **1990**, *68*, 412–422. [CrossRef]
37. Rehwald, W. The study of structural phase transitions by means of ultrasonic experiments. *Adv. Phys.* **1973**, *22*, 721–755. [CrossRef]
38. Tsukada, S.; Kojima, S. Broadband light scattering of two relaxation processes in relaxor ferroelectric $0.93Pb(Zn_{1/3}Nb_{2/3})O_3$-$0.07PbTiO_3$ single crystals. *Phys. Rev. B* **2008**, *78*, 144106. [CrossRef]
39. Ko, J.-H.; Kojima, S.; Koo, T.-Y.; Jung, J.H.; Won, C.J.; Hur, N.J. Elastic softening and central peaks in $BaTiO_3$ single crystals above the cubic-tetragonal phase-transition temperature. *Appl. Phys. Lett.* **2008**, *93*, 102905. [CrossRef]
40. Onoda-Yamamuro, N.; Matsuo, T.; Suga, H. Calorimetric and IR spectroscopic studies of phase transitions in methylammonium trihalogenoplumbates (II). *J. Phys. Chem. Solids* **1990**, *51*, 1383–1395. [CrossRef]

41. Onoda-Yamamuro, N.; Yamamuro, O.; Matsuo, T.; Suga, H. p-T Phase relations of $CH_3NH_3PbX_3$ (X = Cl, Br, I) crystals. *J. Phys. Chem. Solids* **1992**, *53*, 277–281. [CrossRef]
42. Ko, J.-H.; Kojima, S.; Bokov, A.A.; Ye, Z.-G. Light scattering study of acoustic phonon modes and central peaks in $Pb[(Mg_{1/3}Nb_{2/3})_{0.45}Ti_{0.55}]O_3$ single crystals. *Appl. Phys. Lett.* **2007**, *91*, 252909. [CrossRef]

materials

MDPI

Article

Influence of Halides on Elastic and Vibrational Properties of Mixed-Halide Perovskite Systems Studied by Brillouin and Raman Scattering

Furqanul Hassan Naqvi, Syed Bilal Junaid and Jae-Hyeon Ko *

School of Nano Convergence Technology, Nano Convergence Technology Center, Hallym University, Chuncheon 24252, Republic of Korea; furqanhassan05@gmail.com (F.H.N.); syedbilaljunaid@yahoo.com (S.B.J.)
* Correspondence: hwangko@hallym.ac.kr

Abstract: The relationship between halogen content and the elastic/vibrational properties of $MAPbBr_{3-x}Cl_x$ mixed crystals (x = 1.5, 2, 2.5, and 3) with MA = $CH_3NH_3^+$ has been studied using Brillouin and Raman spectroscopy at room temperature. The longitudinal and transverse sound velocities, the absorption coefficients and the two elastic constants C_{11} and C_{44} could be obtained and compared for the four mixed-halide perovskites. In particular, the elastic constants of the mixed crystals have been determined for the first time. A quasi-linear increase in the sound velocity and the elastic constant C_{11} with increasing chlorine content was observed for the longitudinal acoustic waves. C_{44} was insensitive to the Cl content and very low, indicating a low elasticity to shear stress in mixed perovskites regardless of the Cl content. The acoustic absorption of the LA mode increased with increasing heterogeneity in the mixed system, especially for the intermediate composition where the Br and Cl ratio was 1:1. In addition, a significant decrease in the Raman-mode frequency of the low-frequency lattice modes and the rotational and torsional modes of the MA cations was observed with decreasing Cl content. It clearly showed that the changes in the elastic properties as the halide composition changes were correlated with the lattice vibrations. The present findings may facilitate a deeper understanding of the complex interplay between halogen substitution, vibrational spectra and elastic properties, and may also pave the way for optimizing the operation of perovskite-based photovoltaic and optoelectronic devices by tailoring their chemical composition.

Keywords: mixed lead halide perovskites; $MAPbBr_{3-x}Cl_x$; Raman spectroscopy; Brillouin spectroscopy

Citation: Naqvi, F.H.; Junaid, S.B.; Ko, J.-H. Influence of Halides on Elastic and Vibrational Properties of Mixed-Halide Perovskite Systems Studied by Brillouin and Raman Scattering. *Materials* **2023**, *16*, 3986. https://doi.org/10.3390/ma16113986

Academic Editor: Marc Lamy de la Chapelle

Received: 29 April 2023
Revised: 24 May 2023
Accepted: 24 May 2023
Published: 26 May 2023

1. Introduction

Mixed-lead-halide perovskites have emerged as a promising group of materials for various optoelectronic applications, particularly in the field of photovoltaics [1]. These materials exhibit unique properties such as bandgap tunability, high carrier mobilities and long carrier lifetimes, making them highly attractive for solar cell applications [2–4]. Perovskite-based solar cells have rapidly improved their efficiency from just 3.8% in 2009 to over 25% in recent years, rivalling conventional silicon-based solar cells [5,6]. Their ability to continuously tune the band gap, combined with high carrier lifetimes and mobilities, solution processability and improved stability, make mixed-halide perovskites an attractive candidate for next-generation solar cell technologies [7,8]. In addition, mixed perovskite materials offer versatility through their compositional flexibility, allowing researchers to tailor their properties according to the desired application. Despite the numerous advantageous properties of perovskites, their long-term stability remains a significant limitation that requires further investigation [9].

Brillouin and Raman spectroscopy are crucial techniques for the investigation of perovskite materials, as they provide a comprehensive analysis of structural, vibrational and mechanical properties [10,11]. Raman spectroscopy provides critical insight into the crystal structure and phase transitions, as well as the vibrational modes that affect the performance of perovskite devices. Meanwhile, room-temperature Brillouin spectroscopy is essential for determining the elastic properties that could optimize perovskite materials for device stability and performance [12]. Furthermore, elastic constants are fundamental physical properties that are useful barometers for checking the reliability of theoretical calculations. By combining these techniques, we can comprehensively analyze mixed perovskite materials and optimize their properties, leading to better device performance.

In this study, we present a room-temperature Brillouin and Raman spectroscopic investigation of mixed-lead-halide perovskites, $MAPbBr_{3-x}Cl_x$ with x = 1.5 to 3, to gain insight into their vibrational properties and structural stability. We study the variations in optical-mode and acoustic-mode frequencies as a function of halide composition, where the chlorine content increases from 1.5 to 3 at the X-site anion in the ABX_3 perovskite structure. Previous studies have investigated the effect of chlorine substitution at the X site on $MAPbI_{3-x}Cl_x$ systems. A recent study demonstrated an increase in the photovoltaic performance and stability of $MAPbI_{3-x}Cl_x$ mixed perovskites with increasing Cl content [13]. Another study observed increased photoexcitation lifetimes at higher Cl contents [14]. Density functional theory (DFT) modelling by Silvia et al. showed that the incorporation of Cl into the solar-cell device improved the charge collection efficiency [15,16]. Comparing two systems, $MAPbI_{3-x}Cl_x$ and $MAPbI_{3-x}Br_x$, Tombe et al., discovered that Cl-based mixed perovskites exhibited superior film morphology and photovoltaic efficiency [17]. In addition, Zheng et al. highlighted that the charge diffusion properties of $MAPbBr_{3-x}Cl_x$ are superior to those of the $MAPbI_{3-x}Cl_x$ system [18]. While there are numerous reports on $MAPbI_{3-x}Cl_x$ systems, the $MAPbBr_{3-x}Cl_x$ system remains underexplored. A limited number of studies have investigated the stability, vibrational and photovoltaic properties for this system [19–23]. Furthermore, information on the elastic properties of these materials remains scarce in the existing literature.

In the present work, we have used Brillouin spectroscopy to assess the elastic properties of these materials and to investigate the influence of Cl addition at the X site. The acoustic phonon velocities and the corresponding elastic constants were found to increase with Cl content, suggesting a more rigid and stable lattice structure. We further correlated the behavior of the acoustic phonon modes with the optical phonon modes probed by Raman spectroscopy. The measurement range was extended to a lower frequency range compared to our previous Raman study [21]. Low-frequency Raman modes corresponding to lattice vibrations showed an increase in wavenumber with higher Cl content. This observation indicated a change in the bond strength within the lattice due to the replacement of weaker Pb-Br bonds by stronger Pb-Cl bonds upon addition of Cl.

2. Experimental Methods

2.1. Chemicals

Lead chloride ($PbCl_2$, 99.999%), diethyl ether (HPLC grade, 99.9%), lead bromide ($PbBr_2$, >98%), hydrochloric acid (HCl, 37%, ACS reagent), hydrobromic acid (48%, ACS reagent), ethanol (anhydrous 99.5%), dimethyl sulfoxide (DMSO, anhydrous 99.9%), *N,N*-dimethylformamide (DMF, anhydrous 99.8%) and methylamine (CH_3NH_2, 40% in water) were purchased from Sigma Aldrich (Saint Louis, MO, USA). Methylammonium chloride (MACl) and methylammonium bromide (MABr) were synthesized using above precursors, details of which are mentioned in our previous study [10].

2.2. $MAPbBr_{3-x}Cl_x$ Single-Crystal Synthesis

The single crystals of $MAPbBr_{3-x}Cl_x$ with varying degrees of chlorine substitution (x = 1.5, 2, 2.5 and 3) were grown using the solvent evaporation method. This method involved mixing equimolar ratios of MACl and MABr with $PbCl_2$ and $PbBr_2$, respectively, in their respective solvents. The solvent used was a blend of *N,N*-dimethylformamide (DMF) and dimethyl sulfoxide (DMSO) with different volume ratios for various compositions. After complete dissolution of the powders in solvent, the solution was filtered by using a 0.2 μm syringe filter, and the filtrate was kept at constant temperature (80–100 °C) for one or two days until the crystals started to appear. The crystals were then washed with dichloromethane and dried at 60 °C for one night in an oven.

2.3. X-ray Diffraction (XRD)

Powder X-ray diffraction (XRD) patterns were obtained at room temperature using a high-resolution powder X-ray diffractometer (PANalytical X'pert Pro MPD, Malvern, UK). Cu Kβ radiation (λ = 1.5406 Å) was employed to probe the samples within a scan range of $10° < 2\Theta < 60°$ at a scan rate of 3° per minute. Crystalline powders were prepared for the measurements by crushing single crystals.

2.4. Brillouin Spectroscopy

A tandem, multi-pass Fabry-Perot interferometer (TFP-2, JRS Co., Zürich, Switzerland) was utilized to record room-temperature Brillouin spectra, with an excitation source of wavelength 532 nm. Measurements were conducted using backscattering geometry, facilitated by a modified microscope (BH-2, Olympus, Tokyo, Japan) in the backscattering geometry $c(a, a + b)\bar{c}$, where a, b and c represent the cubic axes. In this geometry, the scattered light was concurrently collected along the same path as that of the incident light.

2.5. Raman Spectroscopy

Raman measurements were conducted using a standard Raman spectrometer (LabRam HR800, Horiba Co., Kyoto, Japan). Room-temperature spectra were obtained in a frequency range spanning from 10 to 3500 cm^{-1}. The Raman spectrometer was equipped with a low-frequency notch filter, enabling the lowest frequency limit of 10 cm^{-1}. A diode-pumped solid-state laser with a wavelength of 532 nm was employed to excite the single crystals. All measurements were performed utilizing an optical microscope (BX41, Olympus, Tokyo, Japan) with a 50× magnification objective lens in backscattering geometry. Prior to recording measurements, a silicon standard sample exhibiting a single peak at 520 cm^{-1} was used for calibrating the Raman spectrometer.

3. Results and Discussion

Figure 1 illustrates the variation of unit-cell volume and calculated density as a function of composition in $MAPbBr_{3-x}Cl_x$. It shows a monotonic decrease with increasing Cl content (x), highlighting a strong correlation between structural changes induced by Cl substitution and their direct effect on material properties. The reduction in unit-cell volume can be attributed to stronger electrostatic forces resulting from tighter atomic packing when larger Br ions are replaced by smaller Cl ions within the perovskite structure. This can be explained by the higher electronegativity of chlorine (3.16) compared to bromine (2.96), resulting in a stronger covalent character within Pb-Cl bonds compared to Pb-Br bonds. The absolute values of these parameters are tabulated in Table 1. All mixed crystals examined in this study have a cubic phase ($Pm\bar{3}m$) at room temperature [21].

Table 1. Summary of crystallographic and physical properties for $MAPbBr_{3-x}Cl_x$ with x = 1.5, 2, 2.5 and 3. Values of refractive index for pure components were obtained from ref [24] and interpolated to get values for all mixed compositions.

Composition	Lattice Parameter (Å)	Unit-Cell Volume (Å^3)	Refractive Index	Density (kg/m^3)
$MAPbCl_3$	5.67	182.28	1.90	3147
$MAPbBr_{0.5}Cl_{2.5}$	5.71	186.16	1.96	3278
$MAPbBr_1Cl_2$	5.75	190.10	2.03	3326
$MAPbBr_{1.5}Cl_{1.5}$	5.80	195.11	2.10	3506

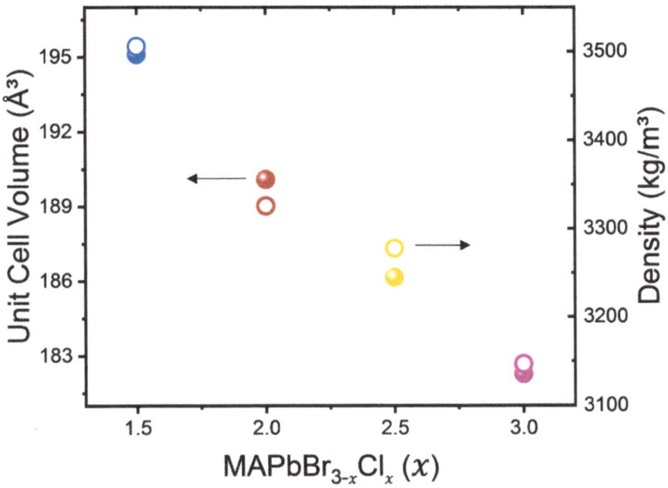

Figure 1. Unit-cell volume (spheres) and density (circles) of $MAPbBr_{3-x}Cl_x$ with x = 1.5, 2, 2.5 and 3.

Figure 2 shows the room temperature Brillouin spectra of $MAPbBr_{3-x}Cl_x$ single crystals and the corresponding fitting lines. The Brillouin spectra consist of two doublets corresponding to the longitudinal (LA) and transverse (TA) acoustic modes. No central peak was observed at room temperature for all compositions. The mode frequencies shift to higher values with increasing Cl content, indicating changes in the elastic properties of the perovskite lattice due to the incorporation of Cl ions. The higher Cl content in the mixed crystal may lead to stronger interatomic interactions, resulting in a stiffer crystal structure, higher elastic moduli and faster acoustic wave propagation. Similar behavior has been observed in previous studies of mixed-halide perovskites, which have shown that the incorporation of halide ions with different ionic radii leads to significant changes in the elastic properties of the materials [12,25].

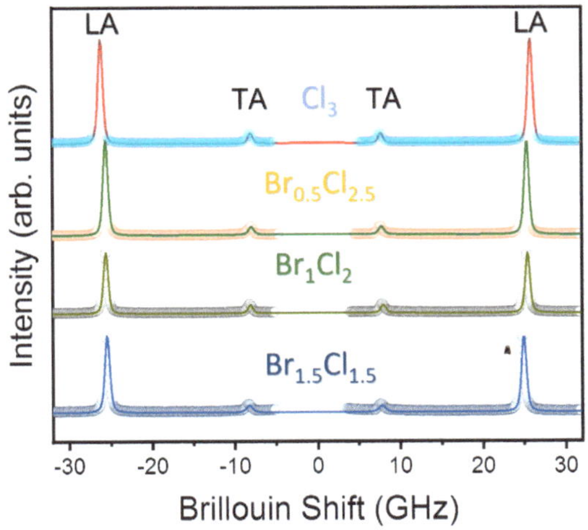

Figure 2. Room-temperature Brillouin spectra of the four mixed-halide perovskites $MAPbBr_{3-x}Cl_x$ with x = 1.5, 2, 2.5 and 3 along with best fitted lines using Voigt function.

The acoustic-mode frequencies and the full width at half maximum (FWHM) of the Brillouin doublets were determined by fitting the experimental spectra with the Voigt function, which is a convolution of the instrumental Gaussian and Lorentzian response functions. Considering the backscattering geometry, where the incident and scattered light are antiparallel, the relationship between the acoustic-mode frequency and the sound velocity can be expressed by the following equation:

$$V = \frac{\lambda \nu_B}{2n} \quad (1)$$

where V denotes the sound velocity, n is the refractive index, λ (=532 nm) is the wavelength of the light source and ν_B is the acoustic-mode frequency determined from experimental spectra by the fitting procedure. The refractive indices of the mixed compositions were estimated by interpolating the experimental values of the pure components, $MAPbCl_3$ (n = 1.90) and $MAPbBr_3$ (n = 2.30), as shown in Table 1 [24]. The above equation can be used to derive the sound velocities of the LA and TA modes propagating along the cubic [100] direction. According to the Brillouin selection rule, the LA and the TA modes for this scattering geometry are associated with the elastic constants C_{11} and C_{44}, respectively. These elastic constants can then be calculated from the sound velocities and the density (ρ) of the crystal as follows:

$$C_{ij} = \rho V^2 \quad (2)$$

The densities for mixed compositions were also calculated from the lattice parameters obtained and the chemical formula. The densities and refractive indices of all compositions are tabulated in Table 1. Figure 3 shows the calculated LA and TA sound velocities as a function of composition and Figure 4 shows the derived elastic constants C_{11} and C_{44}. It is observed that both LA and TA sound velocities, as well as C_{11} and C_{44}, show a monotonic increase with increasing Cl content. A similar trend was previously observed by Kabakova et al. [12]. Among the two acoustic waves, the changes in V of the LA waves and the related C_{11} are much larger than those of the TA waves and C_{44}.

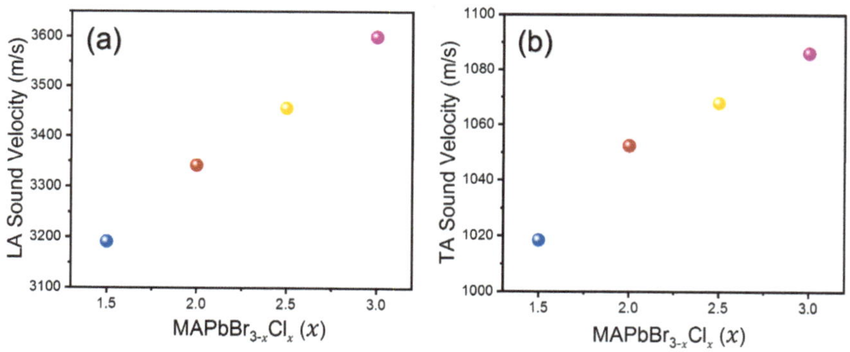

Figure 3. (**a**) Longitudinal acoustic (LA) and (**b**) transverse acoustic sound velocities as a function of composition x in $MAPbBr_{3-x}Cl_x$.

A key factor contributing to the increase in sound velocity is the reduction in the lattice constant due to the incorporation of the smaller Cl ions, as evidenced by the XRD patterns and lattice parameters discussed above. A stiffer lattice, characterized by higher moduli of elasticity, allows faster propagation of sound waves through the material. Consequently, the observed increase in sound velocity with Cl substitution is expected and supports the hypothesis that Cl incorporation plays a crucial role in modifying the mechanical and acoustic properties of the $MAPbBr_{3-x}Cl_x$ system.

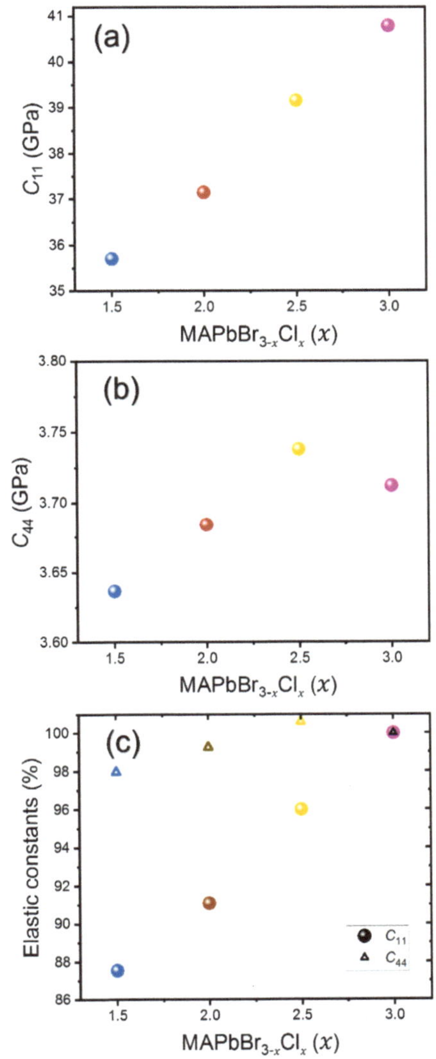

Figure 4. Two elastic constants, (**a**) C_{11} and (**b**) C_{44}, and (**c**) their relative changes with respect to the values of $MAPbCl_3$ as a function of composition x in $MAPbBr_{3-x}Cl_x$.

C_{11} is a measure of the stiffness or modulus associated with the resistance of a material to deformation along a primary crystallographic axis when subjected to stress parallel to the same axis, while C_{44} represents the shear modulus. Table 2 shows a comparison of the sound velocities and the two elastic constants of the mixed-halide perovskites studied in this work. To the best of our knowledge, the elastic constants for mixed perovskites have not been reported in the literature. For $MAPbCl_3$, a theoretical study previously calculated C_{11}(C_{44}) to be 39.5 GPa (2.9 GPa), which is close to our experimentally obtained values [26]. These two moduli play an important role in fully characterizing the mechanical behavior of a cubic crystal structure. As x increases from 1.5 to 3, C_{11} increases from 35.7 GPa to 40.7 GPa, while C_{44} increases from 3.63 GPa to 3.71 GPa. While the C_{44} value is generally low in all the perovskites studied, it becomes even lower with decreasing Cl content, indicating an increased elastic anisotropy and a low resistance to shear stress in mixed perovskites with lower Cl content [27]. Previously, such a difference has also been observed in MA- and FA-based perovskites [11], as well as in Cs-based perovskites [28], and it has been linked to the rotational motions of the cations and the tilting arrangement of the PbX_6 octahedra in these halide perovskites [29]. In addition, the observed increase in C_{11} and C_{44} with higher Cl content suggests that, in the $MAPbBr_{3-x}Cl_x$ system, a larger amount of chlorine enhances the structural resilience at the molecular level, including both longitudinal and shear deformations, which can help the device withstand the regular strains caused by the external forces during practical operation. However, it should be noted that the change in C_{11} is much larger (~12.5%) than that in C_{44} (~2%) for the whole composition change, indicating that C_{44} is insensitive to the composition of the $MAPbBr_{3-x}Cl_x$ system.

Table 2. Summary of calculated sound velocities and elastic constants for $MAPbBr_{3-x}Cl_x$ with x = 1.5, 2, 2.5 and 3.

Composition	V_{LA} (m/s)	V_{TA} (m/s)	C_{11} (GPa)	C_{44} (GPa)
$MAPbCl_3$	3599	1086	40.7	3.71
$MAPbBr_{0.5}Cl_{2.5}$	3455	1067	39.1	3.73
$MAPbBr_1Cl_2$	3341	1052	37.1	3.68
$MAPbBr_{1.5}Cl_{1.5}$	3191	1018	35.7	3.63

The absorption coefficient (α) was also calculated from the FWHM (Γ_B) of the LA mode by the following relation:

$$\alpha = \frac{\pi \Gamma_B}{V} \tag{3}$$

Absorption coefficients represent the ability of a material to absorb and lose energy through mechanisms such as thermal vibration and scattering or coupling to other dynamic degrees of freedom [30]. It is therefore a good measure to characterize how acoustic waves are damped in a system during propagation, especially in the gigahertz range in the present case. The calculated absorption coefficients of the LA modes shown in Figure 5 exhibit a decreasing trend with increasing Cl content. The lower absorption/attenuation coefficients observed in materials with higher Cl content suggest that these materials exhibit weaker damping effects or internal friction. The absorption coefficient of the mixed perovskite $MAPbBr_{1.5}Cl_{1.5}$ was found to be the highest among all the compositions studied, which may be due to increased lattice heterogeneity leading to attenuation of acoustic wave propagation. Especially, the heterogeneous distribution of Br and Cl cations in the X site may be enhanced near $MAPbBr_{1.5}Cl_{1.5}$ where the ratio between Cl and Br is 1:1 maximizing the damping of the acoustic waves. The above observations highlight the importance of investigating how compositional variations not only affect mechanical properties, but also influence energy dissipation mechanisms within the $MAPbBr_{3-x}Cl_x$ mixed-halide perovskite system. On the other hand, the absorption coefficients of the TA modes were much higher than those of the LA modes, more than 9×10^5 m^{-1}. In addition, the absorption coefficient of the TA mode was insensitive to the composition, consistent with the fact that the corresponding elastic constant C_{44} is insensitive to the composition. This suggests that the TA modes are more strongly damped in the mixed crystals compared to the LA modes, which are common to all mixed crystals regardless of composition.

Investigation of the bonding mechanism between the halides (Cl and Br) and the surrounding octahedra ($PbCl_6$ and $PbBr_6$) may be necessary, as the elastic properties of halide perovskites have been found to be determined by the strength of the chemical bond between the Pb and halogen atoms [31,32]. Raman spectroscopy serves as a useful tool to obtain important information about the vibrational properties and molecular dynamics within these mixed-halide perovskites [22]. Figure 6 shows the Raman spectra recorded at room temperature for all compositions in the frequency range 10–3500 cm^{-1}. The Raman spectra show several distinct peaks corresponding to vibrational modes of the lattice as well as the internal modes of the MA cation. The exact peak positions for all the observed vibrational modes, calculated by fitting analyses, together with their respective mode assignments, are tabulated in Table 3 based on previous results [23,33,34].

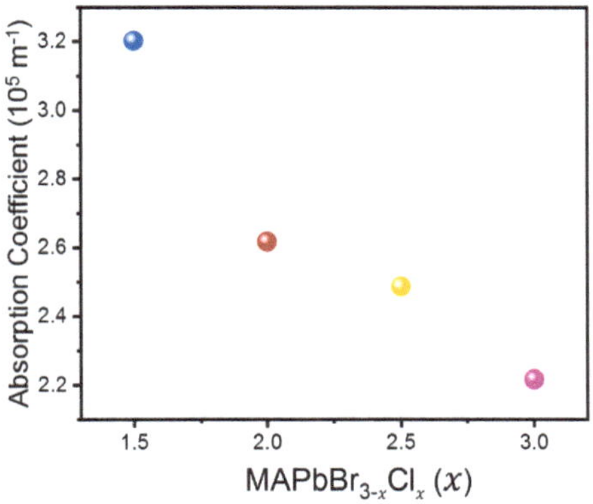

Figure 5. Absorption coefficient of the LA mode as a function of composition x in $MAPbBr_{3-x}Cl_x$.

Table 3. Raman modes along with their respective assignments for $MAPbBr_{3-x}Cl_x$ with x = 1.5, 2, 2.5 and 3.

$Br_{1.5}Cl_{1.5}$	$BrCl_2$	$Br_{0.5}Cl_{2.5}$	Cl_3	Assignment
55	65	73	77	δs (X-Pb-X) [a]
151	161	168	178	ν_s (Pb-X) [a]
226	227	227	236	R of MA+ cation [a]
453	468	479	484	τ (MA) [a,b]
920	923	923	922	ρ (MA) [a,b]
974	975	976	976	ν (C-N) [a,c]
1248	1248	1248	1247	ρ (MA) [a,c]
1426	1426	1426	1426	δ_s (CH_3) [a]
1454	1456	1456	1456	δ_{as} (CH_3) [a]
1483	1484	1485	1484	δ_s (NH_3) [a]
1593	1596	1596	1596	δ_{as} (NH_3) [a]
2829	2830	2830	2830	ν_{as} (C-H) [c]
2895	2896	2897	2897	ν_s (C-H) [c]
2870	2971	2972	2972	ν_{as} (CH_3) [a]
3041	3042	3043	3044	ν_s (NH_3) [a]
3115	3116	3117	3117	ν_{as} (CH_3) [b]
3190	3192	3194	3196	ν_s (NH_3) [b]

δ: bending; ρ: rocking; ν: stretching; s/as: symmetric/asymmetric; τ: torsion; R: rotation. Superscripts [a], [b] and [c] indicate the references [23,33,34], respectively, from which the mode assignment was adapted.

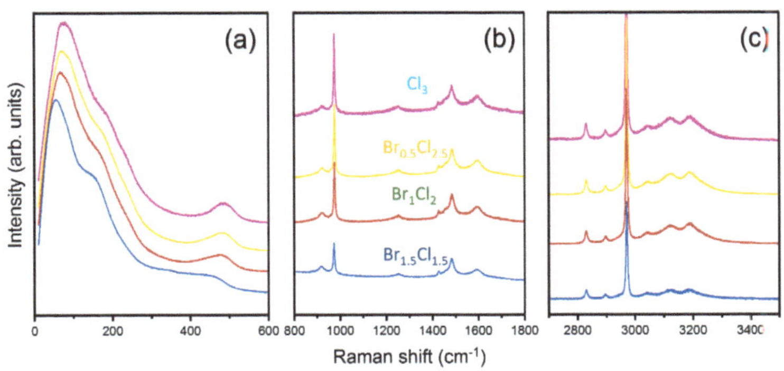

Figure 6. Room-temperature Raman spectra of $MAPbBr_{3-x}Cl_x$ with x = 1.5, 2, 2.5 and 3 in (**a**) 10–600 cm^{-1}, (**b**) 800–1800 cm^{-1} and (**c**) 2700–3500 cm^{-1} ranges.

Figure 6a shows the low-frequency lattice modes below 300 cm^{-1} and the torsional (τ) mode of the MA cation. We observed a significant increase in mode frequencies for the low-frequency lattice modes and the torsional (τ) mode of the MA cation with increasing Cl content. A similar trend was also observed by Xie et al. for a small increase in Cl content where the authors studied $MAPbBr_{3-x}Cl_3$ with $x = 0.25$, 0.33 and 0.5 [22]. Low-frequency lattice modes represent collective vibrations of the crystal lattice, including motions associated with PbX_6 octahedral arrangements (where X refers to Br or Cl). Two low-frequency modes associated with the bending and stretching of the lead-halide octahedra increase from 55 cm^{-1} and 151 cm^{-1} in $MAPbBr_{1.5}Cl_{1.5}$ to 77 cm^{-1} and 178 cm^{-1} in $MAPbCl_3$, respectively. As discussed earlier, the incorporation of smaller Cl ions into the lattice results in a decrease in unit-cell dimensions, leading to stronger electrostatic interactions between neighboring atoms. Consequently, these stronger interactions imply stiffer force constants within the crystal structure, which translates into increased vibrational frequencies of low-frequency lattice modes.

The torsional (τ) mode, which corresponds to an out-of-phase twisting motion around the central nitrogen atom in MA cations, increases from 453 cm^{-1} to 484 cm^{-1} as x increases from 1.5 to 3. This mode is sensitive to hydrogen bonding and the overall interaction strength between the organic moiety and the surrounding halogen atoms [35]. With increasing Cl content, stronger hydrogen bonding may occur due to shorter bonding distances between the amine end (NH_3^+) of the MA cation and the corresponding halogen atoms, ultimately resulting in a shift of the τ mode peak towards higher wavenumbers. These results can be correlated with our Brillouin spectroscopic observations of the elastic properties. The linear translational–rotational coupling of the MA cation has been found to cause a softening of the perovskite lattice, as well as a damping effect on sound waves propagating within the crystal [36]. As the MA cations undergo rotational and translational motions, they impose local strain on the adjacent inorganic lattice. This local strain affects the interatomic forces between neighboring atoms, resulting in variations in the elastic constants and mechanical stiffness of the perovskite lattice.

Figure 6b shows Raman spectra between 800 and 1800 cm^{-1}, including rocking modes of the MA cation and bending of CH_3 and NH_3 groups. Contrary to the behavior of the low-frequency modes, the vibrational modes in this range show no significant change in Raman shifts as a function of composition. Similar behavior was observed for high-frequency modes involving internal bond vibrations such as C-H and N-H stretching within the organic cations, as shown in Figure 6c. The insensitivity of these modes to Cl content implies a minimal influence of halide substitution on the local environment surrounding the organic component and suggests that Cl content primarily affects global crystal-lattice dynamics, while only moderately or negligibly affecting local bond strengths within the organic cations.

The softening of the lattice structure makes the crystal more susceptible to local distortions that can lead to the formation of polarons, which can ultimately have a significant impact on the electronic and optical properties of a material in photovoltaic devices, as they tend to lower the carrier mobility due to their increased effective mass and limited hopping behavior compared to free carriers [27,37,38]. On the other hand, a soft lattice structure can be easily damaged by external shock or vibration, reducing long-term mechanical reliability. Therefore, understanding the elastic behavior of halide perovskites is crucial for the development of advanced and reliable perovskite materials with tailored mechanical and functional properties for use in various optoelectronic and photovoltaic applications.

4. Conclusions

The study of mixed-halide perovskite $MAPbBr_{3-x}Cl_x$ single crystals revealed the relationship between halogen content and the mechanical, acoustic and vibrational properties in this system. X-ray powder diffraction showed a decrease in the lattice parameter with increasing Cl content, which was attributed to the smaller ionic radii of Cl compared to Br. Brillouin spectroscopy showed that higher Cl incorporation resulted in higher sound velocities and elastic moduli, resulting in a stiffer crystal structure that allowed faster propagation of acoustic waves through the material. Raman spectroscopy provided further information on how Cl content affects the vibrational properties and molecular dynamics within the mixed-halide perovskites. Low-frequency lattice modes and torsional modes of the MA cation showed significant increases in wavenumber with increasing Cl content, highlighting stronger electrostatic interactions and hydrogen bonding within the lattice. In contrast, vibrational modes involving rocking, bending and high-frequency internal bond vibrations within the organic component were insensitive to the Cl substitution, suggesting that Cl content primarily affects global crystal-lattice dynamics while only moderately or negligibly affecting local bond vibrations within organic cations.

Elasticity of the perovskite layer is crucial for thin-film technology in perovskite solar cells on flexible substrates, as it prevents cracking and performance degradation. Single crystalline layers are preferrable since the mechanical behavior of polycrystalline layers can be compromised by grain boundaries and defects. Understanding the relationships between compositional variations and the mechanical, acoustic and vibrational properties is crucial for the development of advanced perovskite materials with tailored properties for various optoelectronic and photovoltaic applications. The present results can serve as fundamental physical properties for the incorporation of mixed-halide systems in various optoelectronic applications.

Author Contributions: Methodology, F.H.N. and J.-H.K.; Validation, F.H.N. and S.B.J.; Formal analysis, F.H.N. and S.B.J.; Investigation, F.H.N. and S.B.J.; Resources, J.-H.K.; Data curation, F.H.N. and S.B.J.; Writing—original draft, F.H.N.; Writing—review & editing, J.-H.K.; Visualization, F.H.N.; Supervision, J.-H.K.; Project administration, J.-H.K.; Funding acquisition, J.-H.K. All authors have read and agreed to the published version of the manuscript.

Funding: This work was supported by the Hallym University Research Fund, 2023 (HRF-202301-015).

Institutional Review Board Statement: Not applicable.

Informed Consent Statement: Not applicable.

Data Availability Statement: The data in this study is available upon request to the corresponding author.

Conflicts of Interest: The authors declare no conflict of interest.

References

1. Xu, F.; Zhang, M.; Li, Z.; Yang, X.; Zhu, R. Challenges and Perspectives toward Future Wide-Bandgap Mixed-Halide Perovskite Photovoltaics. *Adv Energy Mater.* **2023**, *13*, 2203911. [CrossRef]
2. Eperon, G.E.; Stranks, S.D.; Menelaou, C.; Johnston, M.B.; Herz, L.M.; Snaith, H.J. Formamidinium Lead Trihalide: A Broadly Tunable Perovskite for Efficient Planar Heterojunction Solar Cells. *Energy Environ. Sci.* **2014**, *7*, 982–988. [CrossRef]
3. Noh, J.H.; Im, S.H.; Heo, J.H.; Mandal, T.N.; Seok, S. Il Chemical Management for Colorful, Efficient, and Stable Inorganic-Organic Hybrid Nanostructured Solar Cells. *Nano Lett.* **2013**, *13*, 1764–1769. [CrossRef] [PubMed]
4. Eperon, G.E.; Burlakov, V.M.; Docampo, P.; Goriely, A.; Snaith, H.J. Morphological Control for High Performance, Solution-Processed Planar Heterojunction Perovskite Solar Cells. *Adv. Funct. Mater.* **2014**, *24*, 151–157. [CrossRef]
5. Kojima, A.; Teshima, K.; Shirai, Y.; Miyasaka, T. Organometal Halide Perovskites as Visible-Light Sensitizers for Photovoltaic Cells. *J. Am. Chem. Soc.* **2009**, *131*, 6050–6051. [CrossRef]
6. NREL. Efficiency Chart. Available online: http://www.nrel.gov/ncpv/images/efficiency_chart.jpg (accessed on 23 April 2023).
7. Li, Z.; Klein, T.R.; Kim, D.H.; Yang, M.; Berry, J.J.; van Hest, M.F.A.M.; Zhu, K. Scalable Fabrication of Perovskite Solar Cells. *Nat. Rev. Mater.* **2018**, *3*, 18017. [CrossRef]
8. Zhao, X.; Lu, H.; Fang, W.-H.; Long, R. Correlated Organic-Inorganic Motion Enhances Stability and Charge Carrier Lifetime in Mixed Halide Perovskites. *Nanoscale* **2022**, *14*, 4644–4653. [CrossRef]
9. Jacak, J.E.; Jacak, W.A. Routes for Metallization of Perovskite Solar Cells. *Materials* **2022**, *15*, 2254. [CrossRef]
10. Naqvi, F.H.; Ko, J.H. Structural Phase Transitions and Thermal Degradation Process of $MAPbCl_3$ Single Crystals Studied by Raman and Brillouin Scattering. *Materials* **2022**, *15*, 8151. [CrossRef]
11. Ferreira, A.C.; Létoublon, A.; Paofai, S.; Raymond, S.; Ecolivet, C.; Rufflé, B.; Cordier, S.; Katan, C.; Saidaminov, M.I.; Zhumekenov, A.A.; et al. Elastic Softness of Hybrid Lead Halide Perovskites. *Phys. Rev. Lett.* **2018**, *121*, 085502. [CrossRef]

12. Kabakova, I.V.; Azuri, I.; Chen, Z.; Nayak, P.K.; Snaith, H.J.; Kronik, L.; Paterson, C.; Bakulin, A.A.; Egger, D.A. The Effect of Ionic Composition on Acoustic Phonon Speeds in Hybrid Perovskites from Brillouin Spectroscopy and Density Functional Theory. *J. Mater. Chem. C* **2018**, *6*, 3861–3868. [CrossRef]
13. Al Katrib, M.; Planes, E.; Perrin, L. Effect of Chlorine Addition on the Performance and Stability of Electrodeposited Mixed Perovskite Solar Cells. *Chem. Mater.* **2022**, *34*, 2218–2230. [CrossRef]
14. Docampo, P.; Hanusch, F.C.; Stranks, S.D.; Döblinger, M.; Feckl, J.M.; Ehrensperger, M.; Minar, N.K.; Johnston, M.B.; Snaith, H.J.; Bein, T. Solution Deposition-Conversion for Planar Heterojunction Mixed Halide Perovskite Solar Cells. *Adv. Energy Mater.* **2014**, *4*, 1400355. [CrossRef]
15. Colella, S.; Mosconi, E.; Pellegrino, G.; Alberti, A.; Guerra, V.L.P.; Masi, S.; Listorti, A.; Rizzo, A.; Condorelli, G.G.; De Angelis, F.; et al. Elusive Presence of Chloride in Mixed Halide Perovskite Solar Cells. *J. Phys. Chem. Lett.* **2014**, *5*, 3532–3538. [CrossRef]
16. Colella, S.; Mosconi, E.; Fedeli, P.; Listorti, A.; Gazza, F.; Orlandi, F.; Ferro, P.; Besagni, T.; Rizzo, A.; Calestani, G.; et al. $MAPbI_{3-x}Cl_x$ Mixed Halide Perovskite for Hybrid Solar Cells: The Role of Chloride as Dopant on the Transport and Structural Properties. *Chem. Mater.* **2013**, *25*, 4613–4618. [CrossRef]
17. Tombe, S.; Adam, G.; Heilbrunner, H.; Apaydin, D.H.; Ulbricht, C.; Sariciftci, N.S.; Arendse, C.J.; Iwuoha, E.; Scharber, M.C. Optical and Electronic Properties of Mixed Halide (X = I, Cl, Br) Methylammonium Lead Perovskite Solar Cells. *J. Mater. Chem. C* **2017**, *5*, 1714–1723. [CrossRef]
18. Zhang, M.; Yu, H.; Lyu, M.; Wang, Q.; Yun, J.H.; Wang, L. Composition-Dependent Photoluminescence Intensity and Prolonged Recombination Lifetime of Perovskite $CH_3NH_3PbBr_{3-x}Cl_x$ films. *Chem. Commun.* **2014**, *50*, 11727–11730. [CrossRef]
19. Doumbia, Y.; Bouich, A.; Soro, D.; Soucase, B.M. Mixed Halide Head Perovskites Thin Films: Stability and Growth Investigation. *Optik* **2022**, *261*, 169222. [CrossRef]
20. Tang, M.-C.; Dang, H.X.; Lee, S.; Barrit, D.; Munir, R.; Wang, K.; Li, R.; Smilgies, D.-M.; De Wolf, S.; Kim, D.-Y.; et al. Wide and Tunable Bandgap $MAPbBr_{3-x}Cl_x$ Hybrid Perovskites with Enhanced Phase Stability: In Situ Investigation and Photovoltaic Devices. *Sol. RRL* **2021**, *5*, 2000718. [CrossRef]
21. Naqvi, F.H.; Ko, J.H.; Kim, T.H.; Ahn, C.W.; Hwang, Y. Disentangled Structural and Vibrational Characteristics of Methylammonium Halide Perovskite $MAPbBr_{3-x}Cl_x$ with x = 0$\sim$3 Studied by X-ray Diffraction and Raman Scattering. *Curr. Appl. Phys.* **2022**, *44*, 150–157. [CrossRef]
22. Xie, L.Q.; Zhang, T.Y.; Chen, L.; Guo, N.; Wang, Y.; Liu, G.K.; Wang, J.R.; Zhou, J.Z.; Yan, J.W.; Zhao, Y.X.; et al. Organic-Inorganic Interactions of Single Crystalline Organolead Halide Perovskites Studied by Raman Spectroscopy. *Phys. Chem. Chem. Phys.* **2016**, *18*, 18112–18118. [CrossRef] [PubMed]
23. Niemann, R.G.; Kontos, A.G.; Palles, D.; Kamitsos, E.I.; Kaltzoglou, A.; Brivio, F.; Falaras, P.; Cameron, P.J. Halogen Effects on Ordering and Bonding of $CH_3NH_3^+$ in $CH_3NH_3PbX_3$ (X = Cl, Br, I) Hybrid Perovskites: A Vibrational Spectroscopic Study. *J. Phys. Chem. C* **2016**, *120*, 2509–2519. [CrossRef]
24. He, C.; Zha, G.; Deng, C.; An, Y.; Mao, R.; Liu, Y.; Lu, Y.; Chen, Z. Refractive Index Dispersion of Organic–Inorganic Hybrid Halide Perovskite $CH_3NH_3PbX_3$ (X = Cl, Br, I) Single Crystals. *Cryst. Res. Technol.* **2019**, *54*, 1900011. [CrossRef]
25. Shahrokhi, S.; Dubajic, M.; Dai, Z.Z.; Bhattacharyya, S.; Mole, R.A.; Rule, K.C.; Bhadbhade, M.; Tian, R.; Mussakhanuly, N.; Guan, X.; et al. Anomalous Structural Evolution and Glassy Lattice in Mixed-Halide Hybrid Perovskites. *Small* **2022**, *18*, 2200847. [CrossRef]
26. Faghihnasiri, M.; Izadifard, M.; Ghazi, M.E. DFT Study of Mechanical Properties and Stability of Cubic Methylammonium Lead Halide Perovskites ($CH_3NH_3PbX_3$, X = I, Br, Cl). *J. Phys. Chem. C* **2017**, *121*, 27059–27070. [CrossRef]
27. Létoublon, A.; Paofai, S.; Rufflé, B.; Bourges, P.; Hehlen, B.; Michel, T.; Ecolivet, C.; Durand, O.; Cordier, S.; Katan, C.; et al. Elastic Constants, Optical Phonons, and Molecular Relaxations in the High Temperature Plastic Phase of the $CH_3NH_3PbBr_3$ Hybrid Perovskite. *J. Phys. Chem. Lett.* **2016**, *7*, 3776–3784. [CrossRef]
28. Li, S.; Zhao, S.; Chu, H.; Gao, Y.; Lv, P.; Wang, V.; Tang, G.; Hong, J. Unraveling the Factors Affecting the Mechanical Properties of Halide Perovskites from First-Principles Calculations. *J. Phys. Chem. C* **2022**, *126*, 4715–4725. [CrossRef]
29. Lomonosov, A.M.; Yan, X.; Sheng, C.; Gusev, V.E.; Ni, C.; Shen, Z. Exceptional Elastic Anisotropy of Hybrid Organic–Inorganic Perovskite $CH_3NH_3PbBr_3$ Measured by Laser Ultrasonic Technique. *Phys. Status Solidi-Rapid Res. Lett.* **2016**, *10*, 606–612. [CrossRef]
30. Lee, J.W.; Naqvi, F.H.; Ko, J.-H.; Kim, T.H.; Ahn, C.W. Acoustic Anomalies and the Critical Slowing-Down Behavior of $MAPbCl_3$ Single Crystals Studied by Brillouin Light Scattering. *Materials* **2022**, *15*, 3692. [CrossRef]
31. Feng, J. Mechanical Properties of Hybrid Organic-Inorganic $CH_3NH_3BX_3$ (B = Sn, Pb; X = Br, I) Perovskites for Solar Cell Absorbers. *APL Mater.* **2014**, *2*, 81801. [CrossRef]
32. Li, Z.; Qin, Y.; Dong, L.; Li, K.; Qiao, Y.; Li, W. Elastic and Electronic Origins of Strain Stabilized Photovoltaic γ-$CsPbI_3$. *Phys. Chem. Chem. Phys.* **2020**, *22*, 12706–12712. [CrossRef] [PubMed]
33. Thu Nguyen, T.T.; Kim, Y.; Bae, S.; Bari, M.; Jung, H.R.; Jo, W.; Kim, Y.H.; Ye, Z.G.; Yoon, S. Raman Scattering Studies of the Structural Phase Transitions in Single-Crystalline $CH_3NH_3PbCl_3$. *J. Phys. Chem. Lett.* **2020**, *11*, 3773–3781. [CrossRef] [PubMed]
34. Leguy, A.M.A.; Goñi, A.R.; Frost, J.M.; Skelton, J.; Brivio, F.; Rodríguez-Martínez, X.; Weber, O.J.; Pallipurath, A.; Alonso, M.I.; Campoy-Quiles, M.; et al. Dynamic Disorder, Phonon Lifetimes, and the Assignment of Modes to the Vibrational Spectra of Methylammonium Lead Halide Perovskites. *Phys. Chem. Chem. Phys.* **2016**, *18*, 27051–27066. [CrossRef]

35. Yin, T.; Fang, Y.; Fan, X.; Zhang, B.; Kuo, J.L.; White, T.J.; Chow, G.M.; Yan, J.; Shen, Z.X. Hydrogen-Bonding Evolution during the Polymorphic Transformations in $CH_3NH_3PbBr_3$: Experiment and Theory. *Chem. Mater.* **2017**, *29*, 5974–5981. [CrossRef]
36. Even, J.; Carignano, M.; Katan, C. Molecular Disorder and Translation/Rotation Coupling in the Plastic Crystal Phase of Hybrid Perovskites. *Nanoscale* **2016**, *8*, 6222–6236. [CrossRef]
37. Nie, W.; Blancon, J.-C.; Neukirch, A.J.; Appavoo, K.; Tsai, H.; Chhowalla, M.; Alam, M.A.; Sfeir, M.Y.; Katan, C.; Even, J.; et al. Light-Activated Photocurrent Degradation and Self-Healing in Perovskite Solar Cells. *Nat. Commun.* **2016**, *7*, 11574. [CrossRef]
38. Neukirch, A.J.; Nie, W.; Blancon, J.-C.; Appavoo, K.; Tsai, H.; Sfeir, M.Y.; Katan, C.; Pedesseau, L.; Even, J.; Crochet, J.J.; et al. Polaron Stabilization by Cooperative Lattice Distortion and Cation Rotations in Hybrid Perovskite Materials. *Nano Lett.* **2016**, *16*, 3809–3816. [CrossRef]

MDPI AG
Grosspeteranlage 5
4052 Basel
Switzerland
Tel.: +41 61 683 77 34

Materials Editorial Office
E-mail: materials@mdpi.com
www.mdpi.com/journal/materials

www.ingramcontent.com/pod-product-compliance
Lightning Source LLC
Chambersburg PA
CBHW041411210926
51850CB00013B/3237